Adaptronics and Smart Structures

Springer

Berlin
Heidelberg
New York
Barcelona
Hong Kong
London
Milan
Paris
Singapore
Tokyo

H. Janocha (Ed.)

Adaptronics and Smart Structures

Basics, Materials, Design, and Applications

With 272 Figures and 24 Tables

 Springer

Prof. Dr.-Ing. habil. H. Janocha
Universität des Saarlandes
Lehrstuhl für Prozessautomatisierung
Gebäude 13
D-66123 Saarbrücken, Germany

Library of Congress Cataloging-in-Publication Data

Adaptronics and smart structures: basics, materials, design, and applications / H. Janocha (editor).
p. cm.
Includes bibliographical references and index.
ISBN 3-540-61484-2 (alk. paper)
1. Smart materials. 2. Smart structures. 3. Actuators -- Materials. 4. Detectors -- Materials.
5. Structural control (Engineering) I. Janocha, Hartmut.
TA418.9S62A35 1999
620.1'1 -- dc21   99-34011

ISBN 3-540-61484-2 Springer-Verlag Berlin Heidelberg New York

© Springer-Verlag Berlin Heidelberg 1999
Printed in Germany

Typesetting: Camera-ready from the editor
Cover design: Erich Kirchner, Heidelberg

SPIN 10543513   57/3144/xo – 5 4 3 2 1 0 – Printed on acid-free paper

Preface

Adaptronics, as we understand it, is the science of structures and systems whose features or behavior can be influenced in a goal-oriented fashion by using multi-functional materials, and which can adapt independently to changing influences of the surroundings and operating conditions. The roots of adaptronics are to be found in automation and control technology, in computer sciences and, in particular, in materials sciences. Facets of each of these areas have to complement and support one another in a symbiotic way in order that adaptronic concepts can be drawn up and successfully implemented. Such concepts and realizations are about to gain a foothold in many areas of our environment.

With this book the editor and the publisher have tackled the task of presenting the 'state of the art' of this both fascinating and demanding technological-scientific field. After an introduction describing the aims and the content of adaptronics, subsequent chapters present the 'scientific pillars' from the viewpoint of the various basic disciplines involved. Thereafter, important components of adaptronic structures and systems, such as actuators and sensors, are described. Other chapters are dedicated to applications of adaptronics in the various technological and biological/medical fields of daily life, and an outlook towards future developments concludes the book.

It is obvious that no single person can master all the specialist knowledge involved in such a detailed and varied field as adaptronics. Thus, we recognize both a necessity and a great opportunity in bringing together, in a fundamental work, the knowledge and the experience of proven experts from across the range of adaptronic disciplines. The editor is proud of the fact that numerous experts from all over the world have supported him in performing this task. To all of these he expresses his gratitude. It will not escape the attention of the reader that, in their nuances, viewpoints about adaptronics may diverge somewhat. But this situation is actually both attractive and stimulating; it is also hardly surprising in view of the fact that adaptronics has only just begun to establish itself as a discipline in its own right.

With this background in mind, the editor and publisher hope that the present book will become a useful source of information and ideas, which a large number of readers can rely on time and again. Perhaps it will help some readers to discover their interest or their vocation to actively and creatively support the field of adaptronics on its path to maturity.

Finally, the editor would like to thank his co-workers Nicole Baqué, Claudia Berlt, Sabine Recktenwald and Albert Schwinn for their untiring help in transferring the manuscripts and figures, which the contributing authors had presented in very varied forms, into a uniform format; he also thanks the publishing house Springer–Verlag for the appealing outward design of the book.

In conclusion the editor wants to assure the critical readership that their constructive comments about the conception, content and presentation of this book are welcome and will be taken into consideration, if possible, in future editions.

Saarbrücken, Germany
May 1999 *H. Janocha*

Table of Contents

Introduction
C. Rogers 1

**1. Adaptronics – a Concept for the Development
of Adaptive and Multifunctional Structures**
D. Neumann 5
  1.1 What Is Adaptronics? 5
  1.2 Examples of Adaptronic Systems 5
  1.3 Multifunctional Elements 8
  1.4 Fields of Technology and Application 9
  1.5 Historical Review 10

2. Concepts of Adaptronic Structures
C.A. Rogers and V. Giurgiutiu 13
  2.1 What Are Adaptronic Structures? 13
  2.2 What Are They Good for? 15
  2.3 What Are They Made of? 18
      2.3.1 Artificial Muscles — Actuators 19
      2.3.2 Artificial Nerves — Sensors 24
      2.3.3 Intelligence — Signal Processing, Communication,
           and Controls 27
  2.4 What Do They Look Like? 30
  2.5 The Future: The New Age of Materials 32

3. Multifunctional Materials — the Basis for Adaptronics
W. Cao 35
  3.1 What Are Functional Materials? 35
  3.2 Basic Principles of Functional Materials 36
      3.2.1 Phase Transitions and Anomalies 36
      3.2.2 Microscopic, Mesoscopic, Macroscopic Phenomena
           and Symmetry 38
      3.2.3 Energy Conversion 43
  3.3 Examples of Functional Materials 46
      3.3.1 Thermally Responsive Materials 46

3.3.2   Materials Responsive to Electric, Magnetic and Stress
Fields   48
3.4   Increased Functionality Through Materials Engineering   51
3.4.1   Morphotropic Phase Boundary   52
3.4.2   Domain Engineering   54
3.5   Functional Composites   55
3.6   Summary   57

4. **Adaptive Control Concepts: Controllers in Adaptronics**
V. Rao and R. Damle   59
4.1   Introduction   59
4.2   Description of the Test Articles   61
4.3   Conventional Model–Reference Adaptive Control Techniques .   62
4.3.1   Experimental Results   64
4.4   Adaptive Control Using Neural Networks   66
4.4.1   Neural Network–Based Model Reference
Adaptive Control   66
4.4.2   Neural Network–Based Optimizing Controller
With On–Line Adaptation   69
4.4.3   Simulation Results   73
4.5   Summary   76

5. **Simulation of Adaptronic Systems**
H. Baier and F. Döngi   79
5.1   Introduction   79
5.2   Basic Elements of System Theory   79
5.2.1   Nonlinear and Linear Systems   79
5.2.2   State–Space Representation   80
5.2.3   Controllability and Observability   80
5.2.4   Stability and Robustness   81
5.2.5   Alternative System Representations   82
5.3   Modeling of Adaptive Structures   82
5.3.1   Basic Equations of Structural Mechanics   84
5.3.2   Constitutive Laws of Adaptive Materials   84
5.3.3   Finite Element Method   85
5.3.4   Equations of Motion   85
5.3.5   Sensor Equations   86
5.3.6   Model Reduction Techniques   86
5.4   Analysis of Adaptronic Systems   87
5.4.1   Stability Analysis   88
5.4.2   Spillover and Robustness   88
5.4.3   Numerical Time Integration   88
5.5   Optimization of Adaptronic Systems   89
5.5.1   Problem Statements   90
5.5.2   Solution Techniques   90

      5.5.3   Practical Examples 92

  5.6   Analysis Tools 94

      5.6.1   Finite Element Analysis Codes 94

      5.6.2   Control Design and Simulation Tools 94

      5.6.3   System Identification Tools 94

      5.6.4   Hardware–in–the–Loop Simulation 95

6. **Actuators in Adaptronics** 99

  6.1   The Role of Actuators in Adaptronic Systems

      H. Janocha 99

      6.1.1   Definition of an Actuator 99

      6.1.2   Actuator as Part of a System 100

      6.1.3   Personal Computers for Process Control 101

      6.1.4   Power Amplifier 101

      6.1.5   Intelligent Actuators 103

      6.1.6   Smart Actuators 104

  6.2   Piezoelectric Actuators

      B. Clephas 106

      6.2.1   Theory 107

      6.2.2   Technical Transducers 110

      6.2.3   Comparison With Other Actuator Types 115

      6.2.4   Example Applications 116

      6.2.5   Outlook 122

  6.3   Magnetostrictive Actuators

      F. Claeyssen 124

      6.3.1   Theory of Magnetostriction

            in Magnetostrictive Devices 125

      6.3.2   Principles and Properties of Various Applications 132

      6.3.3   Summary 143

      6.3.4   Acknowledgement 143

  6.4   Shape Memory Actuators

      J. Hesselbach 143

      6.4.1   Properties of Shape Memory Alloys 144

      6.4.2   Electrical Shape Memory Actuators 149

      6.4.3   Perspectives for Shape Memory Actuators 154

      6.4.4   Innovative Application Examples 156

      6.4.5   Conclusion 160

  6.5   Electrorheological Fluid Actuators

      W.A. Bullough 161

      6.5.1   Limitations to the Concept of Electrorheological Fluids 170

      6.5.2   Future Aims and Present Problems 176

      6.5.3   Summary of Advantages of ER Fluids 179

  6.6   Magnetorheological Fluid Actuators

      J.D. Carlson 180

      6.6.1   Description of MR Fluids 181

       6.6.2 Advantages and Concerns 182
       6.6.3 Ways in Which MR Fluids Are Used 184
       6.6.4 Basic MR Device Design Considerations 185
       6.6.5 Linear MR Fluid Dampers........................ 187
       6.6.6 Rotary MR Fluid Brakes and Clutches 191
       6.6.7 Medical Applications of MR Fluids 194
       6.6.8 Conclusion 195
   6.7 Electrochemical Actuators
       D. zur Megede 195
       6.7.1 Fundamentals................................... 196
       6.7.2 Construction of Reversible Actuators 200
       6.7.3 Possible Uses for ECA 203
       6.7.4 Summary....................................... 206
   6.8 Chemomechanical Actuators
       H. Wurmus and M. Kallenbach..................... 207
       6.8.1 Chemomechanical Energy–Conversion Principles 207
       6.8.2 Some Properties of Conducting Polymers 209
       6.8.3 Polymer Microactuators 212
       6.8.4 Synthetic Polymers as Muscle–Mimetic Actuator 216
   6.9 Microactuators
       H. Janocha 217
       6.9.1 Driving Mechanisms 218
       6.9.2 Ink–Jet Printer Heads 220
       6.9.3 Microvalves.................................... 220
       6.9.4 Micropumps 223
       6.9.5 Microfluid Systems 225
       6.9.6 Other Microactuators 227
       6.9.7 Conclusion and Outlook......................... 229

7. Sensors in Adaptronics 241
   7.1 Sensors in Adaptronics
       J.E. Brignell and N.M. White.......................... 241
       7.1.1 Primary Sensor Defects 241
       7.1.2 A Case in Point: The Load Cell 246
       7.1.3 The Impact of ASICs 247
       7.1.4 Reconfigurable Systems 248
       7.1.5 Communications 251
       7.1.6 Trends 254
   7.2 Fiber Optic Sensors
       W. Habel... 255
       7.2.1 Physical Principle of Fiber Optic Techniques......... 257
       7.2.2 Relevant Types of Fiber Sensors and Sensor Selection . 258
       7.2.3 Integrating and Quasi–Distributed Long–Gauge–Length
             Sensors 259
       7.2.4 Short–Gauge–Length Sensors 265

7.2.5   Sensor Networks 280
7.2.6   Research Problems and Future Prospects 280

8.   **Adaptronic Systems in Engineering** 285
8.1   Adaptronics in Space Missions — An Overview of Benefits
      B.K. Wada 285
      8.1.1   Basic Principles 286
      8.1.2   Application of Adaptronics for Space 287
      8.1.3   Summary 289
8.2   Adaptronic Systems in Aeronautics and Space Travel
      C. Boller 289
      8.2.1   Implications and Initiatives 289
      8.2.2   Condition Monitoring 292
      8.2.3   Shape Control 299
      8.2.4   Vibration Damping 301
      8.2.5   Smart Skins and MEMS 307
      8.2.6   Control 308
8.3   Adaptronic Systems in Automobiles
      H. Janocha 309
      8.3.1   Preamble 309
      8.3.2   Networking of Control Devices 310
      8.3.3   Multiple Usage of Actuators 311
      8.3.4   Integration of Control Units 315
      8.3.5   Self–Diagnostics 318
      8.3.6   Implementation of Functions 319
      8.3.7   Outlook 321
8.4   Adaptronic Systems in Machine and Plant Construction
      H. Janocha 323
      8.4.1   Positioning Systems 323
      8.4.2   Damper Systems 325
      8.4.3   Valves and Valve Systems 328
      8.4.4   Brakes and Clutches 332
      8.4.5   Additional Applications 333
8.5   Adaptronics in Civil Engineering Structures
      G. Hirsch 334
      8.5.1   State of the Art for Active Control of Civil Engineering
              Structures 335
      8.5.2   The Second Generation of Active Control 341
      8.5.3   Application of Active Control
              from Practical Engineering Aspects 342
      8.5.4   Results of Experimental and Full–Scale Tests
              (in Japan and the U.S.) 343
      8.5.5   Conclusions 349
8.6   Vibration Control of Ropeway Carriers
      H. Matsuhisa 350

8.6.1 Dynamic Vibration Absorber 351
8.6.2 Passive Gyroscopic Damper........................ 356
8.6.3 Conclusions and Outlook for Future Research 361

9. **Adaptronic Systems in Biology and Medicine** 371
9.1 The Muscle as a Biological Universal Actuator
     W. Nachtigall 371
     9.1.1 Principles of Construction and Function 372
     9.1.2 Analogies to Muscle Function and Fine Structure 374
     9.1.3 Muscle Contraction 374
     9.1.4 Aspects of Muscle Mechanics 377
     9.1.5 Principal Types of Motion Achievable by a Muscle and
            Its Antagonists................................... 380
     9.1.6 Force and Position of Muscular Levers 382
     9.1.7 Cooperation of Unequal Actuators.................. 383
     9.1.8 Muscles as Actuators in Controlled Systems 386
     9.1.9 Control Loops in Biology: Similarities Within Biology
            and Engineering.................................. 387
9.2 Adaptronic Systems in Medicine
     J.-U. Meyer 391
     9.2.1 Adaptive Implants................................. 392
     9.2.2 Adaptive Diagnostic Systems 397
     9.2.3 Conclusions and Outlook 399

10. **Future Perspectives: Opportunities, Risks
    and Requirements in Adaptronics**
    B. Culshaw... 401
    10.1 Introduction 401
    10.2 Adaptronics — the Generic Enabler? 403
    10.3 Engineering Conservatism: The Concealed Risks 406
    10.4 The Adventurous Engineer: Some of the Benefits 407
    10.5 Opportunities in Adaptronics —
         the Adventurous Conservative?........................... 410
    10.6 Adaptronics Research 413
    10.7 Adaptronics and Education 414
    10.8 Some Conclusions 417

11. **About the Authors** 419

Index 428

Introduction
C. Rogers

For decades, researchers and scientists have used nature as the inspiration for designing advanced robotics and mechanical systems. Today, nature is the inspiration for an entirely new class of material systems: intelligent material systems or adaptronic structures. Using biological analogies and nature's ability to adapt a material's structure, morphology, shape, and properties to accommodate its changing environment and its aging process, modern-day alchemists are designing materials that have functions that allow them to change their shape, monitor their own health, control vibrations, and behave like materials they are not.

Adaptronic structures — also known as smart structures, adaptive systems, intelligent material systems and structures — have been defined in the context of many different paradigms; however, two definitions are prevalent. The first definition is based upon a technology paradigm: 'the integration of actuators, sensors, and controls with a material or structural component'. Multifunctional elements form a complete regulator circuit resulting in a novel structure displaying small complexity, low weight, high functional density, as well as economic efficiency. This definition describes the components that comprise an adaptronic material system, but does not state the goal or objective of the system; nor does it provide guidance toward how to create such a material system. The other definition is based upon a science paradigm. It attempts to capture the essence of biologically inspired materials by addressing the goal of the material system: 'material systems that have intelligence and life features integrated in the microstructure of the material system to reduce mass and energy and produce adaptive functionality'. The science paradigm does not define the type of materials to be utilized. It does not even state definitively that there are sensors, actuators, and controls. Instead, it describes a design philosophy. Biological systems are the result of a continuous process of optimization taking place over millennia. Their basic characteristics — efficiency, functionality, precision, durability, and self–repair — continue to fascinate the designers of man–made structures. The goal of modern-day alchemists is to assist the evolutionary process by designing materials and structures for a continuously changing environment, and to give these structures the tools, e.g. the technologies, to evolve to more refined states. These structures must also be able to share information with the users and designers throughout their useful lifetime.

The vision or guiding analogy of adaptronic structures is nearly universal: learn from nature and the living systems and apply the knowledge in a way that will enable man–made artifacts to have the adaptive features of autopoiesis that we see naturally occurring throughout the living world. This leads to the anatomy of an adaptronic system. Actuators or motors behave like muscles; sensors have the architecture and processing features of nerves and sensory organs; communication and computational networks mimic the motor control system (in the brain, memory, etc.) of biological systems. Although the central analogy is to living systems, it must be emphasized that the deployment of adaptronic structures produces artifacts: they are designed by human beings to achieve human objectives and purposes. Because adaptronic structures are intended to fulfill human desires and purposes, it is attractive to stipulate that the system boundary of adaptronic structures be drawn to include also the human users.

Biological structural systems do not distinguish between materials and structures. The development of natural organisms is an integrated process with multiple functions. The result is a 'cost–effective' and durable living structure whose performance matches adaptively the demands of its surroundings. Likewise, the distinction between adaptronic structures and smart materials is irrelevant. Each of the systems requires a hybrid approach to integrating the technologies that synergistically yield life functions and intelligence. The distinction between material systems and structures can only be defined superficially in terms of the scale of their microstructures. In their endeavors, materials engineers and structural designers have learned from nature that the microstructure of any system governs its behavior — regardless of its size.

What kind of life functions can we expect from adaptronic structures? The survival of biological structures depends on nature's ability to balance the economy of construction and maintenance (metabolic cost) with the needed mechanical properties, such as strength, toughness, resistance to impact, etc. This balance is precisely what we aim for when we attempt to specify material and structural requirements in order to arrive at a design that simultaneously satisfies economic viability and mission–oriented performance. A particularly attractive feature of biological systems is their unique ability to diagnose (through a continuously 'distributed' sensor network) and to repair localized damages to their structure: a clearly desirable attribute for man–made structural systems. Nature's systems have a few general attributes that we can aspire to instill in synthetic material systems. Many of nature's systems can change their properties — shape, color, and load paths — to account for damage and to permit repair. They can also manage the graceful retirement of their aged components and subsystems. Engineers and scientists have made tremendous advances in developing a plethora of devices that mimic a few of nature's capabilities. However, little has been accomplished towards the integration of such life functions into engineered systems that are able to learn, grow, survive, and age with grace and simplicity.

Today, researchers are concentrating on aspects of adaptronic structures that may seem rudimentary when compared with the living world, but will have enormous consequences on the engineered systems of the future. Controlling the movement of an arm is a wonderful example of the seemingly effortless task that we all perform each day and that has been so difficult for engineers to mimic. Though such a task seems trivial at first, it requires the interaction of a multitude of scientific and engineering disciplines. A conventional engineering approach results in complicated machinery riddled with breakdowns and prone to defects; in contrast, an adaptronic solution is one that borrows directly from the biological world. In the description of a human arm and the mechanism by which it operates, the concept of co–contraction of antagonist muscles is always present. How would the adaptronic cantilever beam provide this same type of action? Materials that behave much as muscles do during contraction are used in many adaptronic structures and are called induced strain actuators. The principle is that when energy is applied to the actuators, they attempt to contract and work against any load that is applied to them. The actuators are typically bonded to the surface of a structure, or embedded within the material. This means that the artificial muscles must now work against the inherent structural impedance of the component, just as human muscles are parallel to the skeletal structure or bone. However, whereas the arm has discrete joints about which rotation occurs, the adaptronic structure may be a continuum, thereby necessitating a more distributed actuation system. Rotation of a cantilever beam, for example, will occur not by using a joint but by inducing bending of the structure by means of actuators placed on the top and bottom surfaces of the beam, close to the clamped–end and excited 180° out–of–phase with the other.

One of the things we see so clearly when we evaluate the function of the human arm is its multifunctional utility. It can perform 'robotic' tasks, perform vibration control and sense damage. It can repair certain types and levels of damage and it possesses temperature sensing and many tactile capabilities. Multifunctionality is also one of the objectives in providing structural systems with active control features. Multifunctionality, however, can be considered in two forms. One is the ability of the system to perform more than one task simultaneously; the second is a scenario in which the components engineered into a material system are used for different functions at different stages in its life cycle. One of the features that has intrigued the scientific community is the possibility of providing modern material systems with a 'birth–to–retirement health policy'. 'Birth–to–retirement' refers to the capability of using sensors to control the manufacturing process by which an engineered artifact is created to ensure quality, and to optimize the process itself. With this network of sensors, the health of the system can be monitored throughout its lifetime. In addition, with the use of the actuator system, it can also be used to control structural damage or to modify the structure's behavior or performance in order to elude the results of damage. These sensors are then used to monitor the aging process and determine when the artifact should be repaired or graciously retired.

The success of engineering endeavors is generally based on the ability of a structure or component to perform the function intended in an uncertain environment without failure. If the structure or part fails within his expected lifetime, so has the engineer. If it does not fail, success is implied. The current method of engineering design is articulated so beautifully by Prof. Petroski in his book, *To Engineer is Human*. A well–skilled designer develops a worst–case scenario for the environment in which the product will be used, for the quality of the materials that will be utilized in the product, and for the its actual use and abuse during service. This worst–case scenario approach yields a system with redundancies and large factors of safety to ensure survival in an insufficiently known environment. This design approach — also termed 'defense in depth' — has an obvious flaw: it is unable to foresee all possible future contingencies. In spite of using more natural resources than would be required, consuming more energy to produce and maintain, and expending enormous human effort trying to look into the 'crystal ball', real life failure and catastrophes cannot always be averted. There is no actual way to foresee all the uses and abuses of a long–life engineering structure. In contrast, a adaptronic approach, like the natural growth process, will produce structures that respond to their environment and modify themselves to cater for the new demands and requirements. This is the way for the future: replace one–time design and construction with a through–the–lifetime adaptability.

1. Adaptronics
– a Concept for the Development of Adaptive and Multifunctional Structures

D. Neumann

1.1 What Is Adaptronics?

In German–speaking areas 'adaptronics' is the comprehensive generic term for disciplines that, on an international level, are known by names such as 'smart materials', 'smart structures', 'intelligent systems', etc. The technical term adaptronics (Adaptronik) was created by the VDI Technology Centre and was submitted as a proposed name to a body of experts. Within the scope of a workshop, fourteen experts from the fields of research, development and technology management agreed on the introduction of this new technical term, along with the pertinent definition and delimitation. This was the origin of the term 'adaptronics'.

The term adaptronics designates a system (and its development process) wherein all functional elements of a conventional regulator circuit are existent and at least one element is applied in a multifunctional way. The conformity with a regulator circuit guarantees that the structure shows autonomic adaptive characteristics and can thus adapt itself to different conditions. The limits to the classic control circuit, where normally each single function is achieved through a separately built component, are fixed by the use of multifunctional elements (functional materials). These elements are decisive for making such a technically utilizable system less complex.

An adaptronic system thus is characterized by adaptability and multifunctionality. The aim is to combine the greatest possible number of application–specific functions in one single element and, if appropriate, in one specific material (see Fig. 1.1).

1.2 Examples of Adaptronic Systems

A prime example of an adaptronic system is spectacles equipped with photochromic glass. A photochromic glass which, in dependence on the external ambient brightness, darkens or lets move light through in a self–regulating manner, combines all necessary application–specific functions. It not only covers all three elements of a regulator circuit — the sensor, the actuator and the controlling unit — but also covers the shaping and optical functions as further interesting material properties. This example shows that it is

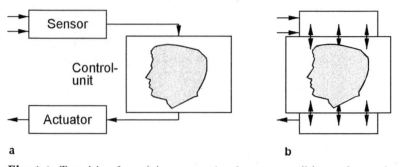

a   b

Fig. 1.1. Transition from **(a)** a conventional system to **(b)** an adaptronic system

possible to successfully combine all functional components of a system into one single element, in this case even into one material; further external components are no longer required. The spectacles' glass represents a complete functional unit.

Examples for adaptronic systems with a more distinct visionary character are window panes whose transparency automatically regulates itself or can be adjusted within seconds by pressing a button; and hydroplanes whose aerodynamic profile adapts itself to the prevailing flight conditions.

Taking an adaptronic shock absorber as an example, Fig.1.2 shows four different levels of creating an adaptronic system. On the basic level it is first necessary to produce materials that have both suitable passive qualities and application–specific functional qualities. Depending on the specific application, passive qualities can be of a mechanical, chemical, thermal, optical or electric nature. For instance, required characteristic features can be resistance to high and/or low temperatures, high mechanical stability, light–transmitting capacity, or good electrical conduction. Functional qualities can be structural changes, changes in the dynamic or static features, or in the chemical, electric, thermal or optical properties. They can, among other things, manifest themselves in a change of transparency depending on the luminous intensity, in a voltage–dependent change in viscosity, or in a temperature–dependent change in dimension or shape.

The example of an adaptronic shock absorber shows how the electrorheological fluid is simultaneously used as a 'classic' absorber fluid and as an actuator (if necessary, additionally as a sensor). This use is made possible by the capacity of such fluids to change their viscosity to a vast extent in less than a second when they are influenced by an electric field.

Functional qualities can, however, only be used in terms of adaptronics if there is success in combining the specific release phenomena with the respective desired functions. What is therefore required in the conception of multifunctional elements (level 2) is the release and specific use of the material–inherent options. For this purpose it is necessary to make use of release phenomena of a physical, chemical or biological nature on material in such a way that, as necessary, several effects can be combined by taking well–directed action. For example, the application of electrorheological fluids in an adaptronic shock absorber requires the production of an electric field, as well as the recording of a motion–dependent, variable intensity of current (i.e., use of the sensor effect). Hence, the multifunctional element does not exclusively consist of the electrorheological fluid but necessarily also of an electric voltage and field–producing electrodes.

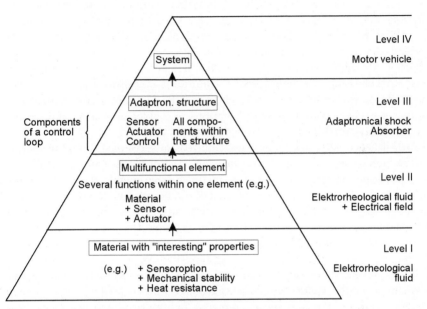

Fig. 1.2. Adaptronic: link between material and system

At the structural level, multifunctional elements must be supplemented to form a complete regulator circuit, always aiming at building up a structure that is marked by minor complexity, low weight, high functional density, and economic efficiency. The successful achievement of this objective will normally depend on the degree to which the functional density is already in existence within the individual elements forming the structural components. In an ideal case — as in case of photochromic glass — all application–specific functions exist in one single element. The outcome will, however, not always be successful. For instance, the multifunctional element existing for the construction of an adaptronic shock absorber must be supplemented by a controlling mechanism, as well as by the structural components required to produce the electric field.

The system level — in the present example the entire motorvehicle — calls for the need to conceptualize during the creation of the adaptronic structure. For instance, the structural shape and damping characteristic of a shock absorber must harmonize with the overall design of a moving gear. Here again, the aim is to optimize the functionality of the entire system.

1.3 Multifunctional Elements

Functional materials constitute the essential basis of all adaptronic systems. The made–to–measure production of functional materials, wherein several functions are interlinked at a molecular level, is therefore of special importance. The more application–specific functions are combined in one single element, the bigger is the advantage in terms of an adaptronic system optimization. Multifunctionality can, however, not be a characteristic feature of an isolated element, but should always manifest itself by meeting user–specific requirements within a system interrelationships. Thus the same element can produce a decisive compression of functions in a given case (A), while it can be completely worthless in a given case (B).

Multifunctionality is by no means required to be coupled to highly sophisticated functional materials. Sometimes amazingly simple concepts lead to a problem–adjusted solution. It is, for instance, conceivable that a gas–filled balloon regulates the volume flow in a fluid flow tube in a temperature–dependent manner. The gas expands with rising temperature, whereupon the balloon reduces the uncovered tubular cross–section. If the temperature decreases, the volume flow is increased along with a smaller balloon cross–section.

This example shows that no limits are set to the user's creativity. Mechanically simple solutions are often advantageous compared with high–technology concepts: they are not only more often reasonably priced, but also frequently marked out by greater functional safety. Made–to–measure solutions, however, can in most cases not fulfill their function without high technology concepts of material scientists.

Materials represent the essential basis for all multifunctional effects. The conception of multifunctional elements is therefore mainly based on the made–to–measure production of functional materials, wherein several functions are interlinked at a molecular level. However, the fact that this is not sufficient in all cases is clearly shown by taking adaptronic shock absorbers as an example, because some effects can only be produced if several materials are combined in suitable interconnected layers or other compounds.

Functional materials, which are characterized by a high potential of functional and application options, are amongst others: shape memory elements; bimetals; electrorheological, magnetorheological, thixotropic, and rheopex fluids; piezoelectric elements; electrostrictors; magnetostrictors; chemochromic, electrochromic, hydrochromic, photochromic, and thermochromic elements ; and functional gels.

1.4 Fields of Technology and Application

The foregoing explanations show that a basis for adaptronic structures is created in numerous different disciplines of science. The range of applications covers various physical, but also chemical and biological technologies (see Fig. 1.3). What prove to be especially user–relevant here are the often interdisciplinary interactions, such as the physical reaction to a specific chemical stimulus or the reaction of micro–organisms to a modification of physical and/or chemical environmental parameters. Scientific disciplines, such as biophysics, biochemistry, and physical or biophysical chemistry, are of special importance here.

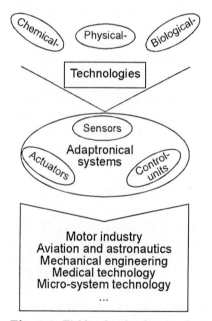

Fig. 1.3. Fields of technology and application

The scope of the application of adaptronic structures or systems can be restricted as the spectrum of influential scientific disciplines. Almost each scientific field covers applications, whose technical benefit and business management utility can be improved by realizing adaptronic concepts. While the need for efficient multifunctional materials certainly originates in the high–technology area, the scope of application is by no means exclusively confined to this field. For example, multifunctional adjusting elements of shape memory alloys are successfully applied for the automatic control of ventilation flaps in greenhouses.

However, even products resulting from highly specialized materials are only partially needed for the realization of efficient adaptronic concepts. Simple adaptive systems, with a minimal number of elements in motion, are of special importance in a surrounding field, where the protection against short-falls is a decisive factor and where little or no well–trained staff are available for the removal of technically complex problems. The broad range of applications covers a number of areas where adaptronic concepts have been intensively pursued and partially have already been translated into concrete action. The specific interest shown in a particular line of business is a result of special security and performance requirements. In this context the fields of aviation and astronautics hold a key position, as both aforementioned aspects are of special importance here. These fields of technology have always been marked by a generally high level of innovation, particularly as a result of their significant financial resources.

1.5 Historical Review

Adaptronics as an overall concept for the development of adaptronic structures and systems is still a young discipline, which was only able to establish itself a few years ago. On the other hand, the research in the fields of multifunctional materials and multifunctional elements, which are the basic elements of the elements of adaptronics, started much earlier. The origins of adaptronics — under a different name — go back to the early 1980s. Early progress came from the arms research sector, especially from various air forces.

In the early eighties, government–sponsored efforts were made in the United States to interlink functions, for instance integrate headlights in the outside plating of combat aircraft. This type of integration not only aimed at the optimization of functions but also at the reduction of weight. This 'Smart Skin' Program lasted nearly one decade, up to the early 1990s. By the mid–eighties, the US airforce likewise had started further adaptronic–oriented programs, which concentrated on the integration of sensor networks in combat aircraft for system supervisor programs. Both the research and application aspects have considerably gained in importance in the United States, although the main fields of application are still aviation and space technology.

In Japan, the driving force behind initial developments was not the military but mainly the civil sector. At first, however, these activities were less concentrated on the conception of systems and rather more on a well–structured and broadly conceived development of multifunctional materials. In 1985, the New Glass Forum came into existence as a program of Japan's Ministry of International Trade and Industry (MITI), the tasks of which included the development of sensor materials with different evaluation options — for example by changing the optical, mechanical and/or chemical conduction properties of the materials. In 1987, the New Glass Forum was dismissed from MITI and a New Glass Association was established in its place. This Association was joined by more than 200 enterprises from different sectors of industry and trade. From July 1987 through November 1989, far–reaching interdisciplinary discussions and harmonizations among scientists working in numerous different areas of reserach took place under the leadership of the state–supported Council for Aeronautics, Electronics and Other Advanced Sciences. The participants came from various sectors, such as medicine, pharmacy, engineering sciences, physics, biology and chemistry, as well as electronics and computer science. The general aim was to formulate and adopt a program for the development of made–to–measure functional materials. In 1989, a comprehensive report was delivered to the Science and Technology Agency (STA), which formed the basis for further promotional activities. Although, in Japan, the expenditures for research activities are largely borne by private enterprises, governmental institutions such as the MITI or the STA exert a significant coordinating influence, pointing the way ahead, despite their comparatively small funds for promotional measures. Within the scope of the 'Basic Technologies for Future Industries' project organized by MITI, the partial project, named 'High Performance Materials', was initiated in 1989 and was carried out up to 1996.

The first German activities in terms of an integrated approach to adaptronics were initiated in the late 1980s in the areas of aviation and space technology. The main topic within the scope of the experiments, which were initially almost exclusively carried out by the big research institutes and large groups of companies, was active vibration suppression. The interest and activities of public institutions started in 1990. The German Federal Minister for Research and Technology entrusted the VDI Technology Centre in Düsseldorf with the coordination of this topic, and initial discussions and harmonization planning took place in 1991. In autumn of that year, the VDI Technology Centre was up and running, and soon formed an expert workshop in which fourteen reputable specialists from the fields of research and development participated. Within the scope of this event, the term 'adaptronics' was introduced and clearly defined within the German language.

In 1992, the first government funded projects were incorporated by the German Ministry for Research and Technology in its 'Materials Research' program. These projects initially concentrated on the improvement of pure material functions. However, it quickly proved necessary to enlarge the basic area of materials and to develop integrated concepts for multifunctional adaptive structures or systems in terms of adaptronics. In this context the objective was the application–orientated optimization of functional materials and their functional integration in a system.

In the spring of 1993 the Ministry for Research and Technology published a study under the title Technologies of the 21st Century, wherein those technologies and trends were described which offered the best chance for maintaining (or even increasing) the competitiveness of German industry. In this study the field of adaptronics was emphasized as one of eight disciplines that were seen to help ensure economic growth parallel to the protection of existing resources. In the early 1994 the first system– and application–oriented projects were started, all focusing on the damping of vibrations in measurement robots.

In November 1994 a further expert workshop took place in Düsseldorf, on the occasion of which some of the main subjects within the broad and interdisciplinary field of adaptronics were thoroughly analysed. In the experts' opinion during that workshop, the greatest application potential could be found in vibration and noise damping in the automobile and mechanical industries, as well as in the fields of aviation and space technology.

The aim for the future, apart from the promotion of individual pilot projects, is the further state–supported advancement of specific areas of adaptronics, which are marked by significant high–technology and application potential.

Further Reading

1. Culshaw, B., Gardiner, P.T.; McDonach, A.: *Proceedings, First European Conference on Smart Structures and Materials.* IOP Publishing Ltd., Bristol, GB (1992)
2. Martin, W.E.; Drechsler, K.: Smart Materials and Structures — Present State and Future Trends. Technische Niederschrift der Messerschmidt–Bölkow–Blohm GmbH, München (1990).
3. Neumann, D.: *Bausteine "Intelligenter" Technik von Morgen — Funktionswerkstoffe in der Adaptronik.* Wissenschaftliche Buchgesellschaft, Darmstadt (1995).
4. Newnham, R.E.: Smart, Very Smart and Intelligent Materials. *MRS Bulletin,* Volume XVIII, No.4, April (1993).
5. Rogers, C.A.: Intelligent Material Systems — The Dawn of a New Materials Age. Journ. of Intelligent Material Systems and Structures, Vol. 4, Technomic Publishing Company, Lancaster, USA (1993).
6. Science and Technology Agency (Government of Japan): The Concept of Intelligent Materials and the Guidelines on R&D Promotion. Tokyo, Japan (1989).
7. Takagi, T.: A Concept of Intelligent Materials. Journ. of Intelligent Material Systems and Structures, Vol. 1, Technomic Publishing Company, Lancaster, USA (1990).
8. Thomson, B.S.; Gandhi, M.V.: Smart Materials and Structures Technologies. An intelligence report, Technomic Publishing Company, Lancaster, USA (1990).

2. Concepts of Adaptronic Structures

C.A. Rogers and V. Giurgiutiu

2.1 What Are Adaptronic Structures?

Adaptronic structures, sometimes referred to as smart materials or structures, are defined in the literature in the context of many different paradigms; however, two are prevalent. The first definition is based upon a *technology paradigm* : 'the integration of actuators, sensors, and controls with a material or structural component'. Multifunctional elements form a complete regulator circuit resulting in a novel structure displaying minor complexity, low weight, high functional density, as well as economic efficiency. This definition describes the components that comprise an adaptronic material system, but does not state a goal or objective of the system; nor does it provide guidance toward how to create such a material system. The definition based upon a *science paradigm* attempts to capture the essence of biologically inspired materials by addressing the goal of the material system: 'material systems with intelligence and life features integrated in the microstructure of the material system to reduce mass and energy and produce adaptive functionality'. It is important to note that the science paradigm does not define the type of materials to be utilized. It does not even state definitively that there are sensors, actuators, and controls, but instead describes a philosophy of design. Biological structural systems are the result of a continuous process of optimization taking place over millennia. Their basic characteristics of efficiency, functionality, precision, self–repair, and durability continue to fascinate designers of engineering structures today. The goal of the modern day alchemist is to assist the evolutionary process by designing structures while cognizant of the unpredictable environment in which all engineered systems are used, and to give these structures the tools, e.g. the technologies, to evolve to more refined states. The structures must then share this information with the users and designers throughout the life of the system.

The vision or guiding analogy of adaptronic structures is nearly universal: that of learning from nature and living systems and applying that knowledge in such a way as to enable man–made artifacts to have the adaptive features of autopoiesis we see throughout nature. This leads to the description of the anatomy of an adaptronic material system: actuators or motors that behave like muscles; sensors that have the architecture and processing features of nerves and memory; and communication and computational networks that represent the motor control system (brain, etc.) of biological systems. Although the central analogy is to living systems, it must be emphasized that the deployment of adaptronic structures produces artifacts: they are designed by human beings to achieve human ends and purposes. This contrasts with living creatures, which exist only for themselves. Because adaptronic structures are intended to fulfill human desires and purposes, it is useful and desirable to stipulate that the system boundary of adaptronic structures be drawn to include the human user.

Biological structural systems do not distinguish between materials and structures. The design and development of natural organisms is an integrated process in which component functions are multiple, and result in a cost–effective and durable structure whose performance matches the demands brought upon the living system. Likewise, the distinction between adaptronic structures and intelligent structures is irrelevant. Each of the systems require a hybrid approach to integrating the technologies that synergistically yield life functions and intelligence. The distinction between material systems and structures can only be defined superficially in terms of the scale of their micro structures. The lessons learned from nature that guide material engineers and designers in their endeavors is that the microstructure of any system governs its behaviour — regardless of its size.

Adaptronic structures must, as their basic premise, be designed for a given purpose; and, by the transduction of energy, must be able to modify its behavior to create an envelope of utility. As an example, a ladder that is overloaded could use electrical energy to stiffen or strengthen it while alerting the user that the ladder is overloaded. The overload response should also be based upon the actual 'life experience' of the ladder to account for aging or a damaged rung; therefore, the ladder must determine its current state–of–health and use this information as the metric for when the ladder has been overloaded. At some point in time, the ladder will then announce its retirement, as it can no longer perform even minimal tasks.

J. E. Gordon's book, *Structures: or, Why Things Don't Fall Down* [1], contains a chapter entitled 'The Philosophy of Design — or the Shape, the Weight and the Cost', which begins with a quote of John Sheldon (1584–1654): 'Philosophy is nothing but discretion'. The science paradigm/definition of adaptronic structures is a philosophy of design, not a technology. It is true that the concept of adaptronic structures does not define a discipline, as it is 'engineering'. Engineering is 'the art or science of making practical application of the knowledge of pure sciences'. The definition of structural engineering that is given in every issue of *The Structural Engineer* adds that it is the 'science and art of designing and making, with economy and elegance'. As mentioned in the beginning of this essay, the objective is to 'instill intelligence in the microstructure to reduce the mass and energy of the system to perform adaptive functions'. If we follow nature's example, grace, beauty, and art will be the reward. The implications of the definition and discussion given above are simply thus.

— We shall learn from nature the method by which natural systems adapt to their environment and the aging process, and how the natural systems of muscles, nerves, and motor control are architected to produce adaptive life functions in engineered artifacts.
— The material systems that can be considered the host, e.g. the skeletal form, of the adaptronic structures can range from biomaterials, to polymers, to electronic materials, to structural composites, to large civil engineering constructs. This is a concept that will manifest new structural materials for automobiles and aircraft, as well as guide the design of new artificial organs or limbs.
— The utility of adaptronic structures is not simply to perform functions programmed *a priori* by a designer, but to learn what the appropriate responses are to a wide range of situations.
— Adaptronic structures will have not only adaptive features to interact with the environment, but also the means with which to convey this information to the designer and user.
— This concept is not technology–limited, but knowledge–limited. The need is firmly to develop the methodology by which material systems, components, and structures will be designed to reduce mass and energy needs by incorporating adaptive features and intelligence. New muscle–like materials or nerve–like sensors will enhance the concept and the capabilities, but this concept can be commercialized today with existing materials and devices.

2.2 What Are They Good for?

How does this design philosophy manifest itself? What are adaptronic structures good for? The applications that adaptronic structures will have an impact upon are numerous and varied. They range from speakers and noise–control devices that are a part of curtains or walls, to quiet commuter aircraft, to windows that can change their reflectivity to control room temperature or light level, or even close automatically when it begins to rain. Biomedical applications, similarly as diverse, include new artificial organs such as a pancreas that continuously monitors blood–sugar level and releases the appropriate amount of insulin in real–time, to artificial limbs that not only provide basic motor control but tactile responses as well.

An aircraft wing with no flaps, ailerons, or articulating parts, that instead possesses the ability to change its shape and camber is one of the applications currently being pursued by researchers in academia and industry. Shape control of propellers, helicopter rotor blades, and aircraft wings can provide vehicles of flight and marine adventure with increased efficiency, maneuverability, reduced vibration and noise, and longer life by reducing vibrations that can cause structural fatigue.

We are already seeing the results of research devoted to specific application areas become reality. Researchers at Penn State have developed solutions for the crippled Hubble telescope by using electro–active ceramic materials to adaptively change the shape of the mirrored surface. Conventional precision antenna and telescope surfaces have always required thick–stiff structures to ensure dimensional stability and accuracy. By incorporating sensors and actuators into thin flexible surfaces with active control algorithms, we can now reduce the weight and volume (as compared to the passive counterpart structures) by using energy (typically electrical energy) to assure that a desired shape is maintained. In this case, you might say, we are using brains instead of brawn.

The need for large, precise space structures and the demand for low volume and low mass have provided the motivation for researchers at MIT and JPL (Jet Propulsion Laboratory) to investigate the potential of using adaptronic structures for controlling the vibrations of large flexible structures, while maintaining accuracy to the fraction of a wavelength of light for football field–sized structures. Researchers have developed space truss elements that can be actively controlled to change their length and behave much like a mechanical muscle. By inserting these elements into a truss, researchers are also able to control the shape, vibration, and precision of the structure and, in some cases, monitor the health of the structure and identify components that are damaged.

The 'design' of plants and animals and of the traditional artificial artifacts did not just happen. As a rule, both the shape and the materials of any structure have evolved over a long period of time in a competitive world, and they represent an optimization with regard to the loads that the structure has to carry and with regard to the financial or metabolic cost. We would like to achieve this same sort of optimization in modern technology. A good example of the ability of living organisms to adapt/evolve to changing loading conditions is bone. Living bone continually undergoes processes by which it remodels itself to accommodate the changes in its loading. The time scale on the process may be of the order of months, but can begin within minutes of an impetus.

Today, researchers are trying to mimic the adaptive load–changing characteristics of bone to develop composite materials that are fatigue– and damage–tolerant. Research has demonstrated that induced strain actuators such as shape memory alloy fibers and piezoelectric ceramics can actively reduce strain concentrations from transverse cracks, near holes and notches, and other forms of damage or engineered strain concentrations.

Pictures of the Aloha Airlines aircraft that landed safely in Hawaii after an inflight catastrophe, which made the disabled aircraft look like it had lost its top, educated many and reminded all of us of the mysteries that surround fatigue failures. Fatigue is generally a quiet instrument of failure in which a long life of utility is abruptly ended by means of catastrophic damage. The signs of fatigue are often difficult to detect. The life experiences, which include every joule of energy that the structure has been forced to accept, are never forgotten, and the end is never graceful. However, experiments in which piezoelectric actuators were used to resist the mechanical strains near locations of known high–strain concentration have extended the fatigue life of some mechanical components by over an order of magnitude. Not only can actuators be used to 'muscle' the mechanical strains on a structure, but the sensors can determine the state of health in real–time, look for signs of weakened material or damaged structure, and then redirect the loads around these portions of the structure until remedial action and repair can occur.

The typical engineering solution to designing structures with strain con-centrations, such as a hole, is either to add mass around the hole to provide the same strength as the rest of the structure, or to design the structure based upon the weak link — the hole — and make the rest of the structure the same thickness, for example. This results in more mass than is actually needed. The concept of using energy to support load, to add stiffness, or to increase strength will allow mass to be replaced with energy.

Using energy to increase stiffness is not always the most 'natural' solution. Nature uses strain energy as a damage–control mechanism by designing sys-tems with large strain capabilities as compared with our modern constructs of steel, concrete, and graphite. Control of structural impedance is perhaps the most fundamental and powerful concept of adaptronic structures. By modifying the structural impedance of a system, we can change its vibra-tion and acoustic behavior as well as change its resistance to damage. Large civil engineering structures can control the transmission of motion and the flow of energy into the structure by adaptively controlling the impedance of the structure at its base. Magneto– and electrorheological fluids can be used to change and control the friction of thrust bearings used to support large buildings in areas of high seismic activity. Trusses that use elements that can change their length in a controlled fashion, known as variable geometry trusses (VGTs), can be used to change the impedance of portions of a large structure and thereby control vibrations very efficiently.

Perhaps one of the most mature application areas of adaptronic structures is active structural acoustic control (ASAC). The objective of ASAC is to reduce the sound radiated from a vibrating structure, be it sound inside an aircraft fuselage that is shaken by engines, or the acoustic signature of a submarine. Adaptronic structures, for the first time, have allowed researchers to demonstrate control of low–frequency noise as a result of the introduction of the artificial muscles that we have referred to as induced strain actuators. The typical approach to controlling structural vibrations has been to use shakers that apply a transverse force to the structure. The magnitude of the force is therefore related to the mass of the shaker. If the shaker is very light, the structure very heavy, and the frequency very low, the result is a shaker that only shakes itself and not the structure. However, induced strain actuators, configured like muscles, use the in–plane structural impedance to react against. Hence, they possess the ability to exert forces on a structure at low frequencies while requiring little mass as compared to conventional shakers, often by orders of magnitude, to perform effective control.

ASAC can fundamentally be performed using two approaches. The first is simply to make the structure stop vibrating completely, which will obviously reduce the sound radiation to zero. This is the 'brute force' approach. The adaptronic approach is to control only the radiating modes. Being that not all vibrational modes radiate sound, and our objective is to reduce the mass and energy needs of the system, then the obvious solution is to sense the structural modes that are radiating the noise and use the actuators that are distributed throughout the structure to control only the highly efficient radiating modes. The efficiency of the solution will be based upon the engineer's insight into the fundamental physical phenomena with which the material system is to interact, and the adaptive capabilities of the adaptronic material system.

Perhaps an application that will prove more useful to the scientific community than any other is the use of adaptronic structures in studying complex physical phenomena and fundamental physics. Material systems with capabilities to sense their environments, store detailed information about the state of the material with refined temporal and spatial resolution, and change their apparent material properties can provide the scientific community with the opportunity to perform experimental parametric studies, which can be useful in extracting detailed information about cause and effect, or use neural networks to aid in identifying features or parameters that are most significant in altering the physical phenomena of interest. Opportunities abound for determining the first principles of many complex systems and the physics that govern them, such as those related to noise generated by turbulent boundary layers, or the process of damage of composite materials. The applications that adaptronic structures may influence are numerous, and the commercial implications enormous; but there may be no application more important or more valuable than their utility in releasing the fundamental secrets of science that have been alluded to us thus far.

2.3 What Are They Made of?

Adaptronic structures are hybrid composite systems, whether fluids, gels, or solids. Instilling intelligence into natural systems requires a cornucopia of materials to synergistically contribute to the creation of life functions. Man–made adaptronic systems require no less. Science has produced a palette of most remarkable materials, from shape memory alloys to piezoelectric ceramics and polymers. They can be successfully used to create adaptronic structural systems. A brief description of some of the technologies, materials, devices and concepts follows. Together they may form the next revolutionary step in human development — the adaptive material systems and structures revolution.

2.3.1 Artificial Muscles — Actuators

Zuk and Clark, in the book, *Kinetic Architecture* [2], write, 'Life itself is motion, from the single cell to the most complex organism, man... It is these attributes of motion, mobility, of change, of adaptation that place living things on a higher plateau of evolution than static forms. Indeed, survival of the living species depends on their kinetic abilities: to nourish themselves, to heal themselves, to reproduce themselves, to adapt to changing needs and environments...' It is fitting, therefore, for us to begin with a description of the technologies that allow for the manifestation of '...motion, mobility, and change': the actuators.

Materials that allow an intelligent or smart structure to adapt to its environment are known as actuators. These materials have the ability to change the shape, stiffness, position, natural frequency, damping, friction, fluid flow rate, and other mechanical characteristics of adaptronic structures in response to changes in temperature, electric field, or magnetic field. The most common actuator materials are shape memory alloys, piezoelectric materials, magnetostrictive materials, electrorheological fluids, and magnetorheological fluids.

Shape memory alloys (SMA) undergo solid–to–solid martensitic phase transformations that allow them to exhibit large, recoverable strains [3]. Strains of up to 8% can be reversed by heating the SMA above its phase transformation temperature — a temperature that can be altered by changing the composition of the alloy (see Sect. 6.4). Nickel–titanium alloys, also known as Nitinol (Ni for nickel, Ti for titanium, and NOL for Naval Ordnance Lab), are high–performance shape memory alloy actuator materials exhibiting high corrosion resistance, large recovery strains, and excellent fatigue behavior [4]. Copper–based SMA, such as copper–aluminum–nickel and copper–zinc–aluminum, can also be used for actuation in adaptronic structures. They are roughly one–tenth of the cost of Nitinol, but have a maximum recovery strain of only 4%.

Original applications of shape memory alloys included weldless pipe couplings and blind fasteners (where the backside of a structure is inaccessible). Only recently have SMAs been used as actuators. Japanese engineers began using Nitinol in micromanipulators and robotic actuators that mimic the smooth motions of human muscles [5]. The controlled shape recovery force allows these devices to delicately grasp fragile objects, such as paper cups filled with water. Other applications for Nitinol actuators include engine mounts and suspension systems that control vibration.

Nitinol wires have been embedded in composite materials to form adaptive composite structures with many similarities to muscles. In addition to applying forces or changing the shape of the structure, the Nitinol wires can be used to actively change the modal characteristics of the composite by changing the stiffness or state of stress in the structure, thereby, changing its natural frequency. Photoelastic damage–control experiments have shown that embedded Nitinol actuators can also be used to reduce stress concentrations in notched tensile coupons by creating localized compressive stresses. Polymeric composites with embedded Nitinol wire have been shown [6] to display large bending deformation when activated (Fig. 2.1).

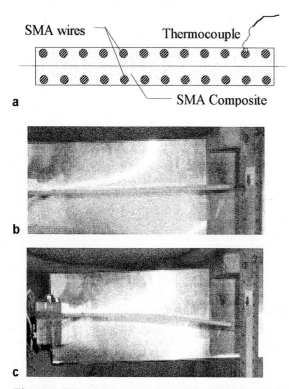

Fig. 2.1. Polymeric composites with embedded Nitinol wires displaying large bending deformation when activated: **(a)** composite beam cross–section; **(b)** beam configuration before activation; **(c)** deflected beam after SMA activation

Like shape memory alloys, piezoelectric actuators can also exert mechanical forces in response to an applied voltage. Rather than undergoing a phase transformation, piezoelectric materials change shape when their electrical dipoles spontaneously align in electric fields, causing deformation of the crystal structure (see Sect. 6.2). Lead zirconate titanate (PZT) [7] and other piezoelectric materials are useful when high–precision or high–speed actuation is necessary. Examples of systems using piezoelectric actuators are:

optical tracking devices, magnetic heads, adaptive optical systems, micropositioners for robots, ink–jet printers, and hifi speakers. Recent research has focused on using PZT actuators with sophisticated control systems in adaptronic structures to perform active acoustic attenuation, active structural damping, and active damage control.

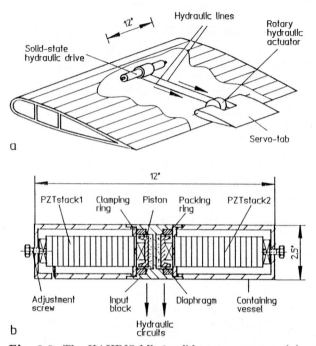

Fig. 2.2. The HAHDIS Mk 2 solid–state actuator: **(a)** proposed implementation for actuating a rotor blade active tab; **(b)** internal construction showing the two hydraulic chambers and the piezoelectric stacks [9]

Magnetostrictive actuator materials are similar to piezoelectric materials, but respond to magnetic, rather than electric, fields. When placed in a magnetic field, the magnetic domains in a magnetostrictor rotate until they are aligned with the field, resulting in expansion of the material. Terfenol–D, which contains the rare earth element terbium, expands by more than 1400 microstrain due to alignment of its magnetic domains [8]. This material has been used in low–frequency, high–power sonar transducers; high–force linear motors; high–torque, low–speed rotating motors; and hydraulic actuators. Terfenol–D is currently being investigated for use in active vibration damping systems (see Sect. 6.3).

Solid–state actuation has found a large field of application in the aerospace industry. The aero–servo–elastic control of vibrations and flutter with the use of solid–state actuated flaps, tabs, vanes, etc. for helicopter rotor blades and aircraft wings is currently being experimented. The challenge in these applications is to realize a capable amplification of the small–amplitude displacement produced by the active material driver of the solid–state actuator.

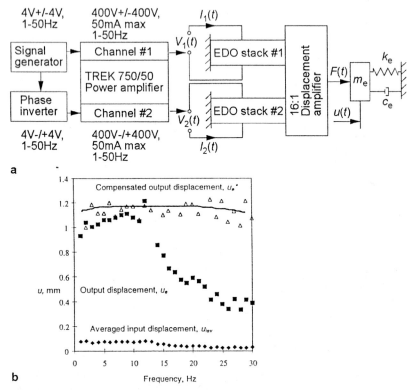

Fig. 2.3. Frequency–response testing of the solid–state proof–of–concept demonstrator HAHDIS Mk 1: **(a)** schematics of the experimental apparatus; **(b)** frequency–response curve [9]

Several concepts have been explored, starting with simple mechanical amplification based on the lever principle and ending with elaborate designs utilizing various non–linear effects of structural deformation. Figures 2.2 and 2.3 illustrate the HAHDIS (Hydraulically–Amplified High–Displacement Induced–Strain) concept based on hydrostatic amplification [9]. Two counteracting active material stacks are placed in a boxer configuration and activate large area diaphragms. Through short hydraulic lines, the volumetric fluid displacements generated by the stacks action operate the opposing chambers of a rotary hydraulic actuator. Since the stack in–and–out operation is directly controlled through electric currents under computer guidance, the servo–tab motion necessary for flutter and vibration control can be positively produced. Experimental validation of this principle was achieved with a proof–of–concept demonstrator, HAHDIS Mk 1, and frequency response curves were obtained over the range 0–30 Hz (Fig. 2.3).

Another displacement amplification principle that has shown remarkable results for angular motion is the LARIS (Large–Amplitude Rotary Induced–Strain) concept [10].

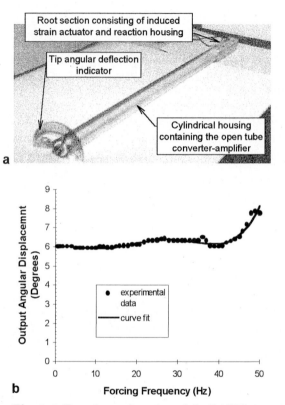

Fig. 2.4. Experimental testing of the LARIS concept: **(a)** the LARIS Mk 2 proof–of–concept demonstrator; **(b)** frequency–response curves up to 50 Hz [11]

The warping–torsion coupling of an open–tube structure is utilized to transform and amplify the small axial displacement produced by the induced–strain actuator and to obtain a sizable angular deflection at the device output The LARIS Mk 2 proof–of–concept demonstrator [11] showed almost flat response up to 50 Hz (Fig. 2.4). This proves the advantages of the proposed concept and its capabilities for implementation in real–life applications.

Fluids can also act as actuators in adaptronic structures. Electrorheological (ER) and magnetorheological (MR) fluids experience reversible changes in rheological properties (viscosity, plasticity, and elasticity) when subjected to electric and magnetic fields, respectively. These fluids contain micron–sized particles that form chains when placed in an electric or magnetic field, resulting in increases in apparent viscosity of up to several orders of magnitude. These fluids can be used to make simple hydraulic valves that contain no moving parts. Other applications include tunable dampers, vibration isolation systems, clutches, brakes and other frictional devices (see Sects. 6.5 and 6.6).

2.3.2 Artificial Nerves — Sensors

One of the critical functions instilled in adaptronic systems and structures is sensing. Damage control, vibration damping, acoustic attenuation, and intelligent processing all require accurate information provided by sensors describing the state of the material system or structure. Sensing capabilities can be given to structures by externally attaching sensors or by incorporating them within the structure during manufacturing. Some of the sensing materials used for this purpose include optical fibers, piezoelectric materials, and 'tagging' particles.

Optical fibers can be used either extrinsically or intrinsically in sensing (see Sect. 7.2). When used extrinsically, the optical fiber does not act as a sensor but merely transmits light. An example of an extrinsic fiber optic sensor is a position sensor, which uses the fiber to collect light from a source. Breaks in the light beam are used to accurately determine the position of a workpiece in robotics applications. Security systems also use this technique to detect intruders. Intrinsic sensing relies on changes in the light transmission characteristics of the optical fiber. The use of optical fibers to perform intrinsic sensing in smart structures was first investigated in 1979 at NASA Langley Research Center. In this early research, optical fibers were used to measure strain in low–temperature composite materials.

Piezoelectric materials can be used for structural health monitoring and Non–destructive Evaluation (NDE). The method is based on the intimate coupling between the electrical impedance of a PZT transducer affixed to the structure and the driving–point mechanical impedance of the structure at the point of contact [12]. Exploratory demonstrations of the electromechanical impedance technique have shown it to be very promising.

Two examples are shown in Fig. 2.5 The vertical–tail–to–fuselage bolted junction of a Piper Model 601P airplane was instrumented with PZT sensor–actuators [13]. In order to monitor the integrity of the connection, PZT crystals were mounted on the fuselage side of the vertical–tail–support brackets, each within one inch of the two securing bolts. A quarter–scale model of a bridge joint was instrumented with bonded PZT sensor–actuators [14]. In both cases, damage was simulated by loosening one or more bolts in the junction area. Incipient local damage was detected from the scalar values of the damage metric.

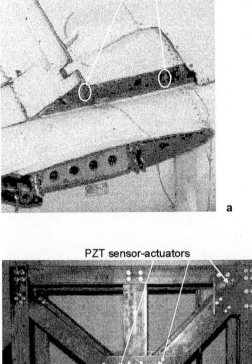

Fig. 2.5. Typical applications of the electromechanical impedance technique: **(a)** light aircraft tail junction [13]. **(b)** 1/4–scale bridge structure [14]

Graphs of the real electrical admittance of a PZT sensor–actuator are shown in Fig. 2.6. Figure 2.6a shows a typical impedance response curve *vs* frequency for near–field (local bolt) damage identification. Figure 2.6b shows the evolution of the damage index. Note that the method is not only capable of identifying damage, but can also distinguish the location of the damage.

The early work in 'smart skins', as it was called then, at NASA Langley provided a catalyst for the development of a variety of fiber optic sensors. Interferometric, refractometric, blackbody, evanescent, modal domain, and time–domain sensors were investigated for use in nondestructive materials evaluation, in–service structural health monitoring, damage detection and evaluation, and composite cure monitoring. Researchers examined experimental objects using fiber optic sensors as magnetic field sensors, deformation and vibration sensors, accelerometers, and sensors in propulsion systems. Resistance to adverse environments and immunity to noise from electrical or magnetic disturbances are among the many advantages of fiber optic sensors.

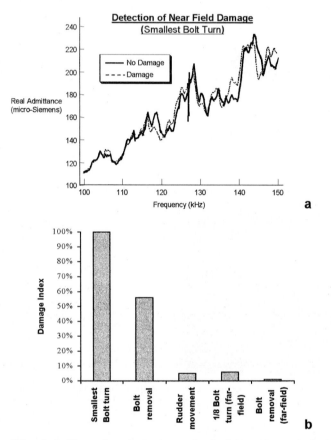

Fig. 2.6. Examples of electromechanical impedance technique: **(a)** detection of near field damage; **(b)** evolution of the damage index [13]

Piezoelectric materials have also found widespread use as sensors in adaptronic structures [15]. Piezoelectric ceramics and polymers produce measurable electrical charges in response to mechanical stress. Because of the brittle nature of ceramics, piezoelectric polymers [16], such as polyvinylidene fluoride (PVDF), are often used for sensing. PVDF can be formed in thin films and bonded to many surfaces. Uniaxial films, which are electrically poled in one direction, can measure stresses along one axis, while biaxial films can measure stresses in a plane.

The sensitivity of PVDF films to pressure changes has been utilized in tactile sensors that can read the Braille alphabet and distinguish different grades of sandpaper. Tactile sensors with ultra–thin (200–300–micron) PVDF films have been proposed for use in robotics. A skin–like sensor that replicates the temperature and pressure sensing capabilities of human skin can be used in different modes to detect edges, corners, and geometric features or to distinguish between different grades of fabric. The pyroelectric effect, which allows piezoelectric polymers to sense temperature, also limits their use to lower temperature ranges.

Piezoelectric composite materials have been developed to overcome the brittleness of piezoelectric ceramics and the temperature limitations of piezoelectric polymers. Flexible composite sensors containing piezoelectric ceramic rods in a polymer–based matrix [17] have been widely used in hydrophones and medical ultrasonic transducers, giving improved sensitivity and mechanical performance over the original piezoelectric ceramics. Polymers containing piezoelectric powders have also been investigated for use as sensing materials. Piezoelectric paint and coatings are being developed that can be applied to complex shapes to provide information about the state of stress and health of the underlying structure.

Passive and active tagging are sensing techniques that involve adding 'tagging' particles to materials. Embedded piezoelectric, magnetostrictive, electrostrictive, or magnetic particles can be used to provide inherent information about the in–process or in–service state of adhesives and polymers [18]. Tagging offers the advantage of distributed in–situ sensing, which is not possible with many types of sensors.

Passive techniques involve sensing the distribution of the particles. An example of a passive technique is adding magnetic particles to an adhesive and using an eddy current probe to detect voids in a tagged adhesive bond. Active tagging involves exciting the sensing particles and measuring the response of the host material (Fig. 2.7).

Applying an alternating magnetic field to a polymer tagged with magnetic particles and measuring the resulting force would be an example of active tagging. Applications of passive and active tagging techniques include characterization of adhesive bonds, cure monitoring, intelligent processing, nondestructive materials evaluation, damage detection, and in–service health monitoring [19].

2.3.3 Intelligence — Signal Processing, Communication, and Controls

Gerard Bricogne is credited with the statement, 'mankind is a catalyzing enzyme for the transition from a carbon–based to a silicon–based intelligence'. It is precisely this reaction that we pursue in the engineering of adaptronic structures. The objective is to include the human 'catalyst' in the process from birth to retirement of the system; to aid the systems creation by means

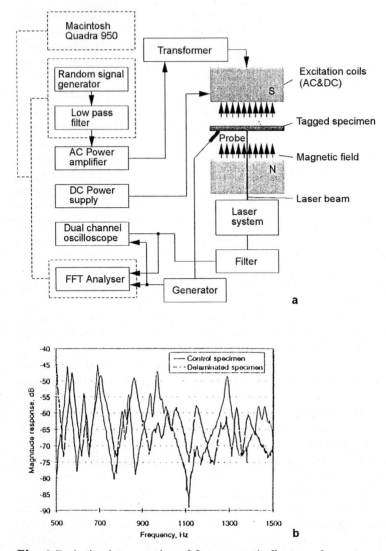

Fig. 2.7. Active interrogation of ferromagnetically tagged smart composites: (a) experimental setup; (b) frequency–response diagrams for control and damaged specimens [19]

of the engineering method; to learn from the system of the 'life experiences' that the system has encountered and the results of such experiences on the health and performance of the system; and to assist in the rejuvenation, repair, and retirement of the system at pragmatic points in the 'life' of the system so as not to allow or force the retirement by catastrophic failure.

The essence of this new design philosophy rests in the manifestation of the most critical of life functions, intelligence. It is this very aspect that has forced the material science, mechanics, and structures communities to debate whether the concept is to create 'adaptive', 'smart', 'very smart', or 'intelligent' material systems. However, the vast majority of the debate to date has been saddled with ignorance and has shown a narrow appreciation for the efforts of the other scientific communities. There is a community of scholars, researchers, engineers, and technicians who have made tremendous strides in developing the genesis of an area called artificial life: an area of study in which the first principles of intelligence are being studied, simulated, and reduced to hardware.

It is a formal endeavor of study where inanimate objects are given the ability to learn, to remember, to forget what is not useful, and to perform functions that ensure life — a field of endeavor that is not just based on theory, but is a laboratory experiment.

'Artificial life, or a–life, is devoted to the creation and study of lifelike organisms and systems built by humans.' This definition comes from the book by Steven Levy entitled, *Artificial Life: the Quest for a New Creation* [20]. The a–life movement is an outgrowth of cybernetics, which Norbert Wiener originally defined as the science of communication and control in animals and machines. The basic premise is that the creation of artificial life forms in the likeness of say, insects, with the mobility and functions of their biological ancestors, allows scientists to evaluate the implications of various theories of control, communication, and learning processes. The ability to learn of one's environment and live within it is a common theme between the a–life and the adaptronic structures communities.

There have been tremendous efforts in developing theories, simulations, and hardware implementations for the control of machines. Modern control approaches, adaptive control, and neural networks are some of the control concepts that today attract so much attention. However, the intelligence features that the adaptronic materials community is trying to create have constraints that the engineering world has never experienced before, but that the biological world seems to accept with simplicity and grace. These constraints arise from the tremendous number of sensors, actuators, and their associated power sources that do not let us use the conventional central processor architecture whereby every piece of sensor and actuator information must be stored and manipulated by silicon.

Nature has used architecture to enhance the behavior of its quite restrictive and far–from–robust material selection; likewise, nature uses architecture to facilitate complex communication, signal processing, and memory using relatively simple devices. The electrochemical devices that we refer to as neurons are not nearly as fast as our modern–day silicon counterparts; however, nature has developed a complex architectural scheme for processing the information from these neurons to allow rather complex tasks to be performed with amazing speed. The key appears to be a hierarchical architecture in which signal processing and the resulting action can take place at levels below and far removed from the central processor, the brain. Removing your hand from a hot stove to prevent damage to the system (in terms of a burn) need only be processed by and organized within the spinal cord; whereas the less automatic behaviors are organized by successively higher centers within the brain. Not until after the action of contracting muscles in the arm and fingers does the information that you have touched a hot surface even reach the brain. This hierarchical approach not only yields control systems that are time–efficient, but yields systems that are fault–tolerant as well. Reliability is a critical factor in reducing energy. A failed system is a tremendous waste of resources and energy; and the control system is as important, if not more important, than the structural components in assuring a system that has a longer lifespan than any one of its components.

The control systems of today can learn, they can change based upon need, they can anticipate a need, and they can correct a mistake. The architecture of control systems will remain an important element in the future manifestations of adaptronic structures, for it is the computational hardware and the processing algorithms that will determine how complex our systems can become — how many sensors we can utilize — and how many actuators we can use to effect change. Will all control systems be neural networks and modeled after biological systems? No. The same paradigm we use to architect the material systems or structures is used to architect the control system – the design that will reduce the mass and energy needs of the system in order to enable it to perform its adaptive functions.

2.4 What Do They Look Like?

Adaptronic structures are first and foremost hybrid material systems. The typical picture that we try to paint is of silicon electronics. The basic tenet of producing, at the micro–level, an array of components that do not serve a single function but that can be interacted with to perform a wide assortment of functions is a concept that allows these systems to have such economy of scale. By creating an embodiment in which several users can use the same device and by which reliability is engineered by redundancy is a simple and cost–effective approach for man–made systems to replicate the adaptive self-preserving features of natural systems.

Adaptronic structures do not look inherently different from the host materials used to support these additional features. The sensors, actuators, and silicon intelligence is reduced to the microstructure, be it nano level for artificial drug delivery systems, micron level for advanced fiber reinforced composites, or meter level for civil engineering constructions. Clearly, each material configuration will manifest its own look for the adaptronic material system. Some may look like fluids, with actuators that cannot be seen by the naked eye but that can manipulate molecules with grace and agility; others may look like materials that are hard and strong and in a moment, upon demand, can behave like jelly just long enough to deflect and absorb energy, as a karate expert reacts to a punch; and yet others may have the mass of small mountains but the perception to become one with nature, so as to ensure the safety of the delicate and intricate human beings they have been designed to protect.

The future of adaptronic structures lies in developing a system with the ability to interface and interact with the network of sensors, actuators, and controls that allows the user/designer/builder to architect a system to perform the function desired with the generic enabling system within the host material. An example can be postulated by focusing on one aspect of the material system, the sensor system. In this scenario, a sensor network is built into the system with many more sensors than are needed by any one application, but by means of adaptive architecture these sensors can be connected together, turned off, or turned on, to create the specific system desired. If a particular sensor fails, the adaptive architecture will replace the failed sensor with the next best alternative and reconfigure the interconnections and the control algorithm to accommodate this change. The sensor network, therefore, could look like the detail of a silicon microchip in which numerous sensors are spread about a polymeric sheet that can be used as the structural ply of a composite laminate. The sensor sheet can be produced by photolithography techniques, which is much like making a photocopy, and such sheets can be massproduced for fractions of a cent per sensor. Similar 'pictures' can be painted for the other components of the system. It seems likely that a system with large arrays of sensors and actuators within a host will require three–dimensional interconnections between the power modulation devices, the control processors, and the sensors and actuators; all via technology that has been developed and refined, once again, by the silicon community.

Adaptronic structures should be as transparent to the user as fiber–reinforced composites are today. Many people drive cars today that have composite components replacing aluminum or other metals; canoers and kayakers expect their boats to be made of lightweight plastic and fibers; and sports equipment from tennis rackets to skis are expected to be similarly advanced. The fact that wood, steel, and aluminum have in many cases been replaced by composites has not changed the way people interact with these products, but the enhancement in performance often times means that a golf ball will travel farther, an airplane will be more efficient, or that a tennis racket will have a larger sweet spot. We can expect no less from adaptronic structures.

2.5 The Future: The New Age of Materials

The endeavor to design adaptronic structures has been called modern–day alchemy. The concept of instilling life functions in inanimate objects and artifacts, such as advanced composite materials and home appliances, seems to be a vision more akin to science fiction than present–day realism. However, as the scientific and engineering communities have begun to develop this area of endeavor, the catalyst was not new compositions of matter but, instead, a new vision of design, a new vision of material function, and a renewed look at the natural world we all live in.

Adaptronic structures may, in the near future, begin to impact our lives by being introduced commercially; but the most lasting impact will be that the philosophy of engineering design will begin to change. Engineers of the future will not have to add mass and cost to a structure to assure safety in structures that are not used as they are intended. Engineers will not have to learn from structural failures but will be able to learn from the 'life experiences' of the structure. Not only will adaptronic structures be of great utility to the consumer but they will have an even more profound influence on science and engineering. They will allow the silent systems we create to inform us, to enlighten us, to educate us of the physics, science, and interaction of the environment on our designs.

J. E. Gordon's book written in 1988 for the Scientific American Library, entitled *Structures and Materials* [21], addresses 'Materials and Structures of the Future'. The last paragraph begins, 'Only recently have materials scientists begun to take the idea of active elasticity seriously. Piezoelectric materials — materials that show strain in response to an electrical impulse — have been known to electronic engineers for a long time, but their structural implications are just now beginning to be considered. If piezoelectric structural materials can be developed — and also reliable means of controlling them — then yet another structural revolution will be upon us. Yet, like many revolutions, it will be a reversion to the past — after all, that is the way in which animals work.'

The adaptronic structures revolution to date has focused upon learning how to use energy as a structural component, how to make structures behave like nature's systems, how to make structures that are 'soft', and how to better utilize the materials around us. New compositions of matter will begin influencing the manifestations of adaptronic structures. Scientists and researchers who are developing new materials, sensory materials, materials with actuator capabilities, energy storage and modulation devices that will allow the integrated system to be autonomous and self–supporting will add fuel to this movement.

Henry Petroski, in his delightful and articulate portrayal of engineering, states that, 'No one wants to learn by mistakes, but we cannot learn enough from success to go beyond the state of the art.' This has been so true; but, as engineers, our need to learn from failures is a result of our inability to learn from structures during their life. In fact, we learn a great deal from the autopsies we perform on structures that have failed, that no longer are able to perform the function they were intended to perform. However, engineering is changing. We will soon have the opportunity to ask structures during their life how they are feeling, where they hurt, have they been abused recently; or better yet, have them identify the abuser. Material systems interacting with their human creator may lead not only to the next materials revolution, but to the next revolution in our understanding of complex physical phenomena.

Will adaptronic structures eliminate all catastrophic failures? No. Not any more than trees will stop falling in hurricane winds or birds will no longer tumble when they hit glass windows. But adaptronic structures will enable man–made inanimate objects to become more 'natural'. Adaptronic structures will be manifestations of the next materials and engineering revolution — the dawn of a new materials age.

References

1. Gordon, J.E.: *Structures, or, Why Things Don't Fall.* Da Capo Press, New York, N.Y. (1981).
2. Zuk; Clark: *Kinetic Architecture.* Van Nostrand Reinhold, New York, N.Y. (1970).
3. Bank, R.: *Shape Memory Effects in Alloys.* Plenum, New York (1975), p. 537.
4. Cross, W.B.; Kariotis, A.H. and Stimler, F.J.: Nitinol Characterization Study. NASA. [contract Rep.] CR NASA CR–1433 (1970).
5. Funakubo, H.: *Shape Memory Alloys.* Gordon & Breach, New York (1987).
6. Giurgiutiu, V.; Rogers, C.A. and Zuidervaart, J.: Design and Preliminary Tests of an SMA Active Composite Tab. SPIE **3041** (1997), pp. 206–215.
7. Jaffe, B.; Roth, R.S. and Marzullo, S.: Piezoelectric Properties of Lead Zirconate–Lead Titanate Solid Solution Ceramics. J. Appl. Phys. **25** (1954), pp. 809–810.
8. Butler, J.L.: Application Manual for the Design of ETREMA Terfenol–D Magnetostrictive Transducers. Tech. Lit. ETREMA Products Inc., Subsidiary of EDGE Technologies, Ames, IA (1988).
9. Giurgiutiu, V.; Rogers, C.A. and Rusovici, R.: Solid–State Actuation of Rotor–Blade Servo–Flap for Active Vibration Control. Journal of Intelligent Material Systems and Structures, **7**, No. 2, March 1996, Technomic Pub. Co., pp. 192–202.
10. Giurgiutiu, V. and Rogers, C.A.: Large–Amplitude Rotary Induced–Strain (LARIS) Actuator. Journal of Intelligent Material Systems and Structures, Technomic Pub. Co., Jan. 1997, pp.41–50.
11. Giurgiutiu, V.; Rogers, C.A. and McNeil, S.: Static and Dynamic Testing of Large–Amplitude Rotary Induced–Strain (LARIS Mk 2) Actuator. Journal of Intelligent Material Systems and Structures, Technomic Pub. Co. (in press) (1997).
12. Giurgiutiu, V. and Rogers, C.A.: January 1996 electromechanical impedance technique for structural health monitoring and non–destructive evaluation. Technical Report No. USC–ME–LAMSS–97–101, Department of Mechanical Engineering, University of South Carolina, Columbia, SC 29208 (1996).

13. Chaudhry, Z.; Joseph, T.; Sun F. and Rogers, C.: Local–Area Health Monitoring of Aircraft via Piezoelectric Actuator/Sensor Patches. Proceedings, SPIE North American Conference on Smart Structures and Materials, SPIE **2443** (1995), pp. 268–276.
14. Ayres, T.; Chaudhry, Z. and Rogers, C.A.: Localized Health Monitoring of Civil Infrastructure via Piezoelectric Actuator/Sensor Patches. SPIE's 1996 Symposium on Smart Structures and Integrated Systems, SPIE **2719** (1996), pp. 123–131.
15. Keilers, C.H.; Chang, F.-K.: Identifying Delamination in Composite Beams Using Built–in Piezoelectrics: Part I — Experiments and Analysis; Part II An Identification Method, Journal of Intelligent Material Systems and Structures, **6** September (1995), pp. 649–672.
16. Lovinger, A.J.: Ferroelectric Polymers. Science, **220** (1983), pp. 1115–1121.
17. Smith: The Role of Piezocomposites in Ultrasonic Transducers. Proc. 1989 IEEE Ultrason. Symp. (1989), pp. 755–766.
18. Quattrone, R. and Berman, J.: Private Communication at the CERL–CI Smart Tagged Composites Meeting, July 17–18, 1977, US Army Corps of Engineers Construction Engineering Research Laboratory, Champagne, IL (1997).
19. Giurgiutiu, V.; Chen, Z.; Lalande, F. and Rogers, C.A.; Quattrone, R. and Berman, J.: Passive and Active Tagging of Reinforced Composite Samples for In–Process and In–Field Non–Destructive Evaluation. Journal of Intelligent Material Systems and Structures, Technomic Pub. Co. **7**, No.6, Nov. 1996, pp. 623–634.
20. Levy, S.: *Artificial Life: the Quest for a New Creation.* Pantheon Books, New York, N.Y. (1992).
21. Gordon, J.E.: *The Science of Structures and Materials.* Scientific American Library, distributed by Freeman, New York (1988).

3. Multifunctional Materials
— the Basis for Adaptronics

W. Cao

Two of the three components in adaptronic structures, i.e., sensors and actuators, are made of single–phase or composite functional materials. In order to design better adaptronic structures, it is necessary to know these functional materials and to understand their functional origin, which will allow us to use them more efficiently and help us to design new and better functional materials following the same physical principles.

3.1 What Are Functional Materials?

Functional materials are materials that can perform certain functions when triggered by environmental stimuli or control signals. The difference between a device and a functional material is that the material will preserve the same functional property when its volume is subdivided. Functional materials may be categorized into two groups: passive and active.

The signature of passive functional materials is the appearance of anomalies, such as maximum, minimum, or singularity, in at least one of the physical quantities. For crystal systems, such extreme behavior is often associated with a structural phase transition and limited in a finite temperature range. The large amplitude change of a particular physical property can then be used to perform certain control functions. Examples of such passive functional materials include positive temperature coefficient materials (PTC) [1], superconducting materials, and partially stabilized tetragonal ZrO_2 [2, 3]. The resistivity of doped $BaTiO_3$ can change more than four orders of magnitude immediately above the paraelectric–ferroelectric phase transition, making it a good material for thermistors. The tetragonal–monoclinic phase transition in ZrO_2 can produce up to 6% volume expansion, which can help to stop crack propagation. There are many passive functional materials that can perform certain functions using their physical anomalies, including voltage dependent resistors (VDRs), carbon fiber–polymer composite near the percolation limit, etc.

Active functional materials are those that can convert energy from one form to the other. Good examples include piezoelectric materials, magnetostrictive materials, electrostrictive materials, and shape memory alloys. These materials can give large responses to external stimuli, but there are no anomalies. The basic energy forms that can be interchanged via functional materials are: thermal energy, electric energy, magnetic energy, and mechanical energy. The energy can be either in a static form such as electrostatic energy inside a capacitor, or in a dynamical form such as electromagnetic waves. Active functional materials are the primary materials used for most of the adaptronic structures. They are sometimes called multifunctional since most of these materials can have several functional properties due to the coupling effects.

While the definition of functional materials is not so stringent, it is worth-while to point out that the property variation must be sufficiently large in amplitude. For example, thermal expansion alone is too small to be utilized for any control purpose; therefore, materials with normal thermal expansion property do not qualify as functional materials. It is very important to understand the fundamental principles that make materials functional, which can help us to use them properly and to create better multifunctional materials.

There are natural functional materials that have been widely used in our daily life. Many composite materials with enhanced functional properties are also created and whose members seem to grow at a fast rate to meet the high demand of adaptronic structures. Since most of the control systems are driven electronically, ferroelectric materials are naturally one of the best functional materials for adaptronics applications. In this chapter, we will use ferroelectric materials as examples to explain some of the fundamental physics.

3.2 Basic Principles of Functional Materials

As defined in Chap. 1, adaptronic structure is designed to perform all three functions: sensing, control and actuation. It is a primitive replica of a biological body. Multifunctional materials are essential components of adaptronic structures in which each component must be able to communicate with the others. Only limited numbers of natural materials can meet the high demand of adaptronics. Therefore, understanding the physical principles of functional materials could help us to engineer composite materials to enhance the functionality or create new functional materials. The quality of functional materials is measured in terms of their 'responsiveness' and 'agility'.

3.2.1 Phase Transitions and Anomalies

High responsiveness is often linked to a stability edge, or a structural phase transition. The commonly referred phase transitions are thermally driven structural instabilities in which ionic displacements occur. A critical transition temperature, T_c, exists at which the crystal structure of the high temperature phase becomes unstable. Below T_c the ions form a new crystal structure with lower symmetry. As a signature of structural phase transitions, one or more of the physical quantities vanish, or become discontinuous. Phase transition is the origin of anomalous response in many crystalline systems.

Fig. 3.1 is an illustration of the ionic displacement pattern in the cubic to tetragonal phase transition in $BaTiO_3$ when cooling the system from a temperature higher than $T_c = 130°C$. At the phase transition temperature, electric dipoles are frozen–in to form the low–temperature tetragonal ferroelectric phase. Fig. 3.1a is the cubic perovskite structure in the paraelectric phase. While cooling through the phase transition, the O–anions move down (note: the top and bottom oxygen atoms shift more than the oxygen atoms on the side faces) and the Ti–cation moves up relative to the Ba frame, as shown in Fig. 3.1b, forming an upward dipole [4]. Associated with the formation of the electric dipole, the unit cell is also elongated in the poling direction, reflecting a strong coupling between the electric dipole formation and structural distortion. The symmetry of the crystal changes from a cubic m3m to tetragonal 4mm, and is accompanied by a dielectric anomaly [5].

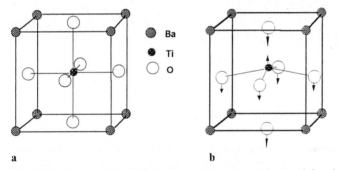

Fig. 3.1. Illustration of the ionic rearrangement in the **(a)** cubic to **(b)** tetragonal ferroelectric phase transition in $BaTiO_3$

Ferroelectric materials are multifunctional materials with many useful functional properties. In addition, the ferroelectric phase transition can provide more useful functional properties when it is chemically engineered to increase its conductivity. In a doped ceramic $BaTiO_3$, the grain boundary can create Schottky barriers that couple with the dielectric anomaly, producing a strong PTC material. PTC resistors have been widely used as thermistors to regulate the temperature in many heating devices.

Anomalies may also be induced by means other than temperature variation. For example, the drastic resistance change in ZnO at a critical electric field level is the basis for varistors that are used for voltage surge protection. The characteristic electric current–voltage curve for a voltage–dependent resistance (VDR) material is shown in Figure 3.2. The varistor has very high

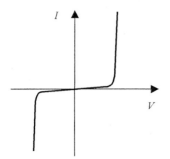

Fig. 3.2. Typical current–voltage curve for a varistor

resistance at low voltage, but becomes a good conductor when the voltage exceeds a critical value.

If a varistor is connected with an electronic device in parallel, it will bypass any large voltage surge to protect this device. The proposed explanation for this anomaly is the creation of paired Schottky barriers at the grain boundaries, as illustrated in Fig. 3.3.

The intergrain layer (IGL) can act as an acceptor to draw electrons from the semiconducting ZnO grains near the IGL region, so that this region will be positively charged. Schottky barriers are then formed at the interface between the grain boundary layer and the grains. The paired Schottky barriers will provide high resistance to current flow in either direction. At low electric field, the barrier for the electron flow will be lowered, but is still too high: only a small fraction of thermally activated electrons will pass through the barriers to provide very low current. At high field level, the electron potential will be raised high enough to allow them passing through the forward biased barrier, and tunnel to the positive charged region of the other grain to produce a surge of current. The reverse–direction electron flow is the same due to the symmetry of the paired Schottky barriers so that the current–voltage curve is antisymmetric.

Passive functional materials are based on anomalies. It is interesting to point out that the requirements for functional materials are very different from those of common materials. These anomalies would be disastrous for many traditional applications since they signify breakdowns and instabilities. The anomalies are, however, essential for constructing adaptronic structures because they can provide clear signals to indicate the operating limits and also respond in large amplitude to repair sudden breakdowns in devices.

3.2.2 Microscopic, Mesoscopic, Macroscopic Phenomena and Symmetry

Most adaptronic structures are used, or are intended to be used, in macroscopic devices. In a single–domain crystal system, macroscopic properties are

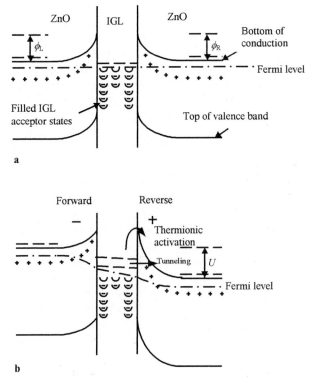

Fig. 3.3. Proposed electronic structure at a junction between semiconducting ZnO grains: **(a)** no voltage applied; **(b)** with applied voltage. (After A.J. Moulson and J.M. Herbert [1])

simply the statistical average of the microscopic properties of each unit cell. For a majority of the cases, however, such a simple average fails because of the nonlocal interactions and the additional mesoscopic structures created at the intermediate length scale, such as domain patterns in single–crystal systems and grain structures in ceramics. These nonlocal interactions and meso–scale structures often produce very strong extra enhancement to the functional properties. Therefore, to grasp the whole picture of these functionalities, one must study the physical principles at different length scales.

Electron band structures control the conductivity and the structural stability at the microscopic level. Based on the electron band structures, inorganic materials can be classified as conductors, semiconductors and insulators. Modification of these band structures through doping, by the introduction of foreign elements into a perfect crystal, could change these band structures so as to regulate the properties of the material and produce conductivity anomalies. The electronic structures also determine the stability of crystal structures. Instabilities may be created in crystal structures at designed temperatures by altering the electronic structures using chemical doping. Many functional materials contain elements of mixedvalences, i.e. the element can have two or more different valences while forming a compound. Doping of these mixedvalence elements, such as transition d–block elements in the periodic table and lanthanide (Eu, Yb, Ce, Pr, Tb, etc), often enhances the functionality of the material [6]. The length scale for this level is unit cell and below.

The next level of dominating factor for the functionalities of materials is the microstructure, such as domains, domain walls, grains, and grain boundaries. In ferroelectric ceramics, for example, contributions to the functional properties from domain wall movement could be as high as 70% of the total effect at room temperature [7]. For shape memory alloys, the superelasticity and shape memory effects are all originated from domain reorientation or the creation and annihilation of domains. Grain boundaries play a key role in the formation of the paired Schottky barriers in PTC and VDR materials. The conduction anomalies in PTC and VDR do not even exist in a single–crystal system. The length scale for these mesoscopic structures is in the range of a few to a few hundred nanometers.

The formation of domains in a phase transition from a high–symmetry phase to a low–symmetry phase is a reflection of the nature trying to recover those lost symmetries. The number of domain states or variants in the low–temperature phase is equal to the ratio of the number of operations in the high and low symmetry groups.

There are 230 space groups and 32 point groups describing the symmetry operations allowed in crystal structures [8]. The point groups refer to those symmetry operations without translation operation, including rotation, mirror reflection and inversion. The representation of these 32 point groups and their graphical representations are listed in Table 3.1 (the shapes reflect the allowed symmetry operations). Although, macroscopically we often treat many systems as isotropic, i.e., having a spherical symmetry, the highest symmetry allowed in a crystal structure is cubic m3m. Structural phase transition is allowed only when the symmetry group of the low temperature phase is the subgroup of the high temperature phase.

At a structural phase transition, there are several equivalent choices (variants) for the high–symmetry phase to deform. For example, two variants exist in a ferroelastic tetragonal 4/mmm to orthorhombic 2/mmm transition, representing the elongated axis in the x– and y–direction, respectively. The situation is illustrated in Fig. 3.4 in terms of the unit cell projection on the x–y plane. Because the two low–temperature variants are energetically degenerate, they have equal chance to form. As a result, there will be a mixture of these two domain states. If the transformation was originated from a single-crystal system, these two kinds of domains can form 90° twins that maintain the atomic coherency across the domain boundaries (domain walls). The do-

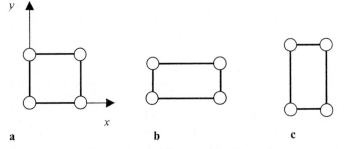

Fig. 3.4. 2–D illustration of the two possible low–temperature states in a tetragonal 4/mmm to orthorhombic 2/mmm ferroelastic phase: **(a)** unit cell of the high–temperature phase, **(b)** orthorhombic phase with the elongation along the x–direction; **(c)** orthorhombic phase with the elongation along the y–direction

main wall orientation can be either in [1 1 0] or [1 $\bar{1}$ 0] for this case. The two sets of twins could also coexist to form more complex domain patterns.

Twinning provides a new functional mechanism for easy shape deformation via the movement of domain walls. If the low–temperature states are polarized, domain wall movement will cause the polar vector to rotate in the region swept by the moving domain wall. This situation is illustrated in Fig. 3.5 for a ferroelectric twin. Under an upward electric field, the domain wall will move to the left. At the same time, the whole region in the righthand side of the wall moves up relative to region I. The dipoles in region II are switched to more favorable positions by the external field, and the global shape change caused by the domain wall movement could be substantial. The switching of these dipoles in region II will give an extrinsic contribution to the dielectric susceptibility, while the shape change will contribute to a macroscopic piezoelectric effect.

Domain walls are a special kind of defect. They create a localized stress gradient or electric (magnetic) field gradient [9] that can have strong interactions with other defects, such as vacancies or charged defects. This interesting nature of domain walls enables us to control the formation of the domain pattern through different chemical doping strategies.

It is a common practice to dope aliovalent elements (non–stoichiometric doping) to create multivalences and vacancies in the material so that domain walls could interact with them and become more mobile. This method has proved to be particularly effective to enhance the mesoscale functionality in ferroelectric materials, the La or Nb doped $Pb(Zr_xTi_{1-x})O_3$ (PZT) has much larger piezoelectric properties than that of the non–doped PZT.

Inhomogeneous stresses produced by local defects may induce local phase transitions above the normal phase transition temperature T_c, causing the material to have mixed low–and high–symmetry phases in certain temperature region. Such a two–phase mixture could be very sensitive to external

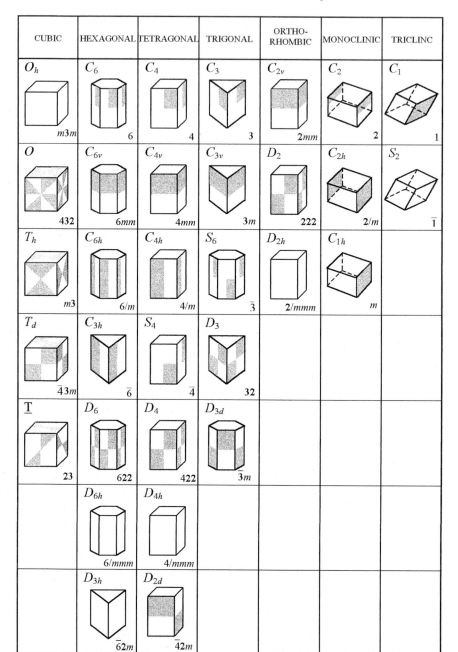

CUBIC	HEXAGONAL	TETRAGONAL	TRIGONAL	ORTHO-RHOMBIC	MONOCLINIC	TRICLINC
O_h — $m3m$	C_6 — 6	C_4 — 4	C_3 — 3	C_{2v} — $2mm$	C_2 — 2	C_1 — 1
O — 432	C_{6v} — $6mm$	C_{4v} — $4mm$	C_{3v} — $3m$	D_2 — 222	C_{2h} — $2/m$	S_2 — $\bar{1}$
T_h — $m3$	C_{6h} — $6/m$	C_{4h} — $4/m$	S_6 — $\bar{3}$	D_{2h} — $2/mmm$	C_{1h} — m	
T_d — $\bar{4}3m$	C_{3h} — $\bar{6}$	S_4 — $\bar{4}$	D_3 — 32			
$\underline{T}$ — 23	D_6 — 622	D_4 — 422	D_{3d} — $\bar{3}m$			
	D_{6h} — $6/mmm$	D_{4h} — $4/mmm$				
	D_{3h} — $\bar{6}2m$	D_{2d} — $\bar{4}2m$				

Table 3.1. The 32 point groups and the symbols of the symmetry groups. The upper–left corner and the lower–right corner in each cell list the Schoenflies and international symbols, respectively

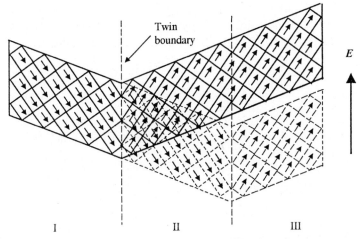

Fig. 3.5. Domain wall movement in a ferroelectric twin structure under an external electric field E

fields or stresses since the phase change becomes barrierless even for a first–order phase transition [10].

The formation of domain structures and the available variants in the low–symmetry phase are dictated by the symmetry of the high–temperature phase. For a macroscopic structure, it is the global symmetry, not the local symmetry, that controls the functionality of the material. Therefore, at the macroscopic level, one can make composite structures of specially designed symmetry to create new functional materials.

3.2.3 Energy Conversion

Energy conversion between different energy forms is the main spirit of many adaptronic structures. Active functional materials must be used for such purposes. For each energy form, there is a set of generalized conjugate variables consisting of a generalized force F and a generalized displacement S. They can be scalars, vectors and tensors. If one kind of generalized force can generate a displacement other than its own conjugate, the material has the ability to convert energy and is an active functional material. Crystal symmetry again determines whether some of the energy conversions are allowed in a particular crystal structure.

There are many phenomena in nature reflecting these energy conversion effects. Fig. 3.6 illustrates the energy conversion phenomena that can occur in a ferroelectric material. There are three energy forms listed: thermal, electrical and mechanical. The generalized force–and–displacement pairs corresponding to these three energy forms are temperature and entropy, electric field and electric displacement, as well as stress and strain. The crossenergy domain coupling could be linear or nonlinear depending on the nature of the material: for example, electric field can generate mechanical strain through linear converse piezoelectric effect and nonlinear electrostrictive effect:

$$S_\lambda = d_{k\lambda} E_k \qquad\qquad (i, j, k = 1, 2, 3; \quad \lambda = 1, 2, 3, 4, 5, 6) \qquad\qquad (3.1)$$

$$S_\lambda = M_{ij\lambda} E_i E_j \qquad\qquad\qquad\qquad\qquad\qquad (3.2)$$

Here the S_λ are the elastic strain components in Voigt notation, $d_{k\lambda}$ are the piezoelectric coefficients, and $M_{ij\lambda}$ are the electrostrictive coefficients.

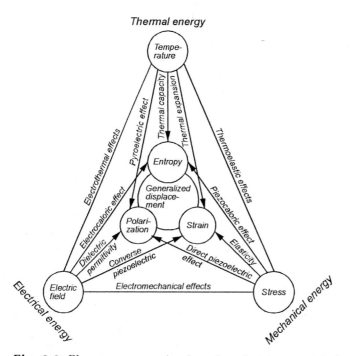

Fig. 3.6. Phenomena occurring in a ferroelectric material that can convert the three forms of energy

Some of the energy conversions can be twoway effects. For example, the piezoelectric effect is actually discovered by the crystal's ability to convert stress into electric charge:

$$D_i = d_{i\lambda} T_\lambda \qquad\qquad (i, j, k = 1, 2, 3; \quad \lambda = 1, 2, 3, 4, 5, 6) \qquad\qquad (3.3)$$

where the T_λ are the stress tensor components in Voigt notation, and the piezoelectric coefficients $d_{i\lambda}$ are the same as those in (3.1). The electrostriction is a one–way effect because of its nonlinear nature. Theoretically speaking, the same M–coefficient as in (3.2) can be used to describe the combined stress and electric field effect on the electric displacement:

$$D_i = 2M_{ij\lambda}E_jT_\lambda \qquad (i,j,k = 1,2,3; \quad \lambda = 1,2,3,4,5,6) \qquad (3.4)$$

However, since the above equation describes a mixed phenomenon for which both the electric field and the stress must be nonzero, pure stress could not generate a charge through this effect when E is zero.

The principle of the cross–energy domain coupling in the atomic scale is illustrated in Fig. 3.7 for a simple two–dimensional lattice. Fig. 3.7a is a binary compound consisting of negative ions sitting at the corners and a positive ion sitting at the center. Assume the square symmetry system goes through a ferroelectric phase transition to become rhombic symmetry; there are two kinds of rearrangements that can reduce the square symmetry to rhombic. One is the formation of dipoles along the diagonals. There are four variants in the low–symmetry phase and the dipoles in different unit cells may or may not be aligned as shown in Fig. 3.7b. The second kind of rearrangement is through a shear deformation of the frame to accommodate the ionic shifts that occurred in Fig. 3.7b. One can see (3.7c) that the dipole formation pushes the frame to deform and in return, the shear deformation of the frame helps to create an ordering of the dipoles. This interdependence between the ordering of dipoles and deformation strain is the fundamental principle of electromechanical coupling.

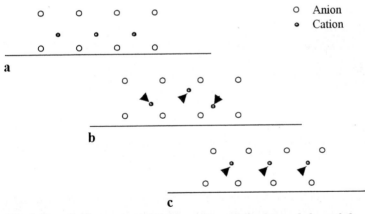

Fig. 3.7. 2–D illustration of the formation of dipoles and shear deformation during a ferroelectric phase transition from square symmetry to rhombic symmetry

3.3 Examples of Functional Materials

In order to make the above concepts correlate to real materials, specific examples are given here to further explain the principles of functional properties. For convenience, we now discuss these functional materials based on their responsive nature.

3.3.1 Thermally Responsive Materials

Thermally responsive functional materials can be produced in the vicinity of phase transitions. For example, the tetragonal–monoclinic phase transition in ZrO_2 can produce as large as 6% volume strain, which can be used for material toughening. Shown in Fig. 3.8 is an enlarged view at a crack tip in a partially stabilized ZrO_2 system. The crack produces tensional stress at the crack tip, which in turn induces the partially stabilized ZrO_2 to transform into monoclinic martensitic phase (darker–shaded interior region). The large volume expansion from the transformation will help to reduce the stress concentration and stop the propagation of the crack. The martensitic phase will form twins to fit the boundary condition, as indicated in the figure.

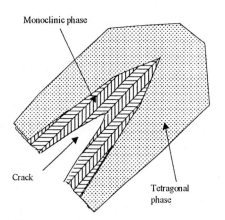

Monoclinic phase

Crack

Tetragonal phase

Fig. 3.8. Magnified view at a crack tip where the partially stabilized ZrO_2 transformed to monoclinic phase. The twin pattern represents the martensite phase

Temperature could also induce large resistivity changes in doped $BaTiO_3$, as mentioned above. The fundamental principle of the PTC material is the coupling of the Schottky barriers at the grain boundaries to the ferroelectric phase transition. The potential barrier similar to that plotted in Fig. 3.3 will be short–circuited in the ferroelectric state due to the presence of charges at the grain boundaries that terminate the polarization. Above the Curie point T_c, the conductivity is proportional to the Boltzmann factor $\exp(\frac{-\phi}{kT})$, with the height of the barrier ϕ approximated to be [1]:

$$\phi = \frac{e^2 N_s^2}{8\varepsilon n} \tag{3.5}$$

where N_s is the surface density of acceptor states near the boundary, e is the electron charge, n is the volume density of donor states in the grain and ε is the dielectric permittivity.

Above the ferroelectric phase transition temperature T_c, the dielectric constant obeys the Curie–Weiss law: $\varepsilon = C/(T - \theta)$, where C is the Curie constant and θ is the Curie–Weiss temperature. (Note: $\theta = T_c$ for a second–order phase transition and $\theta < T_c$ for a first–order phase transition). Thus the resistivity above the transition temperature may be written as [1]:

$$R_{\mathrm{gb}} \propto \exp\left\{\frac{e^2 N_s^2}{8nkC}\left(1 - \frac{\theta}{T}\right)\right\} \qquad T > \theta \tag{3.6}$$

The fast decrease of the permittivity with temperature immediately above T_c drives the resistivity to increase exponentially, producing several orders of magnitude change to the resistivity in a temperature range of a few tens of degrees, making them very sensitive to temperature change.

Shape memory alloys, e.g. Ni–Ti (NITINOL), Ni–Ti–Cu and Ni–Ti–Fe, etc., can recover their original shapes from a large deformation in the martensite phase upon heating back to the austenite phase. This process is demonstrated in Fig. 3.9 by assuming only two variants in the martensite phase. Fig. 3.9a is the high–temperature austenite phase with a perfect rectangular shape. When the system is cooled through the phase transition, a shear deformation occurs and the two martensite variants have equal probability to form. A twin structure is formed between domain states 1 and 2, as shown in Fig. 3.9b; the twinning of the two domains requires no defects and the atomic coherency is preserved. Now, if a tensional stress is applied as shown in Fig. 3.9c, the degeneracy of the two domain states will be lifted so that one type of domain will grow at the expense of the other.

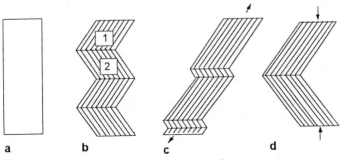

Fig. 3.9. Illustration of shape memory effect

New domains of type 1 may also be generated through a nucleation process to speed up the domain–switching process. This switching may continue until the unfavored domains (type 2, as shown in Fig. 3.9b) are driven out of the system. If the applied stress is compressive, as shown in Fig. 3.9d, some domains will be annihilated; the number of domains in the final state depends on the symmetry of the applied stress. Upon heating, all shapes in Fig. 3.9b–d will go back to the same shape as in Fig. 3.9a. In other words, the shape in the high–temperature phase is 'remembered'. This shape memory effect effectively enhances the elastic limit of the alloy with the help of temperature and the phase transition.

Shape memory effect is directly related to the domain formation and the interaction of domain walls with defects. How accommodating is the martensite phase depends strongly upon the available number of variants. Generally speaking, the more variants, the easier to deform into arbitrary shape in the martensite phase. A cubic–monoclinic transition can generate up to 24 variants; such a martensite can deform into much more complex and elegant shapes than the one shown in Fig. 3.9 without breaking–up the atomic coherency. Because of the drastic change of mechanical strength above and below the martensite phase transition, the shape memory effect can also be used to make shape memory alloy engines that can convert thermal energy into mechanical energy [10].

Another important thermally responsive material is the pyroelectric material that can directly convert thermal energy into electric energy. Pyroelectric effect is a manifestation of the existence of polarization in the material. The change of polarization amplitude with temperature generates electric charge at the sample surface where the polarization terminates. Again, the pyroelectric effect is the strongest near the ferroelectric phase transition temperature. Pyroelectric materials are widely used as infrared sensors, for remote control of electronic devices and for night vision.

3.3.2 Materials Responsive to Electric, Magnetic and Stress Fields

If the adaptronic structure requires temperature stability, active functional materials must be used since they have flat temperature response in a finite temperature range and are controllable with external fields. Most materials in this category are ferroic materials, i.e., ferroelectric, ferromagnetic and ferroelastic materials.

Piezoelectric and electrostrictive materials are materials having the ability to convert electric energy into mechanical energy. The effect is called piezoelectric if the generated surface charge density is linearly proportional to the applied stress. The physical origin of piezoelectricity comes from the noninversion symmetry of the ionic arrangement in crystal systems. The noninversion symmetry makes the anions and cations in a crystal to deform in an asymmetrical fashion under stress, so as to produce a dipole moment. In fact, 20 out of the 21 noncentral symmetric crystal point groups allow piezoelectricity except for the cubic class of 432 (see Table 3.1). The term 'polarization' refers to the volume average of dipole moment and is measured as charge per unit area. For a finite system in static equilibrium, the polarization projection to a surface is equal to the surface charge density.

It is important to recognize that a useful piezoelectric effect is defined macroscopically. Each unit cell has to contribute constructively in order for the macroscopic effect to occur. Global symmetry determines the macroscopic piezoelectric effect. For example, a piezoelectric ceramic material containing randomly oriented crystal grains has no piezoelectric effect even if the symmetry of the unit cell allows piezoelectricity. A net polarization in the material is a sufficient but not a necessary condition for piezoelectricity; for example, quartz is one of the popular piezocrystals without polarization. The existence of polarization, however, does make the piezoelectric effect much more pronounced. Most importantly, the hydrostatic piezoelectric effect belongs uniquely to the polar materials. If the material does have a net polarization, it is also pyroelectric.

Fig. 3.10a shows the polarization arrangement in a ceramic system (note: domains were not explicitly drawn in here; the arrows only represent the net polarization in each grain). The macroscopic piezoelectric effect is zero due to the cancellation of oppositely polarized grains. If the ceramic material is ferroelectric, it can be made piezoelectric by aligning the polarization of different grains using an external electric field through a domain–switching process. A net polarization may be produced along the field direction, as illustrated in Fig. 3.10b. The most popular piezoelectric ceramic, PZT, is produced based on this principle. Because the electromechanical characteristic of the piezoelectric effect is a two–way effect, piezoelectric materials can be used for both sensing and actuation functions.

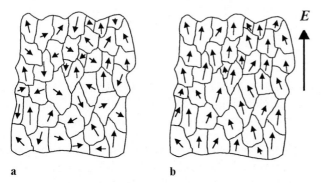

a   b

Fig. 3.10. Polarization distribution in a polar ceramic system: **(a)** random orientation; **(b)** after poling by an electric field $\boldsymbol{E}$

Electrostriction can generate mechanical deformation that is independent of the polarity of the electric field. It exists in almost all materials but is usually very small in effect. However, it could be very large in electrostrictive materials, such as lead magnesium niobate (PMN) systems [11]. The non-linearity is often used to advantage since it can produce tunable functional properties. Because the effect is one–way, as analyzed above, electrostrictive materials are better for actuator applications. Unlike the piezoelectric effect, electrostriction can exist in systems even with center symmetry.

Magnetic field is similar to electric field in many aspects, but it has its own distinct nature. Materials responding to a magnetic field are another important category of functional materials since the change of magnetic properties can easily be converted into electric signals, and vice versa. Magnetic moment of ions is produced by the spins of unpaired electrons. These magnetic moments are randomly oriented in the paramagnetic phase. The system becomes ferromagnetic through an order–disorder phase transition that can align these moments. Unlike the ferroelectric case, the amplitude of each magnetic moment is fixed and the coupling to the ionic structure is much weaker compared with ferroelectric materials. The ordering may appear in antiferromagnetic, ferrimagnetic or ferromagnetic forms as illustrated in Fig. 3.11.

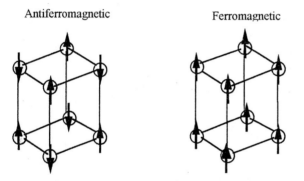

Fig. 3.11. Spin arrangements in an antiferromagnetic and a ferromagnetic system

In a ferrimagnetic state, the spins are only partially aligned or with different amplitude in an antiparallel configuration. If the magnetic spins are strongly coupled to the ionic structure, the system will show magnetoelastic effect similar to the case analyzed in Fig. 3.7. Piezomagnetic effect is allowed in terms of crystal symmetry in many systems; however, it is usually too small to be useful for any control purpose. The nonlinear effect, magnetostriction, however, can be quite large in certain systems. For example, $Tb_{0.3}Dy_{0.7}Fe_2$ (Terfenol–D), can generate a strain level of 10^{-3} at room temperature [12]. Above 700°C, when the spins are randomly oriented, the binary alloy is paramagnetic; at a temperature below 700°C, it transforms to a rhombohedral ferrimagnetic phase with the spins parallel to the <111>. The strong anti-ferromagnetic coupling between the spins of iron and the rare–earth atoms prevents the spins from aligning perfectly in one direction. At room temperature, $Tb_{0.3}Dy_{0.7}Fe_2$ goes through another phase transition that changes the orientation of the spins to <100>, so that the system becomes a tetragonal ferrimagnetic. The crystal has a cubic symmetry with a C15 structure in which the rare–earth atoms form a diamond–like lattice. All three cases are shown in Fig. 3.12, which is the cross–section of [100] and [110]. The arrows indicate the spin orientation in each of the three states.

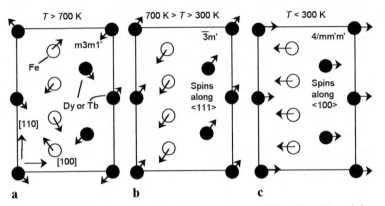

Fig. 3.12. Spin orientations in all three states of $Tb_{0.3}Dy_{0.7}Fe_2$: **(a)** paramagnetic phase; **(b)** rhombohedral ferrimagnetic phase; **(c)** tetragonal ferrimagnetic phase. (After R.E. Newnham [13])

3.4 Increased Functionality Through Materials Engineering

Different strategies have been developed to engineer better functional materials. At the microscopic level, chemical mixing of different materials may create new materials or shift the transition to desired operating temperatures. Experience tells us that better functional materials are mostly mixed compounds and solid solutions. At the mesoscopic level, defects can be introduced to influence the domain structures and grain boundary properties. Many doping strategies have been developed that can greatly enhance the mobility of domain walls and change the characteristics of the conductivity. At the macroscopic level, new composites can be made through structural engineering. These composites can be multifunctional and also more durable compared with single–phase materials.

3.4.1 Morphotropic Phase Boundary

Some isostructural compounds can be atomistically mixed to form a solid solution with enhanced functional properties. A solid solution without solubility gap is called a complete solid solution, in which two or more compounds can be mixed in any proportion to form a single–phase material. The PZT, $Pb(Zr_xTi_{1-x})O_3$, is a good example of such a complete solid solution system. As shown in the phase diagram given in Fig. 3.13, the solid solution $Pb(Zr_xTi_{1-x})O_3$ is cubic above 500°C and will go through a ferroelectric phase transition to become either rhombohedral or tetragonal ferroelectric depending on the composition. The nearly vertical line in the middle of the phase diagram specifies a compositional boundary separating the tetragonal and rhombohedral symmetry phases. This boundary is the morphotropic phase boundary (MPB) at which the two structural phases have degenerate energy. At room temperature, this composition corresponds to a Ti/Zr ratio of 48/52.

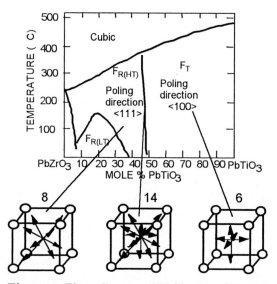

Fig. 3.13. Phase diagram of $Pb(Zr_xTi_{1-x})O_3$ and the illustration of available variants in the ferroelectric phase

The reason for this MPB composition having superior functional properties is due to the fact that there are more variants in the ferroelectric phase. As mentioned above, the symmetry change plays a vital role in determining the functionality of materials. Generally speaking, the more variants can be generated at the phase transition, the better is the performance of the functional materials. For the PZT case, the high–temperature phase has cubic symmetry m3m, which contains 48 symmetry operations. The tetragonal

4mm and the rhombohedral 3m symmetry groups in the ferroelectric phase contain eight and six symmetry operations respectively, so that the number of variants for the tetragonal phase is six and for the rhombohedral phase is eight. This situation is illustrated in Fig. 3.13. Depending on the composition, the dipoles formed at the phase transition in each unit cell may point to any of the six faces to form the tetragonal phase or any of the eight corners to form the rhombohedral phase. At the MPB composition, the two low–temperature phases are degenerate, so that all 14 variants are accessible. This provides a unique situation that gives the most variants in the low–temperature ferroelectric phase.

Another example of a similar situation is given by the Terfenol–D, discussed above. The $TbFe_2$–$DyFe_2$ binary alloy is a resemblance of the PZT solid solution system. There exists a similar compositional boundary between the rhombohedral and tetragonal phases. At room temperature, the best magnetostrictive effect is given by the alloy with a Tb/Dy ratio of 30/70, which falls on the phase boundary line indicated by the arrow in Fig. 3.14.

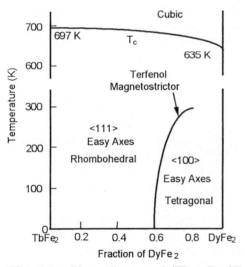

Fig. 3.14. Phase diagram of $(Tb_{1-x}Dy_x)Fe_2$. At room temperature the largest magnetostrictive effect occurs at the $x = 0.7$ composition which is on the phase boundary indicated by the arrow

It is very interesting to see that in both cases, the best composition is at the boundary of two competing structural phases. Due to the degeneracy of the two structural phases at the MPB, the system cannot 'decide' which structure to take at the phase transition. This 'uncertainty' creates high responsiveness to external fields, and hence enhances the functional ability of the material.

Design philosophy: introducing instabilities into the system to create more responsive materials.

3.4.2 Domain Engineering

The next level in material engineering is to manipulate domain structures to increase the extrinsic effects. Aliovalent doping (charged point defects) in a ferroelectric system can create strong localized forces to either facilitate or hinder domain wall movement. This method has been used to improve the piezoelectric properties in soft and hard PZT systems, as mentioned above. Defects, including dislocations and point defects, can also create a local environment to accommodate the stress field generated by the formation of domain walls. For shape memory alloys, defect alignment by the domain walls can provide a reverse shape memory effect. Aligning these defects may take a few rounds of thermal cycle through the phase transition, i.e. the alloy must be trained to remember the exact locations of the walls formed in the martensite phase. This training process is to manipulate domain formation using existing defects.

The second type of domain structure manipulation is to create disorder in an ordered system using physical means other than chemical doping. Single–crystal $Pb(Zn_{1/3}Nb_{2/3})O_3$–$PbTiO_3$ (PZN–PT) solid solution system has recently created some excitement in the transducer and actuator community because it demonstrates an over 90% electromechanical coupling coefficient (compared with 68% for PZT) and a very large piezoelectric coefficient (three times of that of PZT) [14]. The material had been discovered in 1969 [15], but did not generate enough attention because the material, which has 3m symmetry at PT level of 8% or less, cannot maintain remnant polarization along the three axes: the majority of field–aligned dipoles reverse themselves after the external poling field is removed. The key to making the material more attractive now is using misorientational poling. The crystal symmetry of the ferroelectric phase is 3m with the dipoles in each unit cell pointing to the <111> (along body diagonals). It was found that the system could sustain large polarization if the poling field is applied along <100> (one of the normal directions of cubic faces). After poling, each unit cell has a dipole moment along four of the <111> directions in the upper half space, as shown in Fig. 3.15. But the components perpendicular to the poling direction are randomly oriented so that the global symmetry (macroscopic average) is pseudo–tetragonal. Strong elastic interaction among neighboring cells helps to stabilize the poled dipole configuration. The misorientational poling produces a new global symmetry and creates a state that has higher energy than the ground state — and that is therefore much more responsive.

Design Philosophy: Create order in a disordered system, such as aligning random defects in martensite so as to produce reverse shape memory effect, or create disorder in an ordered system, such as misorientational poling of a single–crystal PZN–PT so as to increase responsiveness.

$E \parallel [001]$

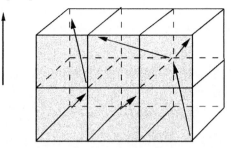

Fig. 3.15. Illustration of misorientational poling in PZN–PT single–crystal system. The field is applied along [001] and the dipoles in each unit cell are pointing to the four upper corners along body diagonals

3.5 Functional Composites

'Composite engineering' is the process of putting several materials together in a certain configuration in order to create improved or new functionalities. Composite engineering allows us to use nonfunctional material to enhance functional material, and to use different functional materials to make new functional or multifunctional composite materials.

The constructive enhancement concept in a composite is analogous to a ballet dance, in which the male dancer does not walk on his toes, although his strength can help the female dancer to stand on her toes much longer by supporting part of her weight. On some occasions, he also lifts her into the air to help her 'fly' to a new height. The physics is to redistribute the load and to reach a new horizon by adding the height of both dancers. Shown in Fig. 3.16 is a 1–3 piezoelectric composite with PZT ceramic rods embedded in a polymer resin. This structure is widely used for medical ultrasonic transducers because the polymer helps to reduce the acoustic impedance mismatch between human body and the PZT so that energy transmission is more efficient. This composite structure also gives a high figure of merit for hydrophone applications [16].

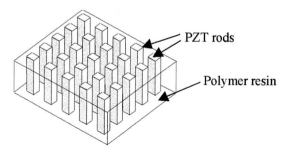

PZT rods

Polymer resin

Fig. 3.16. Piezoelectric PZT–polymer composite

The strength of hydrostatic piezoelectric effect is characterized by the piezoelectric coefficient $d_h = d_{33} + 2d_{31}$. Here d_{33} represents the ability of the material to generate charge on the surface normal to the polarization under a normal stress, and d_{31} measures the ability of the material to generate charge on the same surface by a stress perpendicular to the poling direction. Under a constant electric field, the relationship between the electric displacement D and the hydrostatic pressure p_h is given by $D = d_h p_h$. The d_h value is usually small due to the opposite signs of d_{33} and d_{31}. For PZTs, the d_h value is 20–$60 \cdot 10^{-12}$ C/N, which is an order of magnitude smaller than d_{33}. The mechanism for enhancing the hydrostatic effect in the 1–3 composite is to transfer the normal pressure on the polymer surface to the ceramic rods via shear coupling at the ceramic–polymer interface [17]. This coupling effectively amplifies the pressure on the ceramic rods along the poling direction while leaving the lateral pressure unchanged. As a result, the effective $\overline{d}_h$ value is enlarged through the enhanced effective $\overline{d}_{33}$.

The flextensional 'moonie' structure shown in Fig. 3.17 is another good example of using redirecting force strategy. Through a metal cap, the normal pressure applied to the top and bottom surfaces of the structure is converted to a force with a large horizontal component acting on the outer ring of the PZT disk. The horizontal component of this force will counter the d_{31} effect and the normal component of this force will enhance the d_{33} effect at the contact area. For a small diameter cavity, the main contribution to d_h will be the effective enhanced $\overline{d}_{33}$, since the d_{31} effect will be cancelled; for a large diameter cavity, the main contribution will be $\overline{d}_{31}$ since the contact ring area becomes very small and the cavity area does not contribute to the effective $\overline{d}_{33}$. In this case, the redirected force in the horizontal direction will be much larger than the force produced by the pressure applied to the side of the PZT disk, so that the effective d_h value could be very large. For actuator applications, the radial contraction of the disk will be converted to a much larger normal displacement at the center region of the metal cap. This displacement will add to the direct d_{33} effect so that the effective $\overline{d}_{33}$ could be increased by an order of magnitude [18].

Design Philosophy: Using nonfunctional materials to enhance the ability of functional materials through redirecting force scheme, and to make multifunctional composites using constructive interaction of functional materials.

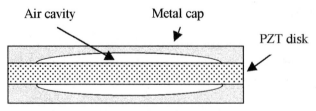

Fig. 3.17. Cross–section of a 'moonie' transducer [18]

3.6 Summary

In this chapter we have discussed some fundamental principles of functional materials using a few examples. Nature has provided us with many functional materials but at the same time puts some limitations on these materials. The objective of materials engineering is to stretch the limits of these functional materials. Better functional materials with multifunctional properties may be developed through innovative engineering at all three levels following those design philosophies. There is no limit to human imagination; we may expect many new functional materials being developed in the near future, especially with the strong drive of adaptronics.

References

1. Moulson, A.J. and Herbert, J.M.: *Electroceramics,* Chapman & Hall, London (1990).
2. Evans, A.G.: 'Advances in Ceramics', Vol. 12, *Science and Technology of Zirconia II,* Ed. Claussen, N.; Rühle, M. and A.H. Heuer, Am. Ceram. Soc., Columbus, OH, pp. 193–212 (1984).
3. Muddle, B.C. and Hannink, R.H.J.: J. Am. Ceram. Soc., **69** [7], pp. 547–555 (1986).
4. Shirane, G.; Pepinsky, R. and Frazer, D.C.: Acta Crystallogr., **9**, pp. 131–140 (1956).
5. Merz, W.J.: Phys. Rev., **76**, pp. 1221–1225 (1949).
6. Wang, Z.L. and Kang, Z.C.: *Functional and Smart Materials,* Plenum Press, New York (1998).
7. Luchaninov, A.G.; Shil'nikov, A.V.; Shuvolov, L.A. and Shipkova, I.JU.: *Ferroelectrics,* **29**, pp. 47–50 (1980).
8. Ashcroft, N.W. and Mermin, N.D.: *Solid State Physics,* Saunders College, Philadelphia (1976).
9. Cao, W. and Cross, L.E.: *Phys. Rev. B,* **44**, pp. 5–12 (1991).
10. Cao, W.; Krumhansl, J.A. and Gooding, R.: *Phys. Rev. B,* **41**, pp. 11.319–11.327 (1990).
11. Newnham, R.E.: Ann. Rev. Mater. Sci., **16**, pp. 47–68 (1986).
12. Clark, A.E.: *Ferromagnetic Materials,* 1, Ed. Wohlfarth, E.P., North Holland, Amsterdam (1980).
13. Newnham, R.E.: MRS Bulletin, **22**, No. 5, pp. 20–34 (1997).
14. Park, S.E. and Shrout, T.: J. Mat. Res. Innov., **1**, pp. 20–25 (1997).
15. Noemura, S.; Takahashi, T. and Yokomizo, Y.: J. Phys., Soc., Japan, **27**, p. 262 (1969).
16. Skinner, D.P.; Newnham, R.E. and Cross, L.E.: Mat. Res. Bull., bf 13, pp. 599–607 (1978).
17. Cao, W.; Zhang, Q. and Cross, L.E.: J. Appl. Phys., **72**, pp. 5814–5821 (1992).
18. Xu, Q.C.; Yoshikawa, S.; Belsick, J.R. and Newnham, R.E.: IEEE Trans. Ultr. Ferr. Freq. Contr., **38**, pp. 634–639 (1991).

4. Adaptive Control Concepts: Controllers in Adaptronics

V. Rao and R. Damle

4.1 Introduction

In recent years, adaptive control of smart structures has become an important component of multidisciplinary research into vibration suppression. The design of controllers for smart structures is a challenging problem because of the presence of nonlinearities in the structural system and actuators, limited availability of control force, and nonavailability of accurate mathematical models. In this study, adaptive control algorithms are being investigated for designing active controllers for smart structures. Both conventional and neural network–based adaptive controllers have been designed and implemented on smart structure test articles. In addition, a neural–network based optimizing control algorithm with on–line adaptation capabilities has been developed that can incorporate nonlinearities in the smart structural system, accommodate the limited control effort and adapt on–line to time–varying dynamical properties. In this algorithm the control signal is computed iteratively while minimizing a Linear Quadratic (LQ) performance index with additional weighting on the control increments.

A central goal of research into adaptive control is to develop control algorithms for time–varying systems, nonlinear systems and systems with unknown parameters [1]–[6]. These controllers have the ability to adjust controller gains for multiple operating points. The adaptive control techniques have been extensively employed for designing controllers for various industrial systems. One of the objectives of this research is to investigate the applicability of adaptive control algorithms for smart structures. When the desired performance of an unknown plant with respect to an input signal can be specified in the form of a linear or a nonlinear differential equation (or difference equation), stable control can be achieved using Model Reference Adaptive Control (MRAC) techniques. The idea behind MRAC is to use the output error between the plant and a specified reference–model to adjust the controller parameters. There are two basic approaches to MRAC. When the controller parameters $\theta(k)$ are directly adjusted to reduce some norm of the output error between the reference model and the plant, it is called direct control. In indirect control, the parameters of the plant are estimated as the elements of a vector $\hat{p}(k)$ at each instant k, and the parameter vector $\theta(k)$ of the controller is chosen assuming that $\hat{p}(k)$ represents the true value of the plant parameter vector $p(k)$. Both the direct control and indirect control algorithms have been implemented on the smart structure, resulting in the following model.

Having successfully implemented conventional MRAC techniques, the next logical step was to try to incorporate the MRAC techniques into a neural network–based adaptive control system. The ability of multilayered neural networks to approximate linear as well as nonlinear functions is well documented and has found extensive application in the area of system identification and adaptive control. The noise–rejection properties of neural networks makes them particularly useful in smart structure applications. Adaptive control schemes require only limited *a priori* knowledge about the system to be controlled. The methodology also involves identification of the plant model, followed by adaptation of the controller parameters based on a continuously updated plant model. These properties of adaptive control methods makes neural networks ideally suited for both identification and control aspects [7]-[11].

A major problem in implementing neural network–based MRAC is translating the output error between the plant and the reference model to an error in the controller output, which can then be used to update the neural controller weights. One recently proposed solution to this problem is based on a constrained iterative inversion of a neural model of the forward dynamics of the plant [12]. This technique predicts the actual and desired output errors to calculate the necessary control signal at the next time instant. The algorithm has shown promise in that it offers a degree of robustness and generates a smooth control. It is from this iterative inversion process that the 'update' method described herein is derived. We use the neural identification model to find the instantaneous derivative of the unknown plant at one instant in time. The derivative is then used iteratively to search the input space of the system to find the input $u^*(k)$ that would have resulted in the correct system output. The control signal error $e_u(k) = u^*(k) - u(k)$ can then be used with a static backpropagation algorithm [10] to update the weights of the neural controller.

For the implementation of MRAC algorithms, we propose to investigate the use of neural networks in order to identify a linear model of a system with the objective of adjusting the parameters of a neural controller to reflect the changes in the plant parameters. This method would be particularly useful when the parameters of the plant change considerably with changes in its operating condition.

A neural network–based Eigensystem Realization Algorithm (ERA) [13] has been utilized to generate a mathematical model of the structural system. For smart structure applications, the size of such networks becomes very large. Therefore, we have developed an adaptive neuron–activation function and an accelerated adaptive learning–rate algorithm, which together significantly reduce the learning time of a neural network. The models obtained by these identification techniques are compared with that obtained from the swept sinewave testing and curve fitting methods [13, 14].

The remainder of this chapter is arranged as follows. A brief description of the two smart structure test articles used to evaluate the adaptive control algorithms is given in Sect. 4.2. Sect. 4.3 includes the outlines of the conventional model–reference adaptive control techniques and their experimental closed–loop performances on the cantilever beam smart structure test article. The neural network–based model–reference adaptive control algorithm and the neural network–based optimizing controller with on–line adaptation have been introduced in Sect. 4.4. The adaptive neuron activation function and an on–line adaptive control algorithm for the neural network–based model–reference adaptive control algorithm are also described in this latter section. The closed–loop performance of adaptive controllers are demonstrated using simulation studies on the two smart structure test article models. Finally, the results of this study are summarized in Sect. 4.5.

4.2 Description of the Test Articles

To demonstrate some of the capabilities of adaptive control using neural networks on smart structures and to determine the limitations imposed by hardware realization, we have designed and fabricated an experimental test article. The smart structure test article was an aluminum cantilever beam with shape memory actuators, strain–gauge sensors, signal–processing circuits and digital controllers. A schematic diagram of the cantilever beam is shown in Fig. 4.1. The system is a single–input, single–output (SISO) system with one actuator and one sensor.

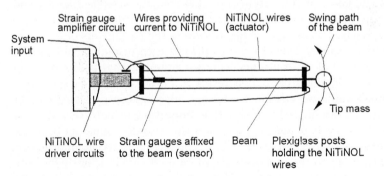

Fig. 4.1. Schematic of cantilever beam test article

The neural network–based control algorithm described in Sect. 4.4 is tested using simulation studies on a cantilever plate system with PZT actuators and PVDF film sensors. A top–view line diagram of the plate structure is shown in Fig. 4.2.

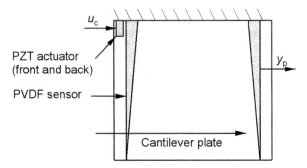

Fig. 4.2. Top view of the plate system

The PVDF film sensors are shaped to measure the displacement and velocity at the free end of the plate [15]. The output of the PVDF film sensor is buffered through a high–pass filter for an output in the range of ± 1 V for a nominal tip displacement of 0.5 inches. The PZT actuators are driven by a high–voltage amplifier such that the control input is in the range of ± 5 V and uses the full linear operating range of the PZT.

4.3 Conventional Model–Reference Adaptive Control Techniques

For many years, there have basically been two distinct methods for finding the solution of the adaptive control problem [2]. These are *direct* and *indirect control methods*. When the controller parameters $\theta(k)$ are directly adjusted to reduce some norm of the output error between the reference model and the plant, this is called direct control or *implicit identification*. In indirect control, also referred to as *explicit identification*, the parameters of the plant are estimated as the elements of a vector $\hat{p}(k)$ at each instant k, and the parameter vector $\theta(k)$ of the controller is chosen assuming that $\hat{p}(k)$ represents the true value of the plant parameter vector p. Figs. 4.3 and 4.4 respectively show the direct and indirect model–reference adaptive control structures for a Linear Time Invariant (LTI) plant. It is important to note that in both cases efforts have to be made to probe the system to determine its behavior because control action is being taken based on the most recent information available. The input to the process is therefore used simultaneously for both identification and control purposes. However, not every estimation scheme followed by a suitable control action will result in optimal or even stable behavior of the overall system; therefore, considerable care must be taken in blending estimation and control schemes in order to achieve the desired performance [2].

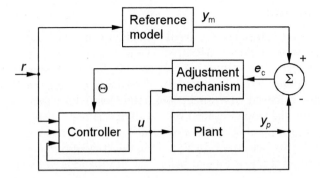

Fig. 4.3. Direct model–reference adaptive control structure

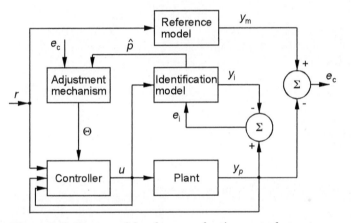

Fig. 4.4. Indirect model–reference adaptive control structure

4.3.1 Experimental Results

Direct Model–Reference Adaptive Control. The first controller implemented on the structure was the direct MRAC shown in Fig. 4.5. This gives
a basis for comparison between direct and indirect control. Fig. 4.6a shows a
plot of the open–loop response envelope, the desired response envelope, and
the closed–loop response achieved. As can be seen, the closed–loop system
adapts to the reference–model response until the deadband is reached (after
approximately 11 s), at which point adaptation is turned off. The deadband is
inherent to the NiTiNOL wire actuators. In Fig. 4.6b, the control parameter
vector $\theta(k)$ has stabilized after about 8 s and before the deadband is reached.

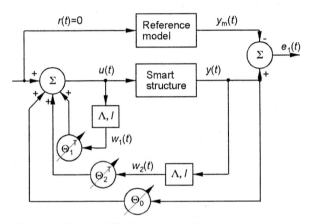

Fig. 4.5. Direct MRAC structure for smart structures

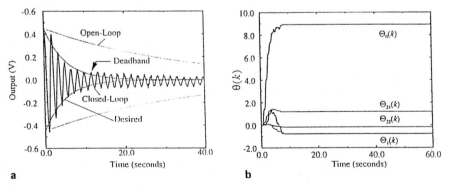

Fig. 4.6. Time–response comparison and evolution for a closed–loop system (direct
MRAC):**(a)** variation of output with time; **(b)** variation of $\theta(k)$ with time

The final values of the controller parameters are given in Table 4.1.

Table 4.1. Final values of the controller gains (direct MRAC)

$\theta(k)$	Final value
$\theta_1(k)$	$\begin{bmatrix} -0.78069 & 0.75716 \end{bmatrix}^T$
$\theta_0(k)$	8.92846
$\theta_2(k)$	$\begin{bmatrix} 1.18814 & -0.16468 \end{bmatrix}^T$

Indirect Model–Reference Adaptive Control. Next, an indirect MRAC was implemented on the structure, as shown in Fig. 4.7. Fig. 4.8a shows a plot of the open–loop response envelope, the desired response envelope, and the losed–loop response and Fig. 4.8b shows the time evolution of the control parameter vector. Again, the parameters converge after about 8s with the deadband reached by 11s.

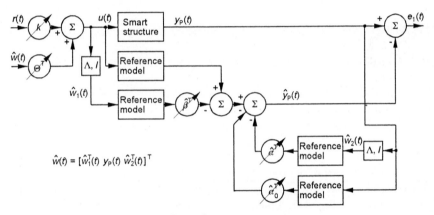

Fig. 4.7. Indirect MRAC regulator for smart structures

The final values of the controller parameters are given in Table 4.2.

Table 4.2. Controller parameters of indirect MRAC

$\theta(k)$	Final Value
$\theta_1(k)$	$\begin{bmatrix} 0.98276 & -1.01170 \end{bmatrix}^T$
$\theta_0(k)$	9.36202
$\theta_2(k)$	$\begin{bmatrix} 1.61385 & -0.14307 \end{bmatrix}^T$

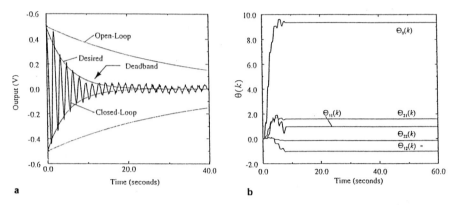

Fig. 4.8. Time–response comparison and evolution for a closed-loop system (indirect MRAC): **(a)** variation of output with time; **(b)** variation of $\boldsymbol{\theta}(k)$ with time

4.4 Adaptive Control Using Neural Networks

4.4.1 Neural Network–Based Model Reference Adaptive Control

After successful implementation of conventional model–reference adaptive controllers on smart structures, the next logical step was to investigate the possibility of using a neural network for adaptive control implementations. The linear and nonlinear mapping properties of neural networks have been extensively utilized in the design of multilayered feedforward neural networks for the implementation of adaptive control algorithms [10].

A schematic diagram of the neural network–based adaptive control technique is shown in Fig. 4.9. A neural network identification model is trained using a static backpropagation algorithm to generate $\hat{\boldsymbol{y}}_{\mathrm{p}}(k+1)$, given past values of $\boldsymbol{y}$ and $\boldsymbol{u}$. The identification error is then used to update the weights of neural identification model. The control error is used to update the weights of the neurocontrollers. Narendra [2] has demonstrated that closed–loop systems may result in unbounded solutions even if the plant is bounded–input and bounded–output stable. In order to avoid such instability, he has suggested that sufficient identification should be made before control is initiated. He has also suggested that the update rate of the identification and controller weights should be chosen carefully. Hoskins et al. [12] have presented a control optimization using a constrained iterative inversion process in order to dynamically search the input space of the identification process. This process provides stability and robustness measures for neural network–based adaptive control systems. We have utilized this technique for designing on–line adaptive algorithms; we have also developed a method for directly deriving a state–variable model using multilayered neural network. These models are useful in generating adaptation data for neural controllers.

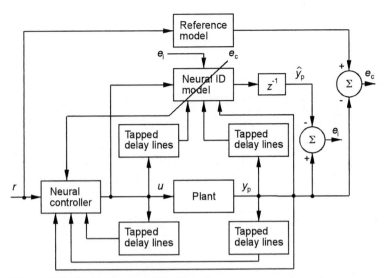

Fig. 4.9. MRAC using neural networks

In the neural network–based adaptive control scheme, a neurocontroller is trained to approximate an inverse model of the plant. We have introduced an adaptive activation function for increasing the training rate of the neural controller, and the proposed function is described in this section.

Adaptive Activation Function. In order to train a neural controller, a multilayered network with linear activation functions was initially considered. During the training process, a large sum–squared error occurred due to the unbounded nature of the linear activation function that caused a floating point overflow. To avoid the floating point overflow we used the hyperbolic tangent activation functions in the hidden layers of the network. The network was unable to identify the forward dynamics of the controller. To overcome this problem, we are proposing an activation function which adapts its shape depending upon the sum–squared error, as shown in Fig. 4.10.

The proposed adaptive activation function is governed by the equation

$$\Gamma(x) = \left(\frac{s+c}{s+1}\right)\tanh\left(\frac{s+1}{s+c}x\right) \tag{4.1}$$

where s is the sum–squared error over the previous time period and c is an arbitrary constant. The transition from a hyperbolic tangent to a linear function is shown in Fig. 4.10

The function has the properties of

$$
\begin{aligned}
\Gamma(x) &\rightarrow \tanh(x) & \text{as } s \gg c, \text{ and} \\
\Gamma(x) &\rightarrow c \cdot \tanh\left(\frac{x}{c}\right) & \text{as } s \ll c
\end{aligned}
\tag{4.2}
$$

When the constant c is chosen large enough, the adaptive activation function can be replaced with a linear activation for implementation with no retraining

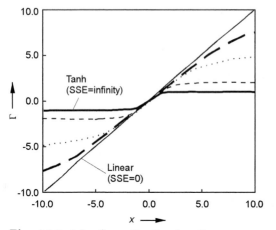

Fig. 4.10. Adaptive activation function

needed. This procedure allows for a one–stage training session of the neural network.

For practical reasons when using the backpropagation training algorithm, it is convenient to be able to express the derivative of an activation function in terms of the activation function itself. The derivative of the adaptive activation function can also be expressed in the form

$$\frac{\mathrm{d}\Gamma(x)}{\mathrm{d}x} = 1 - [\Gamma(x)]^2 \tag{4.3}$$

The proposed activation function was successfully implemented in the training algorithm. The adaptive activation function is also feasible for hardware implementation. Specifically, the Intel i80170 *Electronically Trainable Artificial Neural Network* (ETANN) chip [16] has an external voltage that controls the slope of the activation function. The control level could easily be made a function of the sum–squared error during training and held at the last sum–squared error achieved.

On–Line Adaptive Control Algorithm. A neural network–based model reference adaptive control scheme for nonlinear plants is presented in this section.

Let a system be described by a nonlinear difference equation

$$\boldsymbol{y}_p(k+1) = f[\boldsymbol{Y}_{k,n}(k)] + g[\boldsymbol{U}_{k,m}(k)] \tag{4.4}$$

where f and g are both nonlinear functions in $\boldsymbol{y}$ and $\boldsymbol{u}$, respectively. This model requires two neural networks to identify the plant, one for each nonlinear function, as shown in Fig. 4.11.

For simplicity, let us assume that the function f is linear and g is nonlinear. Then a series parallel neural identification model will have a form

$$\hat{\boldsymbol{y}}_p(k+1) = \hat{f}[\boldsymbol{y}_{k,n}(k)] + N_g[\boldsymbol{u}_{k,m}(k)], \tag{4.5}$$

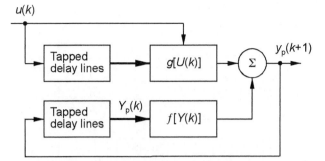

Fig. 4.11. Identification scheme for plant

where the reference model is represented by

$$y_{mm}(k+1) = f[\mathbf{y}_{k,n}(k), \mathbf{R}_{k,m}(k)]. \tag{4.6}$$

The desired control signal $\mathbf{u}(k)$ can be computed by

$$\mathbf{u}(k) = \hat{g}^{-1}[-\hat{f}[\mathbf{y}_{k,n}(k)] + f[\mathbf{y}_{k,n}(k), \mathbf{R}_{k,m}(k)]]. \tag{4.7}$$

The schematic diagram of the model reference adaptive control system is shown in Fig. 4.12.

4.4.2 Neural Network–Based Optimizing Controller With On–Line Adaptation

In this section, a neural network–based design methodology is developed that utilizes the adaptability of neural networks to compensate for the time varying dynamical properties of smart structures. This formulation is designed to be implemented using the ETANN chip and also allows the designer to directly incorporate all the a priori information about the system that may be available. An important feature of this formulation is that it relies only on the experimental input/output data of the system for the design. The ability of neural networks to map nonlinear systems allows this formulation to be extended to incorporate nonlinearity in structural systems.

A functional block diagram of the controller is shown in Fig. 4.13, where the structural system can be represented by

$$\mathbf{y}_p(k+1) = \Phi(\mathbf{y}_p(k), \mathbf{u}_c(k)), \tag{4.8}$$

where Φ can be a linear or a nonlinear function.

The neural network in the controller block diagram has a Model IV architecture with one hidden layer, as shown in Fig. 4.14. It is pretrained to the dynamics of the smart structural system using experimental input/output data. As shown in Fig. 4.15, the input vector to the network consists of $n+1$ samples of the plant input and $m+1$ samples of the plant output. The hidden and output layers have P and 1 neurons respectively.

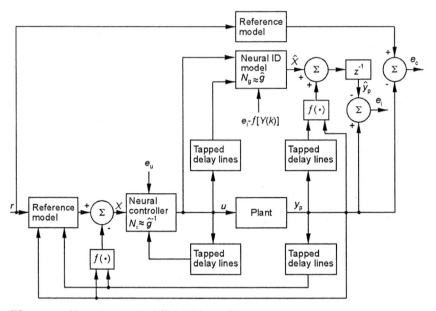

Fig. 4.12. Neural network MRAC block diagram

The activation function of the neurons in the hidden layer is the adaptive activation function (4.1). Models II and III are alternative neural network architectures that can be used to model a dynamical system. Model III is similar to Model IV except for the additional external adder and separate network for the plant input and output parts. Model III can be used to implement high–order dynamical system models using hardware neural networks like the ETANN.

The feedforward equation of the network in Fig. 4.15 can be written as follows. Defining

$$W1_1 = \begin{bmatrix} W1_{11} \\ W1_{21} \\ \cdots \\ W1_{P1} \end{bmatrix}, \quad W1_2 = \begin{bmatrix} W1_{12} & \cdots & W1_{1N} \\ W1_{22} & \cdots & \cdots \\ \cdots & \cdots & \cdots \\ W1 & \cdots & W1_{PN} \end{bmatrix}, \quad \text{and}$$

$$u_2 = \begin{bmatrix} u_c(k-1) \\ \cdots \\ u_c(k-n) \\ y_p(k) \\ \cdots \\ y_p(k-m) \end{bmatrix}, \tag{4.9}$$

the outputs of each of the layers can be written as

$$z = \Gamma(W1_1 \cdot u_c(k) + W1_2 u_2) \quad \text{and} \tag{4.10}$$

$$y_{nn}(k+1) = \Gamma(W2 \cdot \Gamma(W1_1 \cdot u_c(k) + W1_2 \cdot u_2)) \tag{4.11}$$

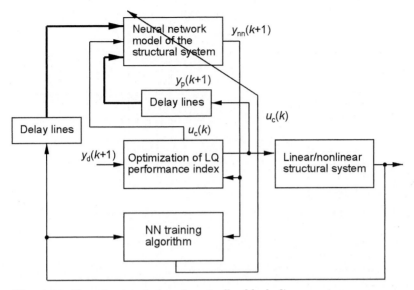

Fig. 4.13. Neural network–based controller block diagram

In the optimization block, the control input applied to the smart structural system is obtained by minimizing a generalized Linear Quadratic (LQ) performance index with weights on the control moves. The performance index is given by

$$\underset{u_c(k)}{\text{Min} J} = \frac{1}{2} E^T Q E + \frac{1}{2} \Delta u^T R \Delta u \tag{4.12}$$

under the constraint given (4.11). The error E is given by $E = y_{nn}(k+1) - y_d(k+1)$ and the control move Δu is given by $\Delta u = u_c(k) - u_c(k-1)$. Q is symmetric positive semi–definite and R is symmetric positive definite. The desired output y_d can either be a constant (regulator problem) or varying (tracking problem). The existence of weights on the control moves alleviates the problem of requiring large sampling times when non–minimum phase zeros exist in plants even in linear unconstrained optimization [17].

In addition to the constraint given by (4.11), any a priori information about the system or the sensors and actuators can be incorporated as additional constraints. Some of the commonly known constraints, such as control effort limits, actuator bandwidth limits and structural bandwidth limits, can be described by

$$
\begin{array}{ccccc}
\Delta u_L & \leq & \Delta u & \leq & \Delta u_H & \text{Actuator bandwidth limits} \\
u_L & \leq & u_c(k) & \leq & u_H & \text{Control effort limits} \\
\Delta y_L & \leq & \Delta y(k) & \leq & \Delta y_H & \text{Structural bandwidth limits}
\end{array} \tag{4.13}
$$

Since this study is restricted to a structural system that is operated in its linear region, the adaptive activation functions approximate to a linear

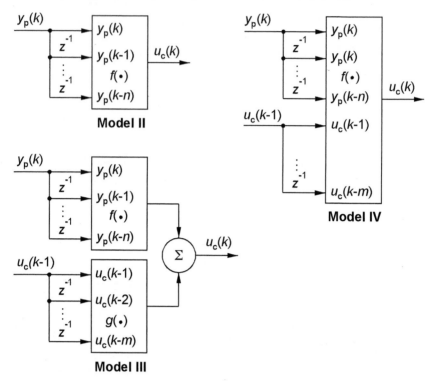

Fig. 4.14. ETANN implementation architectures

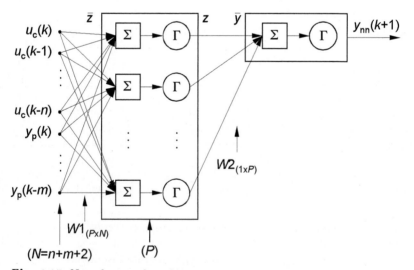

Fig. 4.15. Neural network architecture

function after sufficient training. Therefore the general nonlinear optimization problem given by (4.11) – (4.13) can be simplified for a linear case. After the neural network is sufficiently trained, (4.11) can be written as

$$y_{nn}(k+1) = W_1 \cdot u_c(k) + C_1, \tag{4.14}$$

where $W_1 = W2 \cdot W1_1$ and $C_1 = W2 \cdot W1_2 \cdot u_2$.

Substitution of (4.14) in the error equation above yields

$$
\begin{aligned}
E &= y_p(k+1) - y_d(k+1), \quad \text{or} \\
E &= W_1 \cdot u_c(k) + C_2
\end{aligned}
\tag{4.15}
$$

where $C_2 = C_1 - y_d(k+1)$. The control move equations can then be written as

$$\Delta u = u_c(k) - u_c(k-1) = C_2 - T_1, \tag{4.16}$$

where $T_1 = u_c(k-1)$.

For a single input–single output system, the LQ performance index (4.12) can be written as

$$
\begin{aligned}
J &= \frac{1}{2}QE^2 + \frac{1}{2}R(\Delta u)^2, \quad \text{or} \\
J &= \frac{1}{2}(W_1^2 \cdot Q + R)u_c^2(k) + (2W_1 \cdot C_2 \cdot Q - 2T_1 \cdot R)u_c(k) \\
&\quad + \frac{1}{2}C_2^2 \cdot Q + \frac{1}{2}R \cdot T_1^2.
\end{aligned}
\tag{4.17}
$$

This optimization problem can be solved for $u_c(k)$ using any of the standard optimization algorithms [18].

4.4.3 Simulation Results

Neural Network–Based MRAC. A linear identification model of the cantilever beam test article was obtained and used as N_g in Fig. 4.12. A $3\times10\times5\times1$ neural controller N_c was initially trained off–line using the inverse mapping technique. It was possible to obtain the inverse plant model of the cantilever beam test article because the plant model was at minimum phase. With linear activation functions, the initial sum-squared was of the order of 1.5×10^3, which resulted into an overflow in subsequent time periods. Therefore, the activation functions in the hidden layers were changed to hyperbolic tangent functions. This resulted in a poor sum–squared error performance. Therefore, adaptive neurons described earlier in this Sect. 4.4 were used in the hidden layers.

Once sufficiently trained, the controller N_c was placed in the loop. A neural network–MRAC simulation program was written in C++. This program was used to simulate the response to an initial condition when the reference command signal was 0 for time k, i.e., regulation. Fig. 4.16a shows the open–loop, the desired response of the reference model, and the actual closed–loop plant output. The output of the closed–loop plant goes to zero in about 20s as compared to 80s with open loop. This is adequate control for the limited bandwidth and control effort that can be obtained from the NiTiNOL wire actuators. The control signal provided by N_c is shown in Fig. 4.16b. The control error $X(k) - \hat{X}(k)$ and the control signal error $\Delta u(k) = u^*(k) - u(k)$ are shown in Fig. 4.17. The control signal error is calculated through the 'iterative inversion' process [12].

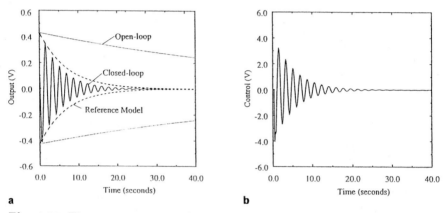

a   **b**

Fig. 4.16. Time–response comparison and evolution for a closed–loop neural network MRAC system: **(a)** variation of output with time; **(b)** variation of control signal with time

Neural Network–Based Optimizing Controller with On–Line Adaptation. To demonstrate the capabilities of the proposed neural network–based controller, a difference equation form of the cantilever plate model was used. With a sampling time of 1/120s, the model is given by

$$
y_p(k+1) = \begin{bmatrix} 2.7013627 \\ -3.6973171 \\ 2.6817469 \\ -9.8675516 \cdot 10^{-1} \\ -1.4358223 \cdot 10^{-3} \\ 6.0667673 \cdot 10^{-4} \\ 6.0650603 \cdot 10^{-4} \\ -1.4234998 \cdot 10^{-3} \end{bmatrix}^T \begin{bmatrix} y_p(k) \\ y_p(k-1) \\ y_p(k-2) \\ y_p(k-3) \\ u_c(k) \\ u_c(k-1) \\ u_c(k-2) \\ u_c(k-3) \end{bmatrix} \tag{4.18}
$$

A three layered feedforward network with $n = 5$ and $m = 7$ is initially trained using the response of the plate for a normally distributed random input. The weights on the error E and the control moves Δu used are $Q = 1$

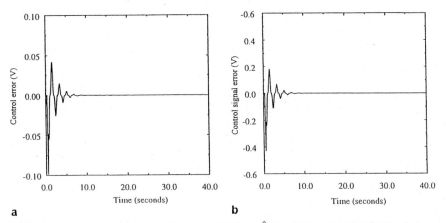

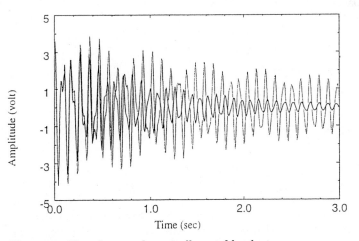

a   **b**

Fig. 4.17. Variation of control error $X(k) - \hat{X}(k)$ **(a)** and control signal error $u^*(k) - u(k)$ **(b)** over time in a closed–loop neural network MRAC system

and $R = 0.001$ respectively. Additional constraints due to the limits on the available control effort and the actuator bandwidth are given as

$$-5 \le u_c(k) \le 5, \quad \text{and}$$
$$-10 \le \Delta u \le 10. \tag{4.19}$$

First the plate system given by (4.18) is controlled as a regulator with the desired output $y_d(k+1) = 0$. The open–loop response of the plate system is compared with the controlled response shown in Fig. 4.18. In this case, on–line adaptation of the neural network was turned off. As shown in Fig. 4.18 the vibrations of the plate are damped out to within $\pm 0.1\,\mathrm{V}$ in about $3\,\mathrm{s}$.

Fig. 4.18. Neural network controller: stable plant

To demonstrate the capability of the controller to adapt on–line to the changes in the structural system, the coefficients of the plate model in (4.18) were altered so as to destabilize the first mode. The model of the plate with a unstable mode is given by

$$y_\mathrm{p}(k+1) = \begin{bmatrix} 2.7104 \\ -3.6973 \\ 2.6817 \\ -9.868 \cdot 10^{-1} \\ -1.4 \cdot 10^{-3} \\ 6.0 \cdot 10^{-4} \\ 6.0 \cdot 10^{-4} \\ -1.4 \cdot 10^{-3} \end{bmatrix}^T \begin{bmatrix} y_\mathrm{p}(k) \\ y_\mathrm{p}(k-1) \\ y_\mathrm{p}(k-2) \\ y_\mathrm{p}(k-3) \\ u_\mathrm{c}(k) \\ u_\mathrm{c}(k-1) \\ u_\mathrm{c}(k-2) \\ u_\mathrm{c}(k-3) \end{bmatrix} \tag{4.20}$$

This change in the plant model was introduced after $0.2\,\mathrm{s}$ in the simulation with on–line adaptation turned on. The closed–loop response of the controller with on–line adaptation is compared to the open–loop response in Fig. 4.19. The controller adapts on–line the unstable dynamics of the plant during normal operation and approaches the desired output level.

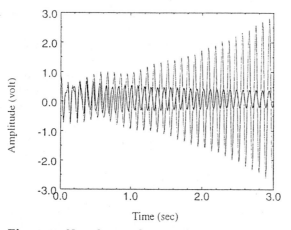

Fig. 4.19. Neural network controller: unstable plant

4.5 Summary

In this study, adaptive control algorithms have been utilized for designing active controllers for smart structure test articles. Adaptive control schemes require only a limited a priori knowledge about the system in order to be controlled. The availability of limited control force and inherent deadband

and saturation effects of shape memory actuators are incorporated in the selection of the reference model. The vibration suppression properties of smart structures were successfully demonstrated by implementing the conventional model reference adaptive controllers on the smart structure test articles. The controller parameters converged to steady state values within 8s for both direct and indirect MRACs.

Various neural network–based adaptive control techniques were discussed in this study. A major problem in implementing neural network–based MRACs is the translation of the output error between the plant and the reference model so as to train the neural controller. A technique called 'iterative inversion', to invert the neural identification model of the plant for calculating neural controller gains, has been used. Due to the real–time computer hardware limitations, the performance of neural network–based adaptive control systems is verified using simulation studies only. These results show that neural–network based MRACs can be designed and implemented on smart structures.

A neural network–based control algorithm based on a LQ performance index which can be implemented using the ETANN chip has been developed. This formulation incorporates a priori information about the structural system. Information such as limits on the control effort and limits on the bandwidths of the sensors and actuators can be incorporated in this formulation. The on–line adaptability property of the ETANN chip–based neural network is also utilized to adapt the controller to time–varying structural systems. The capabilities of this algorithm have been demonstrated on the smart plate system through simulation studies. The ability of neural networks to map nonlinear dynamics as well as linear dynamics makes the control algorithm valid for control of smart structural systems with nonlinearities.

References

1. Aström, K. and Wittenmark, B.: *Adaptive Control.* Addison–Wesley, Reading, MA, 1989, pp. 105–156
2. Narendra, K. and Annaswamy, A.: *Stable Adaptive Control.* Prentice Hall, Englewood Cliffs, NJ, 1989, pp. 21–28, pp. 182–232, pp. 318–345
3. Narendra, K.: Adaptive Control of Dynamical Systems. *Handbook of Intelligent Control: Neural, Fuzzy and Adaptive Approaches.* Editors White, D.; Sofge, D., Van Nostrand Reinhold, New York, NY, 1992
4. Narendra, K.; Duarte, M.: Combined Direct and Indirect Adaptive Control of Plants with a Relative Degree Greater than One, Technical Report #8715., Center for Systems Science, Yale University, New Haven, CT, November 1987
5. Isermann, R.: *Digital Control Systems.* Springer–Verlag, Vol. 1: Fundamentals, Deterministic Control. 2nd rev. ed. 1989, Vol. 2: Stochastic Control, Adaptive Control Multivariable Control, Adaptive Control, Applications. 2nd rev. ed. 1991

6. Rao, V.; Damle, R.; Tebbe, C. and Kern, F.: The Adaptive Control of Smart Structures using Neural Networks. Smart Materials and Structures, 1994, No. 3, pp. 354–366

7. Chen, F.; Khalil, H.K.: Adaptive Control of Nonlinear Systems using Neural Networks — A Dead–Zone Approach. Proceedings of the American Control Conference, 1990, pp. 667–672

8. Chen, F.: Adaptive Control of Nonlinear Systems using Neural Networks. A Ph.D. Dissertation, Dept. of Electrical Engineering, Michigan State University, 1990

9. Tzirkel–Hancock, E.; Fallside, F.: Stable Control of Nonlinear Systems using Neural Networks. Technical Report CUED/F–INFENG/TR.81, July 1991, Cambridge University Engineering Department

10. Narendra, K.; Parthasarathy, K.: Identification and Control of Dynamical Systems Using Neural Networks. IEEE Transactions on Neural Networks, March 1990, pp. 4–27

11. Hoskins, D.A.: Neural Network Based Model–Reference Adaptive Control. Ph. D. Dissertation, University of Washington, UMI Dissertation Services, Ann Arbor, MI, 1990

12. Hoskins, D.A.; Hwang, J.N.; Vagners, J.: Iterative Inversion of Neural Networks and Its Application to Adaptive Control. IEEE Transactions on Neural Networks, March 1992, pp. 292–301

13. Damle, R.; Lashlee, R.: Identification and Robust Control of Smart Structures using Artificial Neural Networks. Submitted to Journal of Smart Structures and Materials.

14. Lashlee, R.; Butler, R.; Rao, V. and Kern, F.: Robust Control of Flexible Structures Using Multiple Shape Memory Alloy Actuators. North American Conference on Smart Structures and Materials, Albuquerque, NM, Feb. 1993

15. Butler, R. and Rao, S.V.: Identification and control of two–dimensional smart structures using distributed sensors. Proceedings of the North American Conference on Smart Structures and Materials, San Diego, CA, March 1995, SPIE **2442**, pp. 58–68

16. *Intel 80170NX Electrically Trainable Analog Neural Network Data Book,* June 1991

17. Garcia, C.E. and Morari, M.: Internal Model Control — Multivariable Control Law Computation and Tuning. Industrial Engineering Chemical Process Design and Development, 1985, **24**, pp. 484–494

18. *OPTIMIZATION TOOLBOX Users Guide,* November 1990, The MathWorks Inc.

5. Simulation of Adaptronic Systems

H. Baier and F. Döngi

5.1 Introduction

The adaptronic systems discussed in this book can be looked upon as dynamic systems with time–varying states subjected to external disturbances. To analyse and simulate such systems, basic ideas of control and system theory can be applied (see Sect. 5.2). The text concentrates on linearized, time–continuous descriptions of adaptive structures. More detailed information can be found in standard textbooks on linear (e.g. [1]) and nonlinear systems (e.g. [2]).

Among the various types of adaptronic systems, adaptive or smart structures are one of the most important groups. The steps to be taken towards a mathematical model of an adaptive structure are outlined in Sect. 5.3.

Once a mathematical model of the adaptronic system is derived and implemented numerically, analyses and simulations have to be carried out to characterize its dynamic behavior. A survey of methods and algorithms is given in Sect. 5.4. Objectives such as stability, performance and robustness, especially in the case of actively controlled structures, are discussed.

The mathematical formulation of the design process of adaptronic systems leads to multidisciplinary and also multicriteria optimization problems, where proper consideration of the interaction of structures and control is essential. Sect. 5.5 combines a discussion of problem statements and solution approaches with some practical examples.

Finally, tools for modeling, analysis, simulation and hardware realization of adaptronic systems are surveyed in Sect. 5.6.

5.2 Basic Elements of System Theory

5.2.1 Nonlinear and Linear Systems

In the first place, all dynamic systems in nature exhibit nonlinear characteristics. Some adaptive materials such as electrostrictive and shape memory alloys imply strongly nonlinear constitutive behavior which requires special effort with respect to modeling and simulation techniques (see for example [3, 4, 5]). However, in many other cases it is possible to derive a linearized

description of the dynamics of an adaptive structure about a chosen state in
the system. Consider the system dynamic behavior described by a set of n
first–order nonlinear differential equations

$$\dot{x} = f(x, u), \tag{5.1}$$

where x and u denote state variables and external influences on the system,
respectively. The overdot symbolizes differentiation with respect to time. A
linearized representation about the equilibrium state $x = 0, u = 0$ is given
by

$$\dot{x} = \left.\frac{\partial}{\partial x} f(x, u)\right|_{x=0, u=0} x + \left.\frac{\partial}{\partial u} f(x, u)\right|_{x=0, u=0} u. \tag{5.2}$$

5.2.2 State–Space Representation

Equation 5.2 represents the state differential equation of the linearized system:

$$\dot{x} = A x + B u \tag{5.3}$$

Combined with the output equation

$$y = C x + D u, \tag{5.4}$$

this description is called the state–space representation of the system. Its
dynamic variables are arrayed in a state vector x $(n \times 1)$. Physical quantities
that exert influences on the system (e.g. actuator forces) are collected in an
input vector u $(p \times 1)$, and measured quantities (e.g. sensor signals) in an
output vector y $(q \times 1)$. In the case of actively controlled adaptronic systems,
the task of control design is to find a suitable input time history $u(t)$ from
a given output time history $y(t)$ such that the system exhibits desirable
dynamic behavior.

A $(n \times n)$ is called the state matrix, B $(n \times p)$ the input matrix, C $(q \times n)$
the output matrix and D $(q \times p)$ the feedthrough matrix of the system. For
many systems $D = 0$ except for cases where the input quantities (actuator
forces and moments) have a direct influence on the sensor measurements. For
example, this happens in the case of active struts based on induced strain
actuators in a truss structure, where sensors measure the displacement, strain
or force in the strut and the strain induced by the actuator directly influences
the sensor signal [6].

5.2.3 Controllability and Observability

The efficiency of actuators and sensors in adaptronic systems can be analysed
using the dual concepts of controllability and observability. To make the basic
ideas more clear, adaptive structures are taken as an example.

The dynamic behavior of structural systems can be characterized in terms of natural frequencies and modes, including possible rigid–body modes of articulated systems. If the natural modes of a system are supposed to be actively controlled using smart actuators and sensors, these elements must be able to influence and sense, respectively, the appropriate modal oscillations. As an example, consider an accelerometer placed at a vibration node of a mode shape. The corresponding mode cannot be detected by this sensor, and hence it is not observable. Analogously, a pin force actuator located in a vibration node of a mode shape is unable to excite this mode, which is then said to be not controllable.

If a state–space description (5.3, 5.4) is given of an adaptronic system its observability and controllability can be determined numerically by various methods. A common way is to compute the eigenvalues of the controllability and observability Gramians

$$P = \int_0^\infty e^{At} \, B \, B^\mathrm{T} \, e^{A^\mathrm{T} t} \, \mathrm{d}t \quad , \quad Q = \int_0^\infty e^{A^\mathrm{T} t} \, C^\mathrm{T} \, C \, e^{At} \, \mathrm{d}t \; . \tag{5.5}$$

P and Q possess real non–negative eigenvalues. Large eigenvalues indicate good controllability and observability, respectively, very small or zero eigenvalues correspond to non–controllable and non–observable states, respectively. Every linear time–invariant system (5.3,5.4) can be transformed into its balanced realization [7]. In this case P equals Q, and the Hankel singular values

$$\sigma_\mathrm{k} = \sqrt{\lambda_\mathrm{k}(PQ)} \, , \tag{5.6}$$

where λ_k denote eigenvalues, can be applied to check for both controllability and observability simultaneously. For lightly–damped adaptive structures, e.g. in space applications, the Hankel singular values can be determined from modal data [8] which makes each numerical application very fast and efficient. Hankel singular values are commonly used for model reduction techniques — see Sect. 5.3.6.

5.2.4 Stability and Robustness

The main objective of a controlled dynamic system is its stability. The notion of stability implies that, after a bounded disturbance, the state variables of the system remain bounded, i.e. they stay within a defined space around a selected state (or even approach this state asymptotically). We use the definition of stability in the sense of Lyapunov:

> The state $x = 0$ is said to be stable, if, for any $R > 0$, there exists $r > 0$, such that if $\|x(t = 0)\| < r$, then $\|x(t)\| < R$ for all $t \geq 0$. It is asymptotically stable if in addition there exists some $\bar{r} > 0$ such that $\|x(0)\| < \bar{r}$ implies that $x(t) \to 0$ as $t \to \infty$.

Controllers for adaptronic systems can be designed based on general proofs of stability, as in the case of colocated dissipative controllers, or based on a mathematical model of the system. In the latter case, it is important to represent the dynamics of a system very accurately because the stability and performance of the controller can only be checked with the mathematical model in the first place. Discrepancies between the dynamic behavior of the mathematical model and the real adaptronic system may lead to loss of performance and even instability when the controller is finally implemented with the real system (see Sect. 5.4.2).

The stability of a controlled dynamic system is said to be robust if the controller designed using a mathematical model stabilizes the real system in spite of modeling errors and/or parameter changes in the adaptronic system. A similar definition holds for the robustness of performance.

5.2.5 Alternative System Representations

An equivalent representation of a state–space system (5.3, 5.4) is the Laplace transform transfer function description,

$$G(s) = C(sI - A)^{-1}B + D \, , \tag{5.7}$$

where s is a complex variable. The elements of matrix $G(s)$ are transfer functions $n(s)/d(s)$ with nominator and denominator polynomials $n(s)$ and $d(s)$, respectively. The common denominator of these transfer functions is the characteristic polynomial of $G(s)$, its roots are equivalent to the eigenvalues of the state matrix A. Poles and zeros of the system are evident if the transfer–function matrix $G(s)$ is transformed into its Smith–MacMillan form [9, p.40 f.]. Variations of the transfer–function matrix representation are the zero–pole–gain and partial fraction models [10, p.510 f.], where the transfer functions in $G(s)$ are factorized into nominator/denominator factors and partial fractions, respectively.

5.3 Modeling of Adaptive Structures

In the larger context of adaptronic systems, this chapter concentrates on adaptive or smart structural systems that consist of the structure as a dynamic system combined with integrated, multifunctional (i.e. load–bearing) actuators and sensors made mainly of smart materials such as piezoelectric or magnetostrictive materials. If these material types are used as actuator and/or sensor materials, a linearization of the system equations may easily be found. From the modeling point of view, the situation is much more complex in the case of electrostrictive materials [3] or shape memory alloys [4, 5] that exhibit highly nonlinear constitutive behavior, or in the case of smart polymer gels [11, 12, 13] that imply coupling between large–displacement mechanics, electro–diffusion processes and chemical reactions. Fig. 5.1 depicts the modeling and simulation process of adaptive structures with cross–references to the appropriate sections in the text.

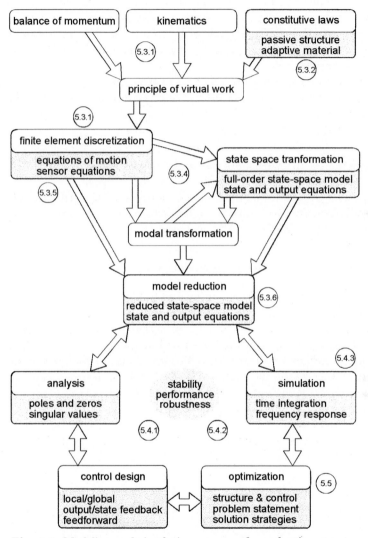

Fig. 5.1. Modeling and simulation process of an adaptive structure

5.3.1 Basic Equations of Structural Mechanics

Consider a linear elastic continuum, which may consist of passive or active, i.e. adaptive, elements. The dynamic equilibrium of the structure can be formulated using the principle of virtual displacements [14, p.121] including inertia loads. To express the internal strain energy in terms of displacement variables, the kinematics of the structure has to be considered. Various types of mechanical structures, e.g. beams, plates, or shells, are defined by kinematic relations and constraints.

Finally, stress σ and strain ϵ are related to each other by the constitutive law of the material. For passive, linear elastic materials the generalized Hooke's law is a valid approximation:

$$\sigma = E\,\epsilon \tag{5.8}$$

Here, E denotes the elasticity tensor of the material. Examples of constitutive laws for adaptive materials are given in the subsequent section.

The equations of dynamic equilibrium, kinematics, and constitutive behavior are combined in the variational formulation on which the discretization using the Finite Element Method (FEM) is based (see Sect. 5.3.3).

5.3.2 Constitutive Laws of Adaptive Materials

In the case of adaptive materials, Hooke's law has to be substituted by a constitutive law that couples the mechanical properties of the material with other physical properties such as electric, magnetic, or thermal entities. Among the large variety of adaptive materials discussed today, piezoelectric and magnetostrictive materials can be described by linearized constitutive laws that are given below. Other widely used material types, such as electrostrictive or shape memory materials, exhibit strongly nonlinear behavior, the modeling of which may become quite demanding (see above).

Piezoelectric Materials. In the constitutive law of piezoelectrics, a coupling between strain ϵ, stress σ, electric displacement $\tilde{D}$ and field $\tilde{E}$ exists as follows:

$$\sigma = E\,(\epsilon - d^{\mathrm{T}}\,\tilde{E}) \tag{5.9}$$
$$\tilde{D} = d\,\sigma + \tilde{\epsilon}\,\tilde{E} \tag{5.10}$$

Here, d and $\tilde{\epsilon}$ are the matrices of piezoelectric coupling and dielectric constants, respectively.

Magnetostrictive Materials. Substituting magnetomechanics for electromechanics, mechanical strain ϵ and stress σ are coupled with magnetic field intensity $\tilde{H}$ and flux density $\tilde{B}$ as follows:

$$\sigma \;=\; E\,(\,\epsilon - d_{\mathrm{m}}^{\mathrm{T}}\,\tilde{H}\,) \tag{5.11}$$
$$\tilde{B} \;=\; d_{\mathrm{m}}^{\mathrm{T}}\,\sigma + \tilde{\mu}^{\mathrm{T}}\,\tilde{H} \tag{5.12}$$

Here, d_{m} and $\tilde{\mu}$ denote the magnetostrictive coupling and free permeability matrices.

5.3.3 Finite Element Method

In the domain of structural mechanics, the Finite Element MethodFinite Element Method (FEM) is a widespread, and powerful tool for numerical analysis of complex structures (see for instance [14]). A large number of commercially available codes exists, as well as those in the public domain.

FE codes based on the principle of virtual displacements model the spatial distribution of displacements using test or interpolation functions scaled with nodal values. In this manner the FEM reduces the continuous formulation of the system dynamics to a discrete set of differential equations for specified nodal degrees of freedom. Full coupling between mechanical and electrical or magnetic properties, respectively, requires introduction of additional degrees of freedom to the system. In most formulations for piezoelectrics, the electric potential is considered at element nodes. Modeling of magnetic fields leads to field intensity degrees of freedom.

Up to the time of writing, only very few generally accessible codes include the fully coupled constitutive laws of adaptive materials (see Sect. 5.6.1). In the literature, however, many formulations exist for piezoelectric materials (e.g. [15, 16, 17]).

As an approximation, the electric or magnetic degrees of freedom can be neglected if the influence of the mechanical properties on these entities is fairly weak. In general, this is the case if large mechanical structures with only small adaptive elements are considered, e.g. shell structures with piezoceramic patches. If a standard FE code is used, piezoelectric or magnetostrictive elements can be modeled with a thermal stress analogy (where coefficients of thermal expansion are substituted by piezoelectric or magnetostrictive coupling coefficients). However, the approximation must not be made if single actuators, such as piezoelectric stacks or magnetostrictive rods, are analysed. From a dynamics point of view, the approximation error can be characterized as an underestimation of system natural frequencies.

5.3.4 Equations of Motion

Application of the FEM to structural dynamics leads to the discrete equations of motion of an adaptive structure:

$$M\,\ddot{q} + D\,\dot{q} + K\,q = F\,u \tag{5.13}$$

Here, M, D, and K denote the mass, damping, and stiffness matrices, respectively. In the case of full coupling for piezoelectric or magnetostrictive material elements in the structure, the vector q of degrees of freedom initially comprises both nodal displacements and electric potentials or magnetic field intensities, respectively. In general, electromagnetic processes are much faster than mechanical vibrations, so that they may be assumed quasi–static in the above equation. As a consequence, electric or magnetic fields only contribute to the stiffness of the system, and a static condensation [14, p.450]) of the corresponding electric or magnetic degrees of freedom can be carried out. Only the mechanical degrees of freedom remain. The full coupling is represented in a electro- or magnetomechanical stiffness matrix. In (5.13) the term $F\,u$ denotes the actuator influence on the structure. The input variables u may represent externally applied actuator voltages (piezoelectrics) or currents (magnetostrictive actuators), and F the corresponding influence matrix. Note that static condensation of the electric or magnetic degrees of freedom leads to changes in F in addition to those in K.

The above equations of motion (5.13) can be transformed into the state equation of a state–space system description (5.3, 5.4) if the displacements and velocities are chosen as state variables:

$$\dot{x} = \begin{bmatrix} \dot{q} \\ \ddot{q} \end{bmatrix} = \begin{bmatrix} 0 & I \\ -M^{-1}K & -M^{-1}D \end{bmatrix} \begin{bmatrix} q \\ \dot{q} \end{bmatrix} + \begin{bmatrix} 0 \\ M^{-1}F \end{bmatrix} u \quad (5.14)$$

$$\dot{x} = A\,x + B\,u \quad (5.15)$$

Alternatively, modal amplitudes and velocities can be chosen as state variables, leading to a decoupling of the state differential equations.

5.3.5 Sensor Equations

An adaptive structure possesses sensor elements in order to gain information about its own state. In the case of actively controlled structural dynamics, sensors may measure a variety of signals, such as accelerations (accelerometers), displacements (Hall sensors, capacitive sensors, laser interferometers, etc.), forces (force transducers), or — typically for adaptive structures — strains or strain velocities (strain gauges, piezoelectric sensors, etc.). Most of these cases can be represented by the following sensor equation:

$$y = \begin{bmatrix} C_1 & C_2 \end{bmatrix} \begin{bmatrix} q \\ \dot{q} \end{bmatrix} + D\,u = C\,x + D\,u \quad (5.16)$$

As mentioned in Sect. 5.2.2, the feedthrough term $D\,u$ becomes important in the case of active struts in truss structures where piezoelectric stacks are placed in series with force transducers.

5.3.6 Model Reduction Techniques

Structural models obtained by using FEM codes are, in general, much too large for application of control design tools. Complex structures are commonly represented by several thousand or tens of thousand nodal degrees of freedom, whereas control design methodologies and analysis tools are restricted to only several tens or hundred degrees of freedom. This discrepancy highlights the reason why a large variety of model reduction techniques has been developed in recent decades, a process which is still continuing.

In the case of model reduction for linear elastic, actively controlled structures, a good survey is given by Craig and Su [8]. It is often advantageous to transform such systems into modal space before reduction, control design, simulation and analysis are carried out. Reduction is then performed by selection of natural modes which

− lie in the frequency range of control;
− are strongly controllable and observable with the chosen actuator and sensor configuration; and
− substantially contribute to undesirable structural motion in case of disturbances.

These criteria may be expressed numerically using Hankel singular values [18, 19], or balanced gains [20] if the system inputs and outputs are chosen appropriately. The method is known as 'balanced reduction'. In the case of lightly–damped adaptive structures, balanced reduction techniques can be applied based on modal data which makes this methodology numerically fast and efficient (see Sect. 5.2.3. Frequency–weighted versions [21, 22] have been developed to account for critical frequency ranges. Another commonly used technique computes modal costs [23]. Methods based on Ritz and Krylov vector projections [24] are advantageous with respect to representation of quasi–static system behavior, but decoupling of the equations of motion is no longer feasible. This implies the risk of severe dynamic spillover (see Sect. 5.4.2). If the control objective is active damping, quasi–static modeling errors are not critical. Therefore, modal representations are often preferred.

5.4 Analysis of Adaptronic Systems

Numerical analysis and simulation of adaptronic systems can be performed in the time and the frequency domain depending on the representation of the system in the state space or as a matrix of transfer functions. Important goals are the assessment of stability, performance and robustness of an adaptronic system. In the case of adaptive structures, performance criteria are often given in terms of allowable static and dynamic errors relating to structural shape if subjected to specified disturbances. Many applications also involve limits in energy consumption and actuator stroke or force, which must be checked in time–history simulations.

5.4.1 Stability Analysis

Every dynamic adaptronic system must be checked for stability in the case of disturbances. For linear elastic adaptive structures, asymptotic stability as defined in Sect. 5.2.4 is guaranteed if the poles (or eigenvalues) of the closed–loop active system lie in the left complex half–plane, i.e. if they have negative real parts. More stringent stability criteria, such as the generalized Nyquist criterion [9, p.59 f.], also consider the zeros of the adaptronic system.

In the case of nonlinear systems that cannot be reduced to a linearized system, stability is much more difficult to assess. Lyapunov's direct method [2, p.57 f.] requires a suitable energy function to be found. Often, only numerical time integration of large–scale, fairly accurate models gives an indication of the dynamic behavior if stability cannot be proven otherwise.

5.4.2 Spillover and Robustness

Structural control systems must be designed using rather small–scale models but are applied in the real structure with a theoretically infinite number of natural modes. Unwanted interaction between the control system and neglected structural modes may occur and lead to instability or loss of performance. This effect is known as spillover [25].
Three different types of spillover can be defined, as follows.

- The actuators influence structural modes that have not been represented in the mathematical model used for control design. This type is known as *control spillover*.
- The sensors produce signals with contributions from neglected structural modes. If this type, known as *observation spillover*, coincides with control spillover in the case of observer–based state feedback control, destabilization of the closed–loop system may be the consequence.
- In case the equations of motion used as a basis for model reduction are not decoupled, coupling terms between selected and neglected degrees of freedom exist. They imply *dynamic spillover*, which may lead to instability of the closed–loop system even if no observer is involved in the design.

The notion of spillover is important in respect of neglected structural modes. Other modeling errors include parametric uncertainties, which are more difficult to model and may have a substantial impact on the stabiliy and performance of the closed–loop system.

5.4.3 Numerical Time Integration

In many cases, stability, performance and robustness are difficult to check with general criteria. Numerical time integration of the state–space model is often used to investigate the dynamic behavior of an adaptronic system.

For the simulation of adaptive structures without control feedback loops, it can be advantageous to use direct time integration schemes to solve the second–order equations of motion (5.13). Examples of widespread algorithms are the Houbolt, Wilson θ, and Newmark schemes [14, p.532 f.]. They exhibit good performance for linear structural dynamics problems. If a large range of structural eigenfrequencies has to be covered, however, very small time steps are required in order to guarantee a stable solution. Modal decoupling of the equations of motion substantially reduces the required computation time and allows for model reduction based on modal selection (see Sect. 5.3.6).

The existence of control feedback loops, especially with actuator, sensor, or observer dynamics, makes the application of direct time integration schemes difficult. Implicit and explicit schemes based on the first–order state differential equation (5.14) are preferred in this case. A large variety of algorithms exist, among them the wellknown Runge–Kutta schemes with modifications for stepsize control (e.g. [26] or [27, p.85]).

5.5 Optimization of Adaptronic Systems

Strictly speaking, optimization of adaptronic systems means the simultaneous consideration of all subsystem aspects (the plant, the controller, actuators and sensors, including their position, etc.) for determination of design variables of these subsystems in a way that optimizes the overall system performance. In order to make the discussion more concrete, again the optimization of adaptive mechanical structures is taken as a representative field of application.

General practice of the related design and optimization process has been and still is sequential so that usually the structural design (the plant) is decided first, and a control system is added subsequently to eliminate or alleviate any undesirable behavior still remaining. Iterations ensue if the control system design is unable to satisfy its share to achieve proper system performance. Though this sequential practice often leads to quite reasonable solutions, in principle it is deficient because it does not explicitly accommodate general overall system objective functions, nor does it enable an explicit trade between structural design variables and control design variables in order to achieve the best system performance and minimum required effort. In addition, adaptive systems and structures in their strict sense require a combined approach from the beginning. So this simultaneous approach is advocated first, followed by some strategies for decomposition e.g. in the case of complex technical systems with difficult if not impossible simultaneous treatment of the subsystems.

5.5.1 Problem Statements

Though there exists a multitude of different possible problem statements, depending on the different technical tasks, a typical design optimization problem with a combined mechanical (m) and control (c) subsystem is the following nonlinear (and usually nonconvex) optimization problem:

$$
\begin{aligned}
&\text{minimize} && f_1(\boldsymbol{v},\boldsymbol{y}) + f_2(\boldsymbol{v},\boldsymbol{y}) + \ldots \\
&\text{such that} && g_k(\boldsymbol{v},\boldsymbol{y}) \geq 0, \\
&&& s_i^m(\boldsymbol{v}^m,\boldsymbol{v}^c,\boldsymbol{y}^m,\boldsymbol{y}^c) = 0 && i = 1,\ldots,q^m \\
&&& s_i^c(\boldsymbol{v}^m,\boldsymbol{v}^c,\boldsymbol{y}^m,\boldsymbol{y}^c) = 0 && i = 1,\ldots,q^c \\
&\text{with} && \boldsymbol{v} = (\boldsymbol{v}^m,\boldsymbol{v}^c)^{\mathrm{T}} \\
&&& \boldsymbol{y} = (\boldsymbol{y}^m,\boldsymbol{y}^c)^{\mathrm{T}}.
\end{aligned}
\tag{5.17}
$$

The design or optimization variables $\boldsymbol{v}^m, \boldsymbol{v}^c$ are to be determined such that a set of objective functions $f_1, f_2, \ldots$ is minimized while constraints $\boldsymbol{g}$ on design variables and system response variables $\boldsymbol{y}^m, \boldsymbol{y}^c$ are to be satisfied. Typical design variables $\boldsymbol{v}^m$ are structural stiffness properties, typical control variables $\boldsymbol{v}^c$ are gain factors and actuator / sensor positions, while objectives are related to structural and control subsystem mass, required power, time integral of response values, etc. Constraints often are put on design variables directly, for example where structural stiffness or actuator forces must not exceed given bounds, and indirectly via constraints on response quantities, for example where displacements or accelerations at specific points on a structure or its eigenfrequencies are bounded.

It should be noted, that the combination of structural and control subsystem masses in the objectives allows an optimal tradeoff between these subsystems. Mathematically, the coupling between the mechanical and control subsystem mainly occurs in the system equations s, where the response quantities $\boldsymbol{y}^m$ (e.g. displacement vector) and $\boldsymbol{y}^c$ (e.g. control forces) are determined depending on (the actual values of) the design variables $\boldsymbol{v}^m$ and $\boldsymbol{v}^c$. These system equations often are the state–space representations as discussed in previous sections, where in the case of adaptive structures the equations of motion and vibration are involved. So all the remarks on modal representation, condensation, etc. again apply, which then also explains the numerical effort involved in the process of finding an iterative solution for the optimization problem.

5.5.2 Solution Techniques

A solution technique for the optimization problem first of all requires an appropriate overall strategy to deal with the coupled problem. In principle, different options are available, such as:

- treating the problem as fully coupled and solving for both the v^m and v^c simultaneously;
- using a decomposition or nested approach, where an optimal structural design with constraints for achieving good controller performance is carried out first, followed by optimal control design with optional side constraints to consider structural requirements, and then eventually followed by optimal structural design, etc.; or
- heuristic decomposition methods.

Of course, treating the problem as fully coupled is the most desirable approach, but it might be difficult to carry out for large complex problems. So a decomposition might be worthwhile, where the subsystems are treated separately without sacrificing too much the overall optimal system performance. Once this overall approach is selected, an improved approximation $v^{(k+1)}$ for the optimum solution as obtained from the kth approximation $v^{(k)}$ is obtained from

$$v^{(k+1)} = v^{(k)} + \Delta v^{(k)}, \tag{5.18}$$

where $\Delta v^{(k)}$ is determined by a nonlinear optimization algorithm such that the objectives in step $k + 1$ are improved and the constraints are (better) satisfied compared with those at the previous step k.

It should be noted that, in principle, at every iteration step, where the evaluation of the objectives and constraints functions is required, the determination of the response quantities y from the system equations s is also necessary. So instead of using the 'full' model via the system equations, often approximation functions are established describing the dependency of the response quantities from the design variables. These are then used during some optimization iterations and are updated from time to time via response quantities (and optionally their derivatives), as obtained from the full model. These then are often used, for instance, for some kind of Taylor series expansions as approximation functions. For the derivatives it is advantageous to use the complete derivatives, which can be formalized by (for example):

$$\begin{bmatrix} I & J_{uc} \\ J_{cu} & I \end{bmatrix} \begin{bmatrix} \frac{dy^m}{dv_i} \\ \frac{dy^c}{dv_i} \end{bmatrix} = \begin{bmatrix} \frac{\partial y^m}{\partial v_i} \\ \frac{\partial y^c}{\partial v_i} \end{bmatrix} \tag{5.19}$$

where $J_{uc} = \partial y^m / \partial y^c$ and similarly. The partial derivatives of the right–hand side of (5.19) are determined from each of the subdisciplines. In structural mechanics these can be obtained by derivations of the finite element equations. This calculation of partial derivatives is implemented in many modern finite element codes.

These complete derivatives are also a key to decomposing the full problem into subproblems via a nested approach as mentioned above. When considering a subproblem, e.g. structural optimization, the linearized contribution from the other subdiscipline control is then also taken into account.

5.5.3 Practical Examples

High–precision reflectors and mirrors should have minimum deviation from their ideal shape even under considerable mechanical and thermal loads. Apart from appropriate design and material selection, this can also be achieved by shape control, with actuators counterbalancing the negative effect of the disturbances. This then means the determination of structural design variables (geometry, stiffness, etc.) and control variables (actuator positions, gain factors, etc.) such that a lightweight reflector has sufficiently accurate shape under simultaneous action of disturbances and controls. Further constraints, for instance on eigenfrequencies or on control force amplitudes, are also to be satisfied. This structure–control design problem could be treated simultaneously or in a nested approach as described above. In the latter, a reasonable albeit suboptimal structural design is determined first, and then actuator positions and control gains are selected to minimize the total shape error. Eventually, a (slight) redesign of the structural subsystem, followed by a redesigned control subsystem, has to be carried out.

Experience with such kinds of problems shows that, typically, an improvement of a factor to five in shape accuracy is obtained compared with a passive system. In the case of simultaneous treatment of the design optimization problem, this can often be achieved with less control energy or structural mass compared with the results from the nested approach.

High–precision instruments in science or optics often require vibration–free positioning and pointing of their sensitive devices. Figure 5.2 shows as an example an optical laser terminal, which transfers high data volumes over very large distances in space. Thus, pointing requirements are in fractions of arcseconds, while microvibrations within the satellite can cause base excitations at the laser terminal. To compensate for this, one concept is to integrate piezoceramic plates into the support structure so as to be used as actuators by a controller.

For design and optimization, a nested approach has been chosen in that example. In a first step, structural stiffness distributions as well as positions of the piezoplates are used as design variables in order to optimize actuator efficiency (maximum transfer functions from actuators to laser pointing angle). Constraints are put on structural mass and on eigenfrequencies. Then design variables of the controller are determined to optimize pointing performance (low vibration amplitude, small settling time). For simulation MATLAB has been chosen, where the structure has been represented by its first five modes in a modal state–space formulation. Results given in Fig. 5.3 for a pulse base excitation show the benefit of the adaptive versus the passive structure. Though some loss is to be observed in the real system because of its various imperfections (e.g. in the actuators), the results are still significantly better than those of a passive design.

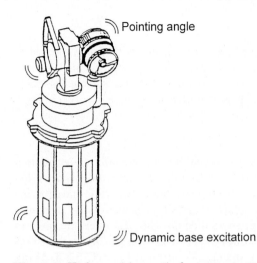

Pointing angle

Dynamic base excitation

Fig. 5.2. High–precision optical communication terminal

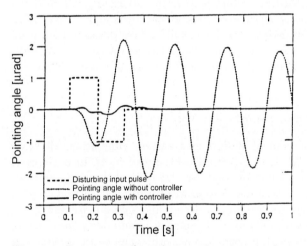

Fig. 5.3. Pointing angle without and with control

5.6 Analysis Tools

This section addresses software and hardware aspects of the simulation and implementation of adaptive structural systems. Note that within the confines of this chapter an extensive overview cannot be given, so that the tools mentioned here reflect the authors' personal choice.

5.6.1 Finite Element Analysis Codes

Among the large number of FE codes available commercially or as public–domain tools, there are still very few that offer constitutive models of smart materials. Fully coupled piezoelectric material laws can be found, for instance, in ABAQUS [28] or ADINA [29]. In industry, standard FE packages for structural analysis are often pre–specified by the customer. In case no suitable constitutive law for smart materials is available in a package, a first approach will be to use a thermal analogy as outlined in Sect. 5.3.3. Many other topics in modeling and simulation of adaptive structures, such as modal analysis or condensation techniques, can be found in standard packages such as NASTRAN [30]. A standard procedure is to set up a large–scale FE model including actuator inputs as thermal or mechanical loads and to read data obtained after modal analysis from the output file. These data are then used to set up a modal state–space representation of the adaptive structure. By this means, actuator and sensor modeling is done partially outside the FE package.

5.6.2 Control Design and Simulation Tools

Among the software for control design and system simulation MATLAB with SIMULINK [31] and MATRIX$_X$ [32] are widespread tools. In particular, MATLAB includes a large variety of toolboxes for control design (standard, non–linear, robust control, etc.) and system identification, (see Sect. 5.6.3). SIMULINK offers the option to graphically design and simulate dynamic systems as block diagrams without any additional programming. These tools, however, are restricted to medium/small–scale problems, so that reduction of large–scale FE models is necessary.

5.6.3 System Identification Tools

Identification of the dynamic behavior of adaptive structures may be performed in the framework of modal testing or in a more control–oriented fashion known as system identification, [33]. In the former case, commercially available software packages by companies, such as LMS or SDRC can be used. They offer a variety of data acquisition and processing capabilities (modal analysis, frequency response functions, etc.) combined with comfortable graphical user interfaces.

Standard system identification techniques are described by Ljung (1987) [34]. Many of these have been implemented in the MATLAB System Identification Toolbox. Algorithms that are specialized in active structural control are the Eigensystem Realization Algorithm (ERA) [35] or the Observer/Kalman Filter Identification algorithm (OKID) [36]. Implementations exist in the NASA System/Observer/Controller Identification Toolbox [37].

5.6.4 Hardware–in–the–Loop Simulation

After design, analysis and simulation, adaptive structural systems are created using data aquisition electronics coupled to the sensors, power amplifiers to provide the electric energy for the actuators, and data processing equipment for control. In special cases, analog feedback loops can be applied between sensors and actuators. More often, control laws have to be implemented digitally using a fast digital signal processor (DSP). Advanced DSPs may be programmed in higher computer languages such as C; the code is transferred to machine language using downloading tools. For less time–critical applications, a PC can be used for data processing.

The German company dSPACE offers turnkey control solutions based on MATLAB / SIMULINK Real–Time Workshop software and various I/O boards combined with Texas Instruments DSPs [38]. This configuration enables the analyst and designer to work in a 'hardware–in–the–loop' configuration in order to test and modify control laws applied to adaptronic systems.

References

1. Kailath, T. (1980): *Linear systems.* Prentice–Hall, Englewood Cliffs, NJ.
2. Slotine, J.–J.E.; Li, W. (1991): *Applied nonlinear control.* Prentice–Hall, Englewood Cliffs, NJ.
3. Blackwood, G.H.; Ealey, M.A. (1993): Electrostrictive behavior in lead magnesium niobate (PMN) actuators. Part 1: materials perspective. Smart Materials and Structures, **2**, pp.124–134.
4. Brinson, L.C.: One–dimensional constitutive behavior of shape memory alloys: thermomechanical derivation with non–constant material functions and redefined martensite internal variable. J. Intelligent Material Systems and Structures, **4**, pp.229–242.
5. Boyd, J.G.; Lagoudas, D.C. (1994): Thermomechanical response of shape memory composites. J. Intelligent Material Systems and Structures, **5**, pp.333–346.
6. Preumont, A.; Dufour, J.–P.; Malékian, C. (1992): Active damping by a local force feedback with piezoelectric actuators. AIAA J. Guidance, Control, and Dynamics, **15**, pp.390–395.
7. Moore, B.C. (1981): Principal component analysis in linear systems: controllability, observability, and model reduction. IEEE Trans. Autom. Contr., **AC–26**, pp.17–32.

8. Craig, R.R. Jr.; Su, T.–J. (1990): A review of model reduction methods for structural control design. Proc. 1st Conf. Dynamics and Control of Flexible Structures in Space, Cranfield, UK.

9. Maciejowski, J.M. (1989): *Multivariable feedback design.* Addison–Wesley, Wokingham, UK.

10. Biran, A.; Breiner, M. (1995): MATLAB *for engineers.* Addison–Wesley, Wokingham, UK.

11. Brock, D.; Lee, W.; Segalman, D.; Witkowski, W. (1994): A dynamic model of a linear actuator based on polymer hydrogel. J. Intelligent Material Systems and Structures, **5**, pp.764–771.

12. Shahinpoor, M. (1994): Continuum electromechanics of ionic polymeric gels as artificial muscles for robotic applications. Smart Materials and Structures, **3**, pp.367–372.

13. Shahinpoor, M. (1995): Micro–electro–mechanics of ionic polymeric gels as electrically controllable artificial muscles. J. Intelligent Material Systems and Structures, **6**, pp.307–314.

14. Bathe, K.–J. (1982): *Finite element procedures in engineering analysis.* Prentice–Hall, Englewood Cliffs, NJ.

15. Ha, S.K.; Keilers, C.; Chang, F.–K. (1992): Finite element analysis of composite structures containing distributed piezoelectric sensors and actuators. AIAA Journal, **30**, pp.772–780.

16. Hwang, W.–S.; Park, H.C. (1993): Finite element modeling of piezoelectric sensors and actuators. AIAA Journal, **31**, pp.930–937.

17. Chandrashekara, K.; Agarwal, A.N. (1993): Active vibration control of laminated composite plates using piezoelectric devices: a finite element approach. J. Intelligent Material Systems and Structures, **4**, pp.496–508.

18. Gregory, C.Z. Jr. (1984): Reduction of large flexible spacecraft models using internal balancing theory. AIAA J. Guidance, Control, and Dynamics, **7**, pp.725–732.

19. Jonckheere, E.A. (1984): Principal component analysis of flexible systems — open–loop case. IEEE Trans. Autom. Contr., **AC–29**, pp.1095–1097.

20. Kabamba, P.T. (1985): Balanced gains and their significance for L^2 model reduction. IEEE Trans. Autom. Contr., **AC–30**, pp.690–693.

21. Al–Saggaf, U.M. (1986): On model reduction and control of discrete time systems. Ph.D. dissertation, Inform. Syst. Lab., Dept. Electr. Eng., Stanford University.

22. Lin, C.–A.; Chiu, T.–Y. (1992): Model reduction via frequency weighted balanced realization. Control — Theory and Advanced Technology, **8**, pp.341–351.

23. Skelton, R.E.; Hughes, P.C. (1980): Modal cost analysis for linear matrix–second–order systems. Trans. ASME, J. Dynamic Systems, Measurement, and Control, **102**, pp.151–158.

24. Su, T.–J.; Craig, R.R. Jr. (1991): Model reduction and control of flexible structures using Krylov vectors. AIAA J. Guidance, Control, and Dynamics, **14**, pp.260–267.

25. Czajkowsky, E.A.; Preumont, A.; Haftka, R.T. (1990): Spillover stabilization of large space structures. AIAA J. Guidance, Control, and Dynamics, **13**, pp.1000–1007.

26. Fehlberg, E. (1970): Klassische Runge–Kutta Formeln 4. und niedriger Ordnung mit Schrittweiten–Kontrolle und ihre Anwendung auf Wärmeleitungsprobleme. Computing, **6**, pp.61–71.

27. Gear, C.W. (1971): *Numerical initial value problems in ordinary differential equations.* Prentice–Hall, Englewood Cliffs, NJ.

28. ABAQUS *Theory Manual Version 5.5* (1995). Hibbitt, Karlsson & Sorenssen, Inc., Pawtucket, RI.

29. Gaudenzi, P.; Bathe, K.-J. (1993): Recent applications of an iterative finite element procedure for the analysis of electroelastic materials. Proc. 4th Int. Conf. on Adaptive Structures, Cologne, pp.59–70.

30. Blakely, K. (1993): *MSC/NASTRAN Basic Dynamic Analysis.* User's Guide, Version 68. The MacNeal–Schwendler Corp., Los Angeles, CA.

31. MATLAB *product catalog (1996).* The Math Works Inc., Natick, MA.

32. MATRIX꜏ *product overview (1996).* Integrated Systems, Inc., Sunnyvale, CA..

33. Juang, J.-N.; Pappa, R.S. (1988): A comparative overview of modal testing and system identification for control of structures. Shock and Vibration Digest, **20**, pp.4–15.

34. Ljung, L. (1987): *System Identification — Theory for the user.* Prentice–Hall, Englewood Cliffs, NJ.

35. Juang, J.-N.; Pappa, R.S. (1985): An eigensystem realization algorithm for modal parameter identification and model reduction. AIAA J. Guidance, Control and Dynamics, **8**, pp.620–627.

36. Juang, J.-N.; Phan, M.; Horta, L.G.; Longman, R.W. (1991): Identification of observer/Kalman filter Markov parameters: theory and experiment. Proc. AIAA Guidance, Control, and Navigation Conf., New Orleans, LA.

37. Juang, J.-N.; Horta, L.G.; Phan, M. (1992): User's Guide for System / Observer / Controller Identification Toolbox. NASA Technical Memorandum 107566.

38. dSPACE Products Overview (1995). dSPACE GmbH, Technologiepark 25, 33100 Paderborn, Germany.

6. Actuators in Adaptronics

6.1 The Role of Actuators in Adaptronic Systems
H. Janocha

Actuators are applied extensively in all spheres of our environment. They can be found in CD players and cameras, washing machines, heating and air–conditioning systems, machining equipment, automobiles, boats and aircraft and even respiratory equipment and artificial limbs [1]. 'New actuators', which are based on the transducer properties of new or improved materials (e.g. solid–state actuators, described in Sects. 6.2 to 6.4 or actuators with electrically controllable fluids, described in Sects. 6.5 and 6.6), are good examples of adaptronic systems. This is most evident in the case of 'smart actuators', as shown in Sect. 6.1.6, which are based on multifunctional materials and which are on the verge of large–scale technical application.

6.1.1 Definition of an Actuator

From a global viewpoint, actuators are the elements that link the information–processing component of electronic control circuitry with a technical or non-technical (such as a biological) process. Actuators can be used to control energy flows or mass/volume flows purposefully. The output quantity of actuators is an energy or power that is often available in the form of mechanical work, namely 'force times displacement'. The input of the actuator is driven electrically — ideally without power loss but at any rate with low losses — with currents or voltages that are compatible with microelectronics whenever possible. It is to be emphasised that particularly the latter–mentioned property distinguishes actuators from 'normal' control elements and gives them system capability — provided that standardised electrical interfaces are used.

The structure of actuators can be described by introducing the elementary functional elements 'energy transducer' and 'energy provider'. The input and output quantities of an energy transducer are energies either of the same type (as is the case with current transformers or torque converters) or of a different nature (as in electromagnetic or piezoelectric transducers). The output of an energy provider is also an energy; however, this energy originates from an auxiliary source and is controlled by an input quantity, as seen in power amplifiers or valves (see Fig. 6.1).

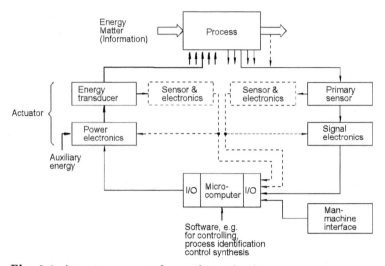

Fig. 6.1. Actuator as part of a regulator circuit

6.1.2 Actuator as Part of a System

Many control tasks in natural and artificial environments can be described by an open control loop: the focus of the loop is composed of processes and procedures in which flows of energy and/or matter are altered purposefully. Actuators intervene in the process in order to achieve this alteration. Their microelectronic compatible input signals are generated by the information–processing component of the electronic control circuitry, which is often decentralised, and applied to the various processes separately, respective of location and function. The control circuitry is predominantly software–controlled and can be implemented using personal computers. Intervention in the process by an operator takes place at a so–called man–machine interface (MMI) — in the simplest case an alphanumeric keyboard and monitor.

In automated processes, the control is performed within a closed loop (compare Fig. 6.1). A key function is the measurement of characteristic process quantities that are then supplied to the control computer following adequate signal pre–processing. The control computer compares the measured values with given reference input values. The differences are used to determine control signals for the actuator or corresponding power electronics according to a control strategy stored in the computer. Provided the process is known in a form usable by the computer — for example in the form of a mathematical model — its process–specific parameters are determined within the computer during an identification cycle. These parameters form the basis for control synthesis running on the computer.

Because of the computing capacity of modern computers, the parameters can follow the state changes of the process constantly. With respect to predefined human goals, the control behavior can then be adapted optimally to the changeable process behavior on–line, and this provides adaptive systems.

6.1.3 Personal Computers for Process Control

The incorporation of actuators and sensors in real–time control concepts — for example on the so–called field or process level within computer integrated manufacturing (CIM) — requires that the computer be capable of processing the necessary user programs intime and almost simultaneously. Normal personal computers with the usual operating systems such as MS–DOS are, as is well known, incapable of fulfilling these requirements. This lack is in contrast to process computers that possess the necessary properties of time sharing, multitasking and interrupt capability directly from the producer. However, possibilities exist to upgrade PCs to micro–processing computers using commercially available hardware and software components.

In most cases, the concept is based on the following distribution of tasks. General computer operations and software development, as well as the visual representation and recording of steps and results, can be performed using a PC/AT permitting use of all typical commercial MS–DOS–based software. The PC functions as the master and assumes the role of the MMI, serving both directions — machine to man and vice–versa. Real–time data collection and processing takes place under a suitable real–time operating system (e.g. RTOS or VX–Works) on a single–board computer operating as the slave.

Another option is made available with multiprocessor systems containing digital signal processors (DSPs). For example, cards with a DSP 32C processor and analog and digital input and output channels are commercially available. These cards are a cost–effective solution offering high computing performance but lacking the benefits of an open bus and a standardised operating system. The communication between the PC and the DSP takes place with the help of special driver software running on the PC. The software enables direct memory access (DMA) of the DSP memory via the ISA bus. Interrupts can be activated in the master processor by a special piece of hardware located on the circuit board.

6.1.4 Power Amplifier

Fig. 6.1 illustrates an actuator as a power transducer connected in series with a power provider (power amplifier); surprisingly, power amplifiers are rarely touched upon in publications about actuators.

Two possible concepts — switching and analog — are implemented nowadays for the power amplifier. Table 6.1 offers a comparative assessment of their important properties with respect to actuators. The table shows that the two concepts are complementary; no one concept offers exclusive advantages.

Table 6.1. Important properties of switching and analog amplifiers for actuators

	Switching Amplifiers	Analog Amplifiers
Power transistor losses	only during switches	continuous up to the rated output power
Re–use of stored field energy	possible in principle	not possible
Residual ripple of output signal	very high	negligibly low
Harmonic distortion of output signal	very great	very little
Electromagnetic compatibility	high–frequency disturbance[1]	very little disturbance
Availability of power transistors	good	less good
Circuit complexity	greater[2]	less

[1] disturbances lower with capacitive loads (piezo transducers)
[2] considerably more complex circuitry for inductive loads (magnetostrictive transducers)

A new circuit concept tries to combine the advantages of the switching with those of the analog amplifier to form a hybrid amplifier. The approach is to design a circuit that regulates the load current in feedback control while feedforward–controlling the transducer voltage — taking a magnetostrictive solid–state transducer for example. In this way the amplifier is adaptable to different loads. For general transient output signals, the current–voltage relationship at the load can be described by a transfer function that can be stored as a model in a microprocessor. The expected voltage can be calculated from the current and vice–versa [2].

In addition to a control component, such a hybrid amplifier for magnetostrictive transducers consists of a switching voltage supply and an analog amplifier for regulating current. Fig. 6.2 illustrates the principle: the hybrid amplifier has a freely programmable digital control unit (a DSP) connected to the final amplifier stage via A/D and D/A converters and special driver transistors. Various control concepts can be implemented by programming the control unit. Control settings gained empirically can also be implemented using the appropriate algorithms. A properly constructed hybrid amplifier can simultaneously

- re–introduce the field energy stored in the transducer into the voltage supply electronics so as to increase the efficiency of the actuator;
- enable high operating frequencies at full output displacement;
- maintain low residual ripple, thereby creating a high transducer precision;
- guarantee low power losses in, and therefore low complexity of the power transistors; and

– control the nonlinearities of the transducer and power amplifier in a stable manner with the help of a suitable controller.

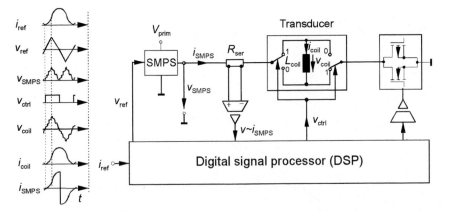

Fig. 6.2. Principle of a hybrid amplifier for magnetostrictive transducers

6.1.5 Intelligent Actuators

As a result of widespread usage, actuators are referred to as 'intelligent' when an integrated microelectronic 'computer intelligence', and possibly sensors, are used to control dynamically the behavior of the actuators. Such 'intelligent' actuators are capable of recognising and automatically correcting deviations from the desired behavior, that result from internal and external imperfections and influences such as hysteresis, temperature fluctuations, reactions from the load, etc. The actuator disturbances are sensed and sent to a fast digital processor that generates the necessary signal correction (see the dashed lines at the left of Fig. 6.1). The procedure will be illustrated briefly for the case of hysteresis in the static output–input response typical of so many transducer types.

In the 1930s, the German physicist Preisach suggested modeling the Weiss domains present in ferromagnetic materials with elementary hysteresis operators having different upward and downward thresholds. A system exhibiting hysteresis can be described by a superposition of an infinite number of these elementary operators. The core of this model is composed of the individual operators in parallel. The local 'memory' of the single operators can be expanded to a global memory for the entire system. In practice, the inverse hysteresis model is stored in a DSP and connected in series with the physical system. Fig. 6.3 shows the result for the case of a magnetostrictive transducer. The output–input response curve of the original system with hysteresis is compared with that of the compensated system. As is apparent from the figure, a computer implementing the Preisach model was able to almost fully compensate magnetic hysteresis and linearize the system response [3].

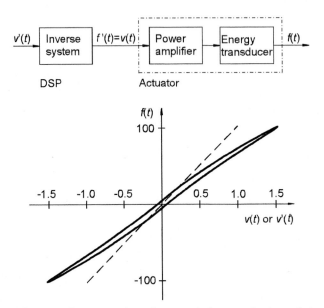

Fig. 6.3. Compensation of magnetic hysteresis through inverse excitation

6.1.6 Smart Actuators

The view of a closed control loop as shown in Fig. 6.1 or the description of 'intelligent' actuators can give the impression that the functions of sensing and positioning are always associated with discrete components or devices. This idea will be overcome by power transducers possessing sensory properties in addition to their exploited properties of actuation (i.e. multifunctional materials). This is precisely the property that characterises 'smart' actuators.

One such transducer material is a shape–memory alloy of nickel and titanium that exhibits a nearly linear dependence of change of shape on electrical resistivity. Electrorheological fluids provide another example: they exhibit a reproducible, and exploitable, relationship between electrical conductivity and mechanical shear loading. Piezoceramics also possess a sensory effect inherent of the material (direct piezoelectric effect = transfer of charge under the influence of a mechanical force).

This behavior is presented in Fig. 6.4. A sinusoidal control voltage is applied to a piezoelectric transducer, resulting in a nearly sinusoidal current. An almost unnoticeable change of current results when the transducer is loaded mechanically. Integration of the difference in current with respect to time —

i.e. determination of the transfer of charge — offers a true representation of the mechanical load without having to implement auxiliary sensors.

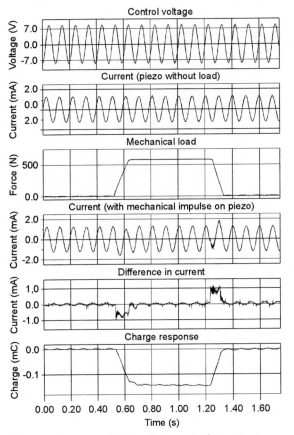

Fig. 6.4. State variables of a smart piezoactuator

An analogous effect is present in magnetostrictive transducers. The magnetostrictive effect may not be reversible in a strict sense, but the magnetic quantities are coupled with the mechanical ones via the magnetoelastic effect, providing sensory properties that can be exploited.

Let us take a closer look at smart actuators using this type of transducer as an example. Solidstate power transducers can generally be described by a three–port network with one electrical and two mechanical ports. One of the mechanical ports represents the transducer output and the other the terminated end. The network model can be reduced to twoports — one electrical and one mechanical — if the mechanical termination is known, e.g. a spring–mass–damper system. Fig. 6.5 illustrates a magnetostrictive power transducer as a two–port network [3, 4].

In a smart magnetostrictive transducer, the current i_{coil} and voltage v_{coil} are measured at the transducer coil and used to calculate the force F_{out} and positioning speed $\dot{s}_{out}$ or displacement s_{out} at the transducer output. The hysteresis in the characteristic curve can be modeled using a suitable Preisach model. Based on the nonlinear transducer and hysteresis models, the parameters are determined for an inverse filter that can compensate for the load–dependent actuator response. In a conventional solution, only the coil current of a magnetostrictive transducer is controlled while the transducer voltage remains ignored; in contrast, a smart actuator makes use of all available signals in order, for example, to compensate for reactions to the mechanical load.

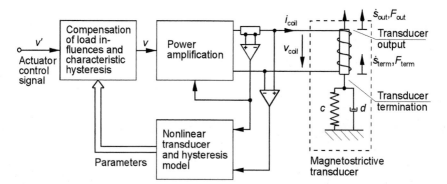

Fig. 6.5. Example of a smart actuator with magnetostrictive power transducer

6.2 Piezoelectric Actuators
B. Clephas

Piezoelectric solid–state transducers are characterized by high forces into the range of kilonewtons, reaction times on the order of a few milliseconds and less, and a positioning accuracy of the order of a few nanometers — but only at low strains of no more than 1 or 2 thousandths. Piezoelectric transducers have been used successfully for many years to solve demanding problems in the field of precision positioning and active vibration control.

The piezoelectric effect is based on the elastic deformation and orientation of electric dipoles in a crystal structure. The effect was first studied scientifically by Jacques and Pierre Curie in a tourmaline crystal in 1888. One fundamental of the piezoelectric effect is the nonsymmetrical structure of the crystal, in which the centres of electrical charge do not coincide and therefore form dipoles. A bundle of several elementary dipoles forms a dipole region known as a domain. The application of an external mechanical force deforms the dipoles, thereby generating a charge at the surface of the crystal: the direct piezoelectric or sensory effect. In contrast, applying an electric field causes a deformation of the dipoles, leading to a constant volume strain of the crystal: the inverse piezoelectric or actuator effect.

Natural piezoelectric materials, of which quartz is the most widely known, are single crystals in which all elementary dipoles possess the same orientation. In contrast, artificially produced ceramics, many of which are based on alloys of lead, zirconate and titanate, are polycrystalline and exhibit a random orientation of the dipole domains. These domains compensate for each other, preventing a charge from occurring at the surface of the ceramic. During the production process, the piezoelectric behavior is impregnated in these ferroelectric ceramics by polarization with an electric field at a high temperature. The orientation of the dipoles is nearly frozen in the ceramic upon cooling. An applied electric field then polarizes the ceramic; the dipole domains orient themselves more in the direction of the field. This reorientation results in an additional strain or elongation of the ceramic.

6.2.1 Theory

Fundamental Equations. In piezoelectric materials, the dielectric charge density D and the mechanical strain S are dependent on the mechanical stress T and the electrical field strength E. By considering the tensors of the mechanical parameters S and T at the surface of the crystal, the linear state equations can be written in vectorial form:

$$D = dT + \varepsilon^{\mathrm{T}} E \qquad\qquad\qquad (6.1)$$

$$S = s^{\mathrm{E}} T + dE \qquad\qquad\qquad (6.2)$$

In this set of equations, the matrix of piezoelectric constants d indicates the strength of the piezoelectric effect. ε^{T} is the matrix of dielectric constants for constant T and s^{E} the elasticity matrix for constant E.

Equation 6.1 describes the direct piezoelectric effect or sensory effect and has been applied for a long time to solve technical problems such as in accelerometers. The inverse piezoelectric effect is recorded in (6.2), which is used to describe actuation: a mechanical stress T causes a strain S (Hooke's Law), which can be controlled by an electric field E. The actuator and sensory effects are coupled so that each piezoelectric transducer can be used either for actuation or for sensing, or for both simultaneously (as in a 'smart' actuator).

Metallic electrodes are affixed to the top and bottom sides of the piezoelectric crystal for applying the electrical driving field. The electrical field parameters D and E are then uniaxial, namely perpendicular to the electrodes. The orientation of the dipoles and the crystal strain occur parallel to the electrical field lines. Due to the constant volume material strain, the crystal experiences shortening perpendicular to the polarization axes, and torsion can also occur within the crystal.

In most transducer constructions, only one of these strain modes is used, either the d_{33} mode in which the crystal elongation is parallel to the electric field (longitudinal effect) or the d_{31} mode in which the change of crystal length takes place perpendicular to the electric field (transversal effect). The d_{31} effect is negative, indicating a reduction in length perpendicular to the crystal and field orientation. In both modes, (6.1) and (6.2) reduce to their scalar forms, and the electromechanical coupling can be represented by the two–port model shown in Fig. 6.6a.

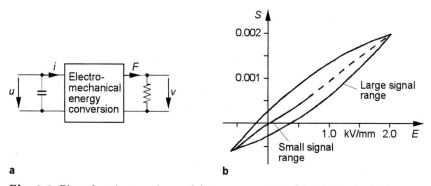

Fig. 6.6. Piezoelectric transducer: **(a)** two–port model [1] **(b)** Strain–field strength response [5]

Based on its construction, a piezoelectric transducer behaves electrically like a capacitor and mechanically like a stiff spring with mass. Without even being connected to external circuitry, the transducer is a system exhibiting two basic natural frequencies of oscillation. One of the natural frequencies is the fundamental oscillation of the spring–mass system, the other the result of the electromechanical coupling of the spring–mass system with the electrical capacitance of the transducer. The two natural frequencies are quite close to one another and, depending upon the transducer geometry, can lie from the audible range up to 200 Megahertz. These natural frequencies can be altered through the interaction with external mechanics or electronics [6].

Fig. 6.6b shows the characteristic strain–field strength response of a PZT ceramic without a mechanical load. The curve clearly verifies that the linear relationships presented in (6.1) and (6.2) are valid only for low–signal operation of the transducer, i.e. for field strengths between about -300 V/mm and +300 V/mm. The coupling factor k_{eff}, which indicates the relationship between the transformed and the stored energies, is also defined for this linear region. The coupling factor is defined by

$$k_{\text{eff}} = \sqrt{d^2/(s^E \varepsilon^T)} \qquad (6.3)$$

and can take on values of up to 0.7: piezoelectric transducers have a high energy density.

Transducers achieve their greatest strains in high–level operation, i.e. up to the field strength $E_{\max}$, which lies between 1500 and 2000 V/mm. Higher values of E can lead to arcing across the electrodes. High negative field strengths (opposing the direction of polarization) result in an irreversible depolarization, which also destroys the ceramic.

Hysteresis, Repolarization, Creep. The relationships described in (6.1) and (6.2) are valid in low–level operation up to about 15% of the maximum permissible field strength. Above this range, the strain–field strength response demonstrates nonlinear behavior with a great degree of hysteresis. The hysteresis is primarily a retarded reorientation of dipole domains, which initially maintain their orientation in the field direction upon reducing its strength.

When the driving field is changed and then held at a constant level, more and more dipoles orient themselves in the direction of the applied field due to mutual influence, leading to an additional strain of the crystal. This so–called 'creep' exhibits a time constant of a few seconds and can result in an additional change in length of a few percent of the maximum displacement in static operation. In contrast, if the piezoelectric ceramic is left a long time without an external electrical field, some of the dipole domains lose their orientation. This effect demands that the ceramic is repolarized in order to achieve the original values in the strain–field strength response curve.

Temperature Effects. The temperature has a large influence on the behavior of piezoelectric materials. The Curie temperature is a crucial value above which the crystals experience a change in structure and lose their piezoelectric properties. The Curie temperature of most piezoelectric ceramics lies between 160°C and 350°C. Depolarisation begins to occur well below the Curie temperature. Therefore the operating temperature of a piezoelectric transducer should be limited to about one–half of the Curie temperature (on the Celsius scale). Changes in temperature also cause changes in the polarization and field strength, an effect known as pyroelectricity. The temperature also influences the dielectric constant ε, the mechanical elasticity constant s and the piezoelectric constant d. The degree of temperature dependence differs from one piezoelectric material to the next.

Electrostriction. Electrostriction is a general term referring to the elastic deformation of a dielectric material under the influence of an electric field. Piezoelectric transducers therefore belong to the category of electrostrictive transducers. In a narrower sense, however, the term 'electrostrictive transducer' is used to describe ceramics such as those based on lead, niobium and magnesium that, in contrast to piezoelectric ceramics, are not polarized but rather exhibit a change of length due to a spontaneous orientation of the dipoles in an electric field. In contrast to piezoelectric transducers, electrostrictive transducers elongate in the presence of both positive and negative electric fields. The actuation forces and reaction times of electrostrictive transducers are comparable to those of piezoelectric transducers; the strain, however, is smaller and the hysteresis much less. The electrostrictive materials used to date are much more temperature–sensitive than piezoelectric materials, and their operating temperature is less than 40°C [7]. Electrostrictive transducers are of less practical importance than piezoelectric transducers, primarily due to their temperature dependence.

Aging. High–level mechanical and electrical loading of piezoelectric materials can lead to microcracking, which can change the material properties or even destroy the transducer. These effects are increased by local concentrations of air or other foreign matter in the ceramic and the drift of ions from the electrodes. Improvements in manufacturing processes and material properties have recently resulted in many ceramics that are capable of achieving 10^9 high–level cycles without substantial damage. Driving the transducers only in unipolar operation also contributes to an extension of their life. However, aging of the material parameters must be taken into account [8].

6.2.2 Technical Transducers

Material. Of the naturally occurring monocrystalline piezoelectric materials, only quartz has been used extensively in electromechanical resonators with high mechanical quality. Artificially manufactured polymers on the basis of polyvinylfluoride (PVDF) are used in piezoelectric films. Mostly sintered lead–zirconate–titanate (PZT) ceramics are used for transducer applications, since these materials exhibit relatively large piezoelectric constants d for up to 0.5×10^{-9} As/N (even as great as 1×10^{-9} As/N in high–level operation). For comparison, in quartz $d = 2 \times 10^{-12}$ As/N. Additional properties of PZT ceramics are summarized in Table 6.2.

PLZT ceramics, which contain lanthanum in addition to lead, zirconate and titanate, exhibit strains twice as great as those of PZT ceramics. However, PLZT ceramics are still in research, as are artificially produced piezoelectric monocrystals, which exhibit piezoelectric constants eight times those of PZT ceramics [10].

Transducer Construction. PZT discs for actuator applications are manufactured with thicknesses of 0.1 mm to 0.5 mm. Displacements of up to 1 μm can be achieved with a single piezoelectric disc of 0.5 mm thickness. And a diameter of 5 mm is sufficient to generate a force of over 1000 N. Greater displacements can be achieved by placing several discs on top of one another to form piezoelectric stacks, as shown in Fig. 6.7.

Displacements of several hundred micrometers can be achieved with such stacks consisting of several hundred individual discs. The use of adhesive to bond the discs to one another can reduce the stiffness of the transducer. Adhering so many individual discs is a manufacturing challenge and can lead to cracks and failure in dynamic operation. So–called multilayer transducers are manufactured by sputtering, together a multiple of metallized single layers.

Table 6.2. Material parameters of various PZT ceramics [9]

Piezoceramic	**PPK**		**11**	**26**	**62**
Curie temperature	ϑ_C	°C	175	220	340
Density	ρ_m	10^3 kg/m^3	8.1	7.7	8.0
Stiffness constant	s_{33}^E	10^{-12}/Pa	19.0	15.6	12.8
	s_{11}^E		15.9	14.7	12.8
Mechanical quality factor	Q_m		70	80	1900
Frequency constants	N_1	m/s		1450	1620
	N_3			1840	2000
Depolarisation pressure	–	10^6 Pa			300
Relative permittivity	$\varepsilon^T/\varepsilon_0$		5000	2500	1200
Dielectric dissipation factor	$\tan\delta$	10^{-3}	20	30	3
Coupling factors	k_{33}		0.72	0.67	0.66
	k_{31}		0.43	0.35	0.20
Piezoelectric charge constant	d_{33}	10^{-12} C/N	680	410	210
	$-d_{31}$		350	190	60
Possible Application	Bending transducer (PPK 11)				
	Stack actuator (PPK 26)				
	Ultrasonic transducer (PPK 62)				

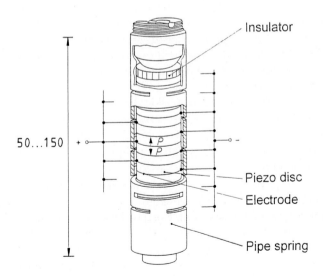

Fig. 6.7. Construction of a piezoelectric stack

An additional means of achieving greater displacements is to make use of d_{31} transducers in a bending mode (Fig. 6.8). A bending transducer can consist of a PZT ceramic affixed to a sprung metal plate (unimorph) or of two bonded ceramic strips, which bend in accordance with the driving voltage applied to each one (bimorph). The displacement achievable at the end of the strip is on the order of 1 mm, but the actuating force less than 1 N.

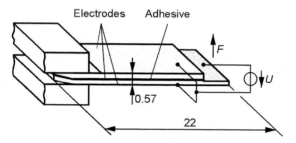

Fig. 6.8. Piezoelectric bending transducer (bimorph) [9]

An additional transducer construction is that of the tube–shaped piezo-electric ceramic with one cylindrical electrode affixed to each of the inner and outer surfaces of the tube. This type of transducer can be designed to make use of either the d_{33} or d_{31} effect, and is therefore optimized in terms of axial or radial displacement — for example, to achieve a clamping function.

Of the great number of further possible constructions of piezoelectric transducers, only the composite transducers for ultrasonic transmitter and receivers will be discussed (Fig. 6.9). Composite transducers operating based on the d_{33} mode consist of small parallel rods embedded in an epoxy matrix. This construction enables the transducer to be formed in order to achieve a focusing or defocusing characteristic (depending on the application requirements). In comparison with transducers based on piezoelectric discs and intended for operation in sonic ranges of frequency, the acoustic impedance of composite transducers is adapted closer to that of the surrounding medium air. This impedance matching enables a considerably greater radiation of energy and receiving sensitivity to be achieved.

Mechanical Properties. Fig. 6.10 shows typical strain–field strength responses of PZT ceramics under different loading conditions.

The achievable strain S_{max} of the transducer is dependent upon the type of mechanical load. A constant mechanical load results in a compressive deformation S_{mech} of the transducer. However, the piezoelectric strain S_{max} remains unchanged. The blocked force, which is the force required to compress the transducer such that $S_{mech} = S_{max}$, is defined as a characteristic parameter. A load proportional to the displacement (spring load) is illustrated in Fig. 6.10b. With this type of load, the achievable displacement is reduced to S'_{max} because of the elastic behavior of the transducer.

PZT-stacks

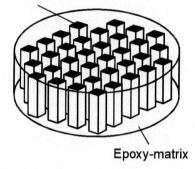

Epoxy-matrix

Fig. 6.9. Schematic construction of a composite transducer [11]

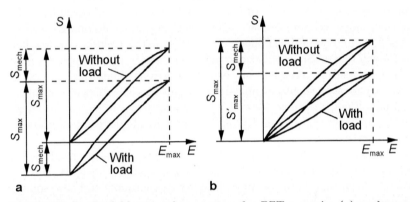

Fig. 6.10. Strain–field strength response of a PZT ceramic: **(a)** under constant mechanical load; **(b)** working against a spring load [1]

Compressive loading of up to 100 MPa is permissible depending on the material. In contrast, very small tension or shearing loads can destroy the ceramic. The ceramic should therefore be mechanically preloaded. Measurements of different transducers have shown that the mechanical stiffness at high–level signals cannot be considered constant but, rather, nonlinear and dependent upon the field strength. This behavior results in a hysteretic relationship between the strain and the force (similar to that of the strain–field strength response). Depending upon the ceramic, the mechanical stress T can have a negligibly low to considerably high influence on the piezoelectric constant d, meaning that the maximum achievable displacement S_{max} is also dependent upon the mechanical prestress.

Operating Voltage. PZT plates with a thickness of 0.5 mm require a driving voltage of 1000 V (high–voltage transducer) in order to achieve a saturation field strength of about 2000 V/mm. Transducers composed of 0.1 mm–thick plates are correspondingly driven at 200 V (low–voltage transducer). To achieve the same maximum displacement as a transducer containing plates of thickness 0.5 mm, a transducer made up of 0.1 mm–thick plates requires five times as many plates. Since the capacitance on a single plate is proportional to the electrode surface area and inversely proportional to its thickness, such a low–voltage transducer has a capacitance about 25 times that of the equivalent high–voltage transducer. Resulting capacitances can be in the range of microfarads, calling for high driving currents and high reactive power in dynamic operation.

Some piezoelectric films with a thickness of 50 μm or less are produced in the form of multilayer transducers. The nominal operating voltage is then below 100 V — resulting, however, in still higher capacitances and correspondingly higher currents in dynamic operation. As a compromise, the common layer thickness is 100 μm, resulting in a driving voltage of 200 V.

Efforts are being made to achieve transducers that can be driven with voltages of 24 V or lower. However, the necessary films with thicknesses of less than 25 μm cannot yet be produced with sufficient purity and surface quality.

Power Amplifiers. Power amplifiers for piezoelectric transducers are predominantly realized as voltage sources by implementing one of two concepts: either Class AB analogue electronics (with low output ripple and simple construction) or switching electronics (with low internal losses and the possibility of energy recovery). One or other amplifier concept can be advantageous depending upon the problem set. Amplifiers which combine the advantages of the individual concepts without exhibiting the respective disadvantages are currently undergoing laboratory development. In such a hybrid amplifier, the fine control of the output voltage is achieved using an analogue Class AB final stage, whose input voltage is given by a preceding switching amplifier. The results so far indicate very low internal losses (see Sect. 6.1). Energy recovery is also possible, i.e. the energy stored in the transducer capacitance can be returned to the source during the discharging cycle.

Displacement Amplifiers. Increasing the transducer displacement is possible using hydraulic or mechanical displacement amplifiers that are based on the principle of a lever. The achievable actuation force is reduced linearly, the stiffness quadratically, with the displacement amplification factor. The dynamic behavior of the complete system is determined predominantly by the displacement amplifier: permissible driving frequencies are considerably lower than those achievable by the transducer alone. Due to the small transducer displacements, even minimal play in the links of the displacement amplifier can lead to a substantial reduction of the maximum achievable output displacement. Displacement amplifiers making use of flexible joints have gained recognition, such as those of hybrid design [12]. On the whole, the use of displacement amplifiers can eliminate some of the specific advantages of piezoelectric transducers.

6.2.3 Comparison With Other Actuator Types

Many of the first applications of piezoelectric actuators involved the substitution of conventional actuators. This approach often resulted in a suboptimal solution. A system adapted especially to a piezoelectric transducer and its special properties makes much better use of its potential. The use of piezoelectric transducers has proven itself particularly in applications such as active vibration control and fine positioning (see Sect. 8.4), in which conventional actuators have reached their limits with respect to actuation force, precision and maximum operating frequency.

Piezoelectric transducers are solid–state transducers, as are electrostrictive (see Sect. 6.2) and magnetostrictive transducers, which, consisting of giant magnetostrictive materials (e.g. Terfenol–D), elongate in the presence of a magnetic field (see Sect. 6.3). Although piezoelectric and magnetostrictive transducers are exchangeable in many applications — with comparable actuation strains and stresses, reaction times and electromechanical efficiencies — piezoelectric transducers have established a greater presence, particularly because of the greater variety of commercially available piezoelectric ceramics and ready–made transducers and their greater availability at considerably lower prices. Various strain effects (e.g. longitudinal, transverse) can also be implemented, whereas magnetostrictive actuators can currently only make use of the d_{33} effect. Piezoelectric transducers can maintain their static displacements nearly without the need for additional electrical energy. In magnetostrictive transducers, a constant current (or permanent magnets) is necessary to achieve an offset magnetization. Magnetostrictive transducers have therefore only found application where their specific advantages over piezoelectric actuators — low hysteresis, high operating temperature, low–voltage driving signal due to possibility of current control and the possibility to separate the driving coil from the magnetostrictive rod — are crucial [13].

6.2.4 Example Applications

As described in Chap. 1, adaptronics is the integration of sensors, actuators and adaptive control within a single material or in a structure containing at least one multifunctional member. Piezoelectric transducers, which can be implemented as sensors and as actuators, can form an essential part of such adaptronic systems. The following characteristic examples make clear the potential of piezoelectric actuators.

Sensors. As presented in (6.1), applying a mechanical force to a piezoelectric transducer results in a dielectric charge density D, which evokes a charge Q at the electrodes. This property can be implemented in the construction of sensors such as accelerometers, in which the inertia of a seismic mass applies a force to a piezoelectric element. Some advantages of sensors based on the piezoelectric effect include their high sensitivity and high cutoff frequency even into the range of MegaHertz [14]. Based on their operating principle, piezoelectric sensors cannot be utilized to measure static forces. The high internal resistance of piezoceramics, however, enables quasistatic measurements with the use of suitable pre–amplifiers (charge amplifiers) (see Chap. 7).

Shakers. Piezoelectric transducers were first implemented to generate vibrations because of their inherent capacity for electromechanical resonance even in the absence of external circuitry. Quartz resonators are the most widely used piezoelectric transducers for frequency stabilization, for example in radio and computer electronics. The need for greater amplitudes or energies of vibration, as is the case in shakers, is satisfied by driving piezoelectric transducers with high–level dynamic signals. The frequency–dependent hysteresis losses can cause such a strong heating of the transducer that the maximum operating frequency must be far below the mechanical resonance of the transducer despite the use of cooling.

Valve Actuation. Piezoelectric transducers find further application as driving elements in quick valves because switching times below a few milliseconds are rarely achievable with conventional forms of actuation (usually electromagnetic drives). For this reason, piezoelectric bending transducers are used in ink–jet printers; in addition to higher operating frequencies, the precision of the position control of the transducers provides for a repeatable and clear printing result.

Additionally, piezoelectric transducers can hold a position with nearly no power loss and therefore require very little energy between switching cycles. This feature is advantageous for battery–supplied applications.

An application in which great development effort is being invested currently is the use of piezoelectrically driven injection valves in combustion engines. High–speed actuation with piezoelectrics enables a pilot injection for optimizing the combustion cycle, which in turn leads (for example) to lower noise levels and reduced fuel consumption. Fig. 6.11 shows a schematic representation of such an injection valve.

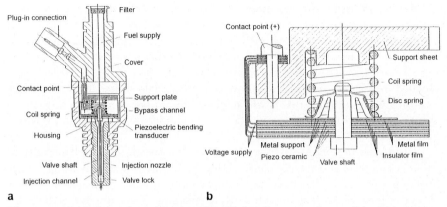

Fig. 6.11. Injection valve with piezoelectric bending transducer: (**a**) axial view; (**b**) detail

Piezoelectric Motors. Piezoelectric transducers can also be used to generate motion in motors. Several concepts are possible.

Ultrasonic motors In ultrasonic motors, piezoelectric elements distributed over an area generate a traveling wave as a result of phaseshifted electrical drive signals. The wave transmits a friction force onto a traveler. Various designs implementing this concept are available. One design, industrially produced in large quantities, is presented in Fig. 6.12 [15].

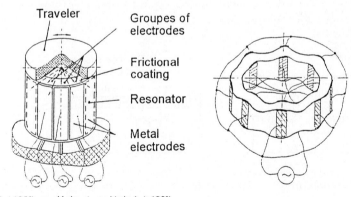

$v_1 = V \sin(\omega t - 120°)$ $v_2 = V \sin \omega t$ $v_2 = V \sin(\omega t + 120°)$

a   **b**

Fig. 6.12. Ultrasonic motor: (**a**) Motor construction; (**b**) deformation of the resonator

Fig. 6.12a shows the schematic construction of an ultrasonic motor with a radially polarized hollow piezoelectric cylinder as resonator. The traveler is held onto a frictional coating with a mechanical prestress. Metal electrodes, whose width corresponds to the length of the ultrasonic wave, supply the active material with the electric signal. That stator has a total of 24 electrodes divided into three groups that are driven by signals shifted by 120°. Three standing waves are generated, which when superimposed result in a traveling wave. The deformation of the piezoelectric cylinder by the standing waves is represented in Fig. 6.12b.

Ultrasonic motors are strong already at low speed. Their advantages include a compact and simple construction, quiet operation, a holding torque free of power, slip under overload (sliding coupling) and a high specific power. A motor from Seiko Instruments with a thickness of 4.5 mm and a diameter of 10 mm operates at a voltage of 3 V with a driving torque of 0.1 mNm at a rotational speed of 6000 min^{-1} [16]. The transmission of force by friction proves disadvantageous, with a strong dependence on aging and temperature; ultrasonic motors can be designed for rotary as well as linear motion. The efficiency of ultrasonic motors can be as high as 45% [16]. They are used, for example, in auxiliary automobile equipment and in the adjustment of autofocus camera lenses.

Micropush motor The operating principle of the micropush motor is similar to that of ultrasonic motors, with exception of the force transmission onto the traveler. Here, the transmission is achieved by impacting which results in very responsive start and stop. One prototype with a rod traveler (Fig. 6.13a) achieves driving speeds between a few μm/s to 0.6 m/s, with a positioning accuracy on the order of a nanometer. Fig. 6.13b demonstrates the micropush concept for rotary motion [17].

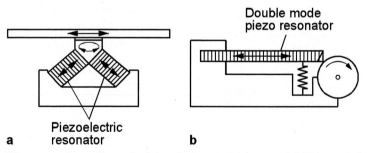

Fig. 6.13. Operating principle of micropush motors: **(a)** linear motion; **(b)** rotary motion

Inchworm Motor Another motor concept making use of piezoelectric transducers is the inchworm motor typically used to generate linear motion, as shown in Fig. 6.14.

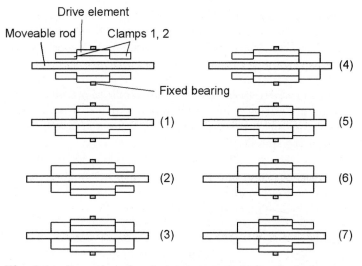

Fig. 6.14. Operation of an inchworm motor [18]

Such an inchworm motor consists of a movable rod as the traveler and a stator with a drive element and two clamping elements. Stacked, bending or, to simplify the construction, hollow cylinder transducers may be implemented as the drive and clamping elements.

A step begins with a closed (1) clamping element 1 forming frictional contact with the rod. The driving element elongates (2), and the clamping element 2 closes (3). Clamp 1 opens (4), and the drive element contracts (5). The original condition of clamping is then assumed (6,7) — the traveler and stator have experienced a total of one step of relative motion.

The traveling distance achieved with such motors is up to 200 mm with steps below one micrometer. Loads of several kilograms can be transported at several millimetres per second. Inchworm motors can achieve a greater positioning accuracy than ultrasonic motors, but require more design effort.

One laboratory inchworm motor makes use of piezoelectrically driven clamps and a magnetostrictive actuator as the drive element. The complementary transducers form a resonant electric circuit in which a repeated exchange of reactive power takes place. Such a hybrid actuator enables the realization of a compact power amplifier that incorporates the transducers themselves. The motor has a maximum step size of $10\,\mu$m, which can be reduced continuously to zero; the natural frequency is $500\,$Hz.

The use of a hybrid actuator is not limited to resonant operation. An appropriate control electronics enables step or triangular signal outputs also to be achieved, thereby opening up further areas of application [19].

Combined Actuation Piezoelectric stacks can be combined with conventional linear drives (e.g. spindle drive) to achieve large displacements in short times and with a high positioning accuracy [20]. Applications include drive systems for linear or rotary tables, as for example used in the photolithographic production of fine structures. Piezoelectric transducers enable positioning accuracy of the order of nanometers and can hold a position with nearly no power loss.

Fig. 6.15 shows a rotary table with a DC–current spindle as primary drive. The angular resolution of position is improved from 0.18 s to 0.015 s through the implementation of a piezoelectric stack adding a further and more precise displacement to that of the spindle.

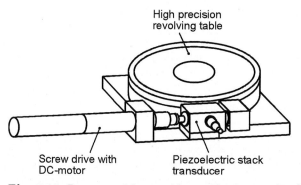

High precision
revolving table

Screw drive with   Piezoelectric stack
DC-motor   transducer

Fig. 6.15. Rotary positioner with combined actuation [12]

In systems making use of combined actuation, piezoelectric transducers are driven with high–level signals. Hysteresis and nonlinearity in the displacement response may not be ignored.

The displacement hysteresis can be compensated for with closed–loop position control. A displacement sensor determines the actual displacement, and the transducer control voltage is adjusted corresponding to the control input. The cost of an additional sensor is typically much greater than the necessary control and signal processing electronics as these are now easy and inexpensive to come by. A further means of compensating hysteresis is to make use of the relationship between the piezoelectric strain and the electrical charge. This relationship is considerably more linear than the displacement–voltage response of piezoelectric transducers. The control electronics must then incorporate a charge amplifier in which the current flowing into the transducer is measured and integrated over time. Small offset voltages in the measurement signal or within the signal processing electronics can lead to large errors in the output signal, particularly in quasistatic operation. A different approach therefore considers the hysteretic behavior already in the pre–control stage. This can be achieved making use of inverted characteristic curves saved in the signal–processing area of the control electronics. A new approach uses a real–time, operator–based mathematical model of the transducer hysteresis, which simultaneously takes creep into account [21].

Distributed Actuators. Mechanical structures such as slabs or frameworks can be made active by affixing or incorporating piezoelectric transducers. Such structures can be used to achieve active forming (e.g. of rotor blades in helicopters) or for active vibration damping (e.g. in an airplane fuselage). Vibrations, such as chatter between the tool and work piece of manufacturing machinery, result in processing uncertainty. The amplitudes of vibration are typically just a few micrometers, but the forces are very high. Due to its narrow–band nature, passive damping is often ineffective in the battle against a large number of natural modes of vibration present in machine structures. The integration of piezoelectric transducers in the structure of machines can be used to decouple or dampen vibrations between the tool and the work piece through the appropriate supply of an opposing energy.

Typically, many single transducers are distributed throughout the structure. Achieving suitable signal processing electronics demands great attention in such actuator systems. The intelligent management of energy also deserves increasing attention in order to achieve a high operating efficiency, to even the load on the power source, or to limit the supply of energy to the maximum allowable for any given operating conditions.

Vibration frequencies in the range audible to humans can often be disturbing. Since passive noise reduction at low frequencies is often not economically feasible, active noise reduction can be achieved using piezoelectric transducers — for example in the frame of a window or in the form of thin–film PZT actuators integrated into the structure. Fig. 6.16 shows such a structure with embedded piezoelectric transducers.

A structure used for active forming consists of several layers of piezoelectric material used alternately for their actuator or sensor properties [23]. Another possibility of combining actuator and sensor properties lies in the inherent sensory effect described in (6.1) above, enabling the use of a single piezoelectric element for both actuation and sensing (a smart actuator). This effect enables the external mechanical loads on the transducer, the displacement, as well as wear and aging, to be detected without the need for additional sensors.

Fig. 6.17 shows an experiment involving active vibrational damping of a cantilever beam [24] with a smart actuator. The piezoelectric transducer is controlled by a computer with a D/A converter. The sensory information is evaluated with the help of a bridge configuration according to the linear relationship of (6.1). The displacement sensor in this setup serves only to verify the achieved damping. The experiment provided an impressive confirmation of the principle of a smart actuator. However, the experiment also demonstrated the need to consider nonlinearity and hysteresis to achieve greater accuracy. Advanced applications therefore took into consideration the voltage dependence of the capacitance as well as the hysteresis in the displacement–voltage response [25]. An operator–based approach, in which the mechanical parameters are reconstructed from a model, is in an advanced stage of

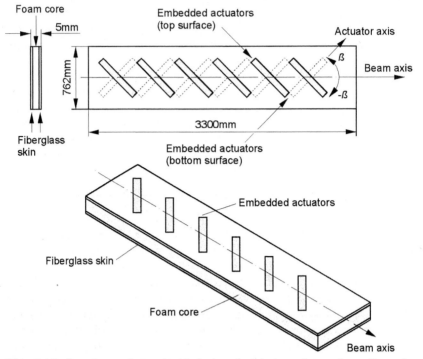

Fig. 6.16. Structure using embedded piezoelectric transducers for active noise reduction and active forming [22]

research as is another method, in which the dependence of the transducer capacitance on the force is exploited (see also Sect. 6.1 and [21]).

Microactuators. The described piezoelectric effects are not limited to use on a macro scale. As a result of the advancement of miniaturization, microstructures are attracting more and more attention. Microstructures can be based on lithographic or anisotropic etching methods and require integrated actuators for driving the movable structures. The actuators, such as thermal and electrostatic drives or solid–state actuators, including piezoelectric bending transducers or thin–film actuators, are simple to produce on a miniaturized scale. Typical applications of microactuators are as drive elements for valves and micropumps, in which a membrane is actuated by a piezoelectric element (see also Sect. 6.9).

6.2.5 Outlook

In comparison with conventional actuators, piezoelectric transducers demonstrate higher energy densities, larger actuating forces, greater positioning accuracy and shorter response times. Piezoelectric transducers are already being used commercially in areas where conventional actuators have reached

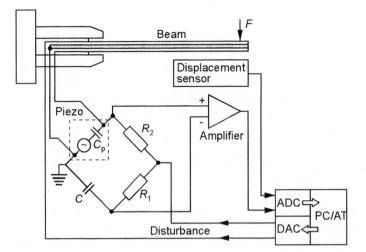

Fig. 6.17. Active vibrational damping with a smart actuator

their limits. However, widening the application of piezoelectric transducers will require improvements in material properties of the piezoceramic itself, for which greater strains and temperature stability are aimed, but also of the electrodes and electrical contacts, the factors representing the greatest manufacturing and material costs. Clear statements about dependability, for example with respect to aging, durability and fail–safe operation, are important. Too few studies have addressed these issues till now.

An additional aspect constraining the use of piezoelectric transducers in actuator applications is the lack of international standards for high–level signal characteristics as they are defined for low–level operation. The user therefore does not have the fundamentals for designing his system, whether with respect to the mechanical process interface or the sizing of the power amplifier. In the case of amplifiers, there is yet a discrepancy between the compact construction of the transducer and the voluminous power electronics. Even when the voltage source can be placed at an uncritical location, small amplifiers are essential for miniaturizing the system as a whole.

Piezoelectric transducers are relatively simple to use. Encompassing both the transducer and the sub–systems in which it is integrated, namely the mechanics and the control electronics including signal processing, a holistic view of the system is required of researchers, developers and designers in order to make use of the advantages and the corresponding potential of piezoelectric transducers.

6.3 Magnetostrictive Actuators
F. Claeyssen

Magnetostriction occurs in the most ferromagnetic materials and leads to many effects [26, 27]. The most useful is referred to as the Joule effect, and is responsible for the expansion (positive magnetostriction) or the contraction (negative) of a rod subjected to a longitudinal static magnetic field. In a given material, this magnetostrain is quadratic and occurs always in the same direction whatever the field direction.

Rare–earth–iron 'Giant' Magnetostrictive Alloys (GMAs), discovered by A.E. Clark [28], feature magnetostrains that are two orders of magnitude larger than Nickel. Among them, $Tb_{0.3}Dy_{0.7}Fe_{1.9}$, often called Terfenol–D, presents at room temperature the best compromise between a large magnetostrain and a low magnetic field. Positive magnetostrains of 1000–2000 ppm obtained with fields of 50–200 kA/m are reported for bulk materials [28, 29]. New composite materials of Feredyn offer an interesting possibility for high–frequency ultrasonic applications [30]. More recently, high magnetostrains (in the range of 500–1000 ppm) have also been obtained in rare–earth–iron thin films [31]. However, these expansion strains are rarely used directly because most applications require a linear behavior. The linearity is obtained by applying a magnetic bias and a mechanical prestress in the active material. Moreover, in the case of applications based on a mechanical resonance, it is a condition of producing 'giant' dynamic strains that their peak–to–peak amplitude is greater than that for the static magnetostrain [32].

The static magnetostrain of the GMAs permits the building of linear actuators offering small displacements (20–200 μm) and large forces (500–5000 N) at low voltage. These linear actuators are constructed to be used 'direct', for instance for micro positioning tools or for damping structures. They can also be used as components of a more complex actuator, such as inchworm motors. Such motors present holding forces/torques that are often much higher than piezoelectric inchworm motors; they also provide good positioning accuracy. Their main disadvantage is a low efficiency, which is due to their static operating conditions. Giant dynamic strains (up to 4000 ppm) can be produced in Terfenol–D linear actuators using the device of mechanical resonance, even against high load working; in such conditions, very large powers and rather good efficiency can be achieved. Using these properties, some magnetostrictive underwater transducers already outperform PZT transducers in the low–frequency domain and receive a great deal of attention. Some research works are being pursued in order to use also mechanical resonance in magnetostrictive motors, aiming at greater mechanical power and a better efficiency than in inchworm motors.

Although there is no large–volume application for magnetostrictive actuators at the moment, some are already used for specific applications in domains such as pumps, micro–positioners, and transducers, and research into other applications is growing. It is likely that we shall also see magnetostriction finding applications in the microactuators domain in the future.

6.3.1 Theory of Magnetostriction in Magnetostrictive Devices

Constitutive Equations. In the most general way, the behavior of magnetostrictive materials is nonlinear [26, 27] and has to be described with nonlinear relations:

$$S = f(\mathsf{T}, \boldsymbol{H})$$
$$\boldsymbol{B} = g(\mathsf{T}, \boldsymbol{H}) \tag{6.4}$$

relating S and T, the tensors of strain and stress, to $\boldsymbol{B}$ and $\boldsymbol{H}$, the vectors of induction and magnetic field. The functions f and g may be obtained by measuring the magnetostriction and the magnetization against the applied field and the external stress [33]. Then functions f and g can be described numerically by an interpolation method [34, 35]. This technic is used in the SANDYS software based on a finite difference method [36] for modeling two–dimensional (2D) structures. Another method could consist in developing f and g as a Fourier series, taking some first–order terms, and such an approach is being applied in the ATILA software, based on a finite element method [37], for modeling the nonlinear behavior of three–dimensional (3D) structures, including electrostrictive materials [38].

However, although magnetostrictive materials are nonlinear, the behavior of most magnetostrictive devices may be rather well described using a linear theory, because the active materials are biased. Experimental results obtained on a high–power transducer (see Sect. 6.3.2) show that linearity can be rather good even with large excitation fields and large dynamic strains.

The bias conditions are defined by the magnetic bias H_0 and the mechanical pre–stress T_0, applied along the magnetostrictive rod axis, which is referred to as the third axis. Then, considering only the variations around this initial bias state, the material behaves in a quasi–linear manner and follows piezomagnetic laws [39]:

$$\begin{aligned} S_i &= s_{ij}^H T_j + d_{ni} \boldsymbol{H}_n \\ \boldsymbol{B}_m &= d_{mj} T_j + \mu_{mn}^T \boldsymbol{H}_n \end{aligned} \quad \begin{vmatrix} i,j=1,\dots,6 \\ m,n=1,\dots,3 \end{vmatrix} \tag{6.5}$$

where s^H, d and μ^T are the tensors of constant–H compliance, piezomagnetic constants and constant–T permeabilities, respectively. They are called the magneto–elastic coefficients. S and T are the tensors of varying strain and stress, $\boldsymbol{B}$ and $\boldsymbol{H}$ are the vectors of varying induction and magnetic field. In the actuators, $\boldsymbol{H}$ is called the excitation field.

The real situation in the material can be reconstructed by adding the bias static situation to the variations. For instance, the real field in the material is the vectored sum of static magnetic bias H_0 and the varying magnetic field $\boldsymbol{H}$. Note also that the values of the coefficients of the materials tensors depend strongly on the bias and the prestress [39, 33]. Complete sets of values for the tensors s^H, d and μ^T and other equivalent tensors of Terfenol–D have already been established [39, 40, 41]. Longitudinal coefficients ('33') and shear coefficients ('15') may be determined using length expansion and shear resonators such as the MB [40] and DCC [42] types (as described in Sect. 6.3.2 and shown in Fig. 6.21). Other coefficients may be found using some special assumptions [39].

Terfenol–D is often used in long rods, subjected to an excitation field parallel to the rod axis. In this case, the simple theory of the longitudinal mode can be applied. Such theory can be used to obtain a preliminary system design, before the use of numerical models to refine it. In such a situation, it is presumed that the transverse excitation fields are negligible ($H_1 = H_2 = 0$). In theory, a pure longitudinal mode ('33' mode) is then obtained starting from the assumption that radial stresses are equal to zero ($T_1 = T_2 = 0$) and that there is no shear effect ($T_4 = T_5 = T_6 = 0$), leading to the following equations:

$$\begin{aligned}
S_1 = S_2 &= s_{13}^H T_3 + d_{31} H_3 \\
S_3 &= s_{33}^H T_3 + d_{33} H_3 \\
B_3 &= d_{33} T_3 + \mu_{33}^T H_3.
\end{aligned} \tag{6.6}$$

The '33'–mode coupling coefficient associated with this mode is given by

$$k_{33}^2 = \frac{d_{33}^2}{s_{33}^H \mu_{33}^T} \tag{6.7}$$

This coefficient represents the capability of the material to convert electric energy into elastic energy. Its value is high in Terfenol–D even with high prestress and bias [33] (see Table 6.3). As will be shown later, the combination of a high coupling, a high prestress and a high bias is required to obtain giant dynamic strains and very high output powers [32].

Table 6.3. Magneto–elastic longitudinal coefficients of Terfenol–D at about 90 kA/m bias versus prestress T_o

T_0	(MPa)	30	35	40	50
Y^H	(GPa)	29	21	23	40
s_{33}^H	(1/GPa)	0.034	0.048	0.043	0.025
Q^H		4.6	3.5	4.3	8.3
μ_{33}^T/μ_0		3.7	4.2	3.8	3.0
Q^T		2.0	1.9	2.2	2.8
d_{33}	(nm/A)	8.0	11.0	9.7	5.0
k_{33}	(%)	63.1	69.3	67.4	52.0

Simplified Theory of Magnetostrictive Linear Actuators. It is interesting to analyse the behavior of linear actuators because most applications are based on such actuators. To simplify the presentation, we can consider an actuator with one end working either free (no load) or against a purely resistive load (R_{load}); the other end of the actuator is clamped. The vibration against this load produces an output power (either mechanical or acoustic), and its behavior is representative of any magnetostrictive device. Most of them can be analysed as whole systems, including a compliance k^H (at constant field), an effective mass M and a mechanical resistance R_m due to internal mechanical losses. The magnetostrictive part is activated by a longitudinal field H_3 produced by a coil driven by an excitation current I. In such a system, all the strain is converted to displacement of the free mass.

Under quasi–static conditions, according to (6.6) and neglecting prestress spring stiffnesses for a first approximation (which gives $T_3 = 0$), the strain S_3 of Terfenol–D in an unloaded actuator is:

$$S_3 = d_{33}H_3. \tag{6.8}$$

A maximum excitation field H_3 equal to the bias H_o can be applied. Higher values lead to a frequency–doubling effect. In this situation, the actuator is field–limited. The heating of the coil is another limitation often encountered in static conditions. A high excitation field needs a high current density in the coil wires, typically in the range of $10\,\text{A/mm}^2$. As it is a rather high value, a significant heating may occur and it is therefore necessary either to use the actuator during short pulses or to cool the coil.

When the unloaded actuator is excited with a constant field amplitude against frequency, a sharp peak is obtained for the induced vibration. A typical example of strain curves (Fig. 6.18) without load or with a load is given by a linear Terfenol–D actuator based on a driver such as MAP (described in Sect. 6.3.2) (A load value of $R_{\text{load}} = 10^4\,\text{kg/s}$ is used in Figs. 6.18 to 6.20, and is denoted by $L = 10^4$). The natural longitudinal vibration mode occurs, and because of the coupling, this mode is magnetically excited. Compared with static strains, the strains at resonance are magnified by a factor called the 'mechanical quality factor' Q_m (the load value being equivalent to $Q_m = 2$) where

$$S_3 = Q_m d_{33}H_3. \tag{6.9}$$

This mechanical quality factor defines the damping of the resonance. When the vibrating end is unloaded, the damping is only due to internal mechanical losses and Q_m is equal to the material mechanical quality factor Q^H. When a load is applied, the resistive part of the load provides an additional damping that reduces the device's mechanical quality factor. Typical values for Q^H in Terfenol–D are in the range of 3 to 20. Consequently, for a very first approximation, the maximum strain S_3 at resonance under such conditions is determined from (6.6) and (6.9), by the stress T_3, since $d_{33}H_3$ is necessarily small compared with $s_{33}^H T_3$. We thus have

$$S_3 = s_{33}^H T_3. \tag{6.10}$$

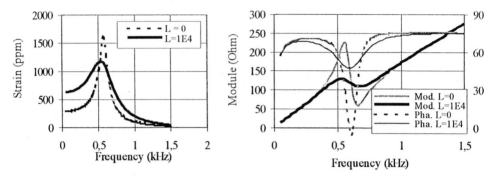

Fig. 6.18. Strain S_3 versus frequency at constant currents, without load and with a load equivalent to $Q_\mathrm{m} = 2$

Fig. 6.19. Module and phase impedance versus frequency, without load and with a load equivalent to $Q_\mathrm{m} = 2$

Without load (or also with a small load), the actuator is limited at resonance by the stress: the dynamic stress level T_3 reaches the prestress value T_o. With use of a high prestress, the maximum dynamic peak–to–peak amplitude of strain may be much larger than the maximum static strain (1600 ppm for this material) [32]. For instance, with $T_\mathrm{o} = 40\,\mathrm{MPa}$, the peak–to–peak strain is about $S_\mathrm{pp} = 2$, $S_3 = 3500$ ppm according to (6.10) and Table 6.3. This high strain is also permitted by the good coupling factor of Terfenol–D at such high prestress, and can be obtained under low load with a low field amplitude $H_3 = 40\,\mathrm{kA/m}$ according to (6.9) and Table 6.3. Intensive research on giant strains is being conducted and has allowed experimentalwork with peak–to–peak strains of 3500 ppm and more (see Sect. 6.3.2).

Due to the strong coupling, the mechanical resonance obtained at constant current is associated with the electrical antiresonance f_a, the maximum impedance (Fig. 6.19). Using a constant voltage, the mechanical resonance would occur at the electrical resonance f_r, the minimum impedance. These resonances determine the effective coupling factor k_eff of the device:

$$k_\mathrm{eff} = \sqrt{1 - (f_\mathrm{a}/f_\mathrm{r})^2}, \tag{6.11}$$

and this factor represents the capability of the device to convert electric energy to elastic energy. As shown below, the output power of a device depends strongly on this factor. In the best theoretical case, it is equal to the material coupling factor; in the best actuators, the measured k_eff may reach 55–60%.

The high–power handling capability of Terfenol–D can be observed by applying a high load. A high load condition is achieved when the mechanical quality factor Q_m of the vibration mode of the system is low (load higher than the optimal load). In this case, the actuator is field–limited, even at resonance. Then the maximum excitation field that can be applied is equal to the bias. It is important to notice that even against such high loads –

and unlike to PZT actuators in a same condition – the maximum strain of Terfenol–D actuators remains very high (Fig. 6.18).

A special case is obtained with an optimal load. Both stress and field limits are reached. This permits production of the absolute maximum power. The optimal load of an actuator can be determined theoretically. Typically (see Table 6.4, Sect. 6.3.2), it leads to a mechanical quality factor in the range of 2 to 3, which also show the ability of Terfenol–D to work against high loads.

The output power can be compared with the electric power through the efficiency (Fig. 6.20). The curve of efficiency against frequency shows that the best way to produce a significant output power with an actuator or a transducer is to work at resonance. A good efficiency ($\geq$50%) may be obtained with a high load ($Q_m \leq 2$).

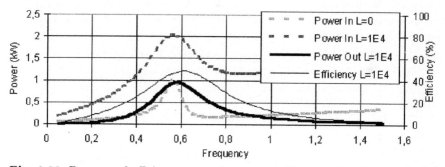

Fig. 6.20. Powers and efficiency versus frequency, without load and with a load L equivalent to $Q_m=2$

The expression of the output power at resonance [39] permits examination of the role of some parameters:

$$P_{\text{output}} = \omega e_m k_{\text{eff}}^2 Q_m (L_{\text{LF}} I^2/2), \tag{6.12}$$

where $L_{\text{LF}} I^2/2$ is the electric energy stored in the low–frequency inductance L_{LF} of the device, $e_m \equiv 1/(1 + R_m/R_{\text{load}})$ is a mechanical efficiency and ω is the resonance pulsation. In general, the pulsation and the load are often prescribed by the application. When it is possible to select the load, a high load is preferred to obtain a high efficiency. The stored electric energy can be increased using higher prestress, bias, and current. However, bias values much greater than $100\,\text{kA/m}$ are difficult to produce with permanent magnets. The effective coupling factor can be optimised by improvement on the basic design.

The maximum force that can be produced by the actuator is the clamped force. This force F is given by G, the force factor (also called the electromechanical conversion factor):

$$F = GI \tag{6.13}$$

with:

$$G = k_{\mathrm{eff}} \sqrt{L_{\mathrm{LF}} k^H} \qquad (6.14)$$

This is also the blocked force of the main mode of the actuator at resonance. So, it is an important parameter for several applications: for example, in both quasi–static and resonant motors it influences strongly the maximum force/torque of the motors.

This simplified theory provides an understanding of some important features of linear magnetostrictive drivers of actuators, transducers, etc. It shows, for instance, that a driver may be limited either by the stress or by the field, and that the strain at resonance may be much larger than that of a static system and yet may require much less field. However, because of the assumptions on the field shape, the strain uniformity and so on, it is not possible to predict accurately the behavior of the device, especially its exact limits. So, without a good knowledge of these limits, it is difficult to use the full potential of the device. That is why a more accurate model is required and has been developed.

Variational Principle of Magnetostrictive Devices. An accurate description of the behavior of magnetostrictive devices implies the taking into account of mechanical, magnetic and electrical states in a consistent way and without any assumption on the device geometry or the solution.

This can be performed, first, by applying Euler's equations at any point of the device thus:

- Newton's law (for the mechanical domain, including magnetostrictive parts);
- magnetic flux conservation (for the magnetic domain, including magnetostrictive parts and the vacuum surrounding the device); and
- Ampere's law (for the electric domain: the currents into the excitation coils).

These equations are local and do not depend on the material's constitutive equations, (6.4) and (6.5), which have additionally to be taken into account.

A second method, called the variational principle, relies on the stationary nature of a quadratic functional L (a kind of energy minimisation) with respect to the state variables. A satisfying variational principle has to be mathematically equivalent to the previous Euler's equations and has to take into account the boundary conditions. The chosen state variables are in this problem the displacement field u, the reduced scalar potential Φ of the magnetic field and the set of currents I in the coils.

For this second approach, it can be shown that a satisfying variational principle is simply an extension of the Hamilton principle [39, 40]. The stationary quantity L is equal to a generalised Lagrangian $\mathcal{L}$ added to a sum of virtual work $\mathcal{W}$. It is calculated on the volume V of the domain, defined by the device and the vacuum around the device, where the field is not zero. Thus

$$L = \mathcal{L} + \mathcal{W}. \tag{6.15}$$

The stationary condition is obtained for small variations δ around the solution which can be written as

$$\delta L = 0. \tag{6.16}$$

The generalized Lagrangian $\mathcal{L}$ is defined by the difference between the kinetic energy E_c and the magnetic enthalpy $\mathcal{H}$.

$$\mathcal{L} = E_c - \mathcal{H} \tag{6.17}$$

The kinetic energy term E_c can be written differently according to the type of analysis (static, harmonic, or transient). For example, in static analysis it is equal to 0, and in harmonic analysis at pulsation ω it is an integral [39, 40] depending on the displacement field u and the local density ρ.

The enthalpy term includes internal mechanical energy, magnetic energy and electric energy defined by the product of fluxes generated by the coils $\mathbf{Q}^V$ and the coil currents I.

$$\mathcal{H} = \int_V (\frac{1}{2}\mathsf{TS} - \frac{1}{2}\boldsymbol{BH})\mathrm{d}V + \boldsymbol{Q}^V I \tag{6.18}$$

Although the notation here is simplified (lower letters have been dropped), this term always takes into account all the components of stress $\boldsymbol{T}$, strain $\boldsymbol{S}$, induction $\boldsymbol{B}$ and field $\boldsymbol{H}$ that are bounded by a strong coupling relation, which can be either nonlinear (6.4) or linear (6.5). The sum of virtual work $\mathcal{W}$ is the sum of the mechanical work done by the prescribed normal forces T_N on the surface ST and the work done by the prescribed induction B_N on the surface SB [39, 40].

Note that all these equations can be used not only for the magnetostrictive materials but also for the whole of a magnetostrictive device, including the air around it. Then the coupling terms between elastic tensors (S, T) and magnetic vectors $(\boldsymbol{H}, \boldsymbol{B})$ are simply zero.

This thermodynamic interpretation of the proposed variational principle ensures for this method the existence and the uniqueness of the derived solution.

Finite Element Method for Magnetostrictive Devices. The main advantage of a variational principle is its global approach (in contrast with Euler's equations, which are local), which permits the problem to be treated by a Finite Element Method (FEM). Thus the resolution of the 3D magnetostrictive problem is equivalent, when discretized using the FEM, to the numerical resolution of a linear system, the unknowns of which are the state variables u, Φ at the nodes of the mesh and the coils flux $\boldsymbol{Q}^V$. For example, the matrix system is as following:

$$[\boldsymbol{K}][U, \Phi, I] = [\boldsymbol{F}, \boldsymbol{Q}, -\boldsymbol{Q}^V] \tag{6.19}$$

The vectors of force $\boldsymbol{F}$ and fluxes $\boldsymbol{Q}$ are equal to zero, but respectively on prescribed displacements u and potentials ϕ that occur on Dirichlet–type boundaries but that are unknowns. In prescribed currents analysis, the set of currents I is known. When this system is solved, the strains from the displacements, the magnetic field from the potentials, and the currents (reduced–scalar potential method) can (for example) be deduced, and thereby the stress field and the induction field (6.4). The inductances are given by the resulting value of vector $\boldsymbol{Q}^V$.

Because of the proposed variational principle and the reduced–scalar–potential formulation [39], strong algorithmic analogies have been recorded with respect to the method implemented within the ATILA software for piezoelectric transducers [37]. This software has therefore been extended to the computation of 2D and 3D magnetostrictive transducers [43] under advantageous conditions. Different types of resolutions have been developed: constant–current quasi–static, harmonic or modal analysis (antiresonances and resonances). The software has been used for the first time in the analysis of magnetostrictive transducers and actuators [41, 44]. It has subsequently been demonstrated that it permits the analysis of piezoelectric and magneto-strictive friction motors [45, 46, 47]. In addition, a 3D transient solver has also been implemented.

6.3.2 Principles and Properties of Various Applications

Linear Actuators and Drivers. Many linear actuators have been built [48]–[55]. For example, Etrema has a wide range of products of different sizes, all of which are adapted for quasi–static use. The 50/6MP, for instance, [48] is based on a 50 mm–long by 6 mm–diameter Terfenol–D rod. It is biased with a field H_o of about 40 kA/m. Low prestress and bias have the advantage of yielding to the highest d_{33} values; consequently, via (6.9), a high static strain S_3 of the unloaded actuator is obtained with a small field H_3. The maximum static strain of 500 ppm, leading to a displacement of 25 μm, can be achieved with a field of about 35 kA/m. It gives a strain of 14 ppm per kA/m, better than that of the MAP actuator of Cedrat Recherche, which offers only 7 ppm per kA/m. On the other hand, the prestress T_0 of the 50/6MP model is lower than 20 Mpa, much smaller than that of MAP. So the maximum dynamic strain of the 50/6MP is limited to about 1000 ppm and is also much smaller than that of MAP, which reaches more than 3000 ppm. This example shows that each actuator should be designed for its specific application.

The design problems of magnetostrictive linear drivers have been addressed at Cedrat through several actuators [41, 56, 47]. These actuators (see Fig. 6.21) are identical except in their bias system. They are all based on one driver and two symmetrical head–masses. Their driver contains a total length of Terfenol–D of 100 mm. The rod diameter is 20 mm. The first actuator, called MB, is biased with a DC current in a coil giving a bias field from 0 to 160 kA/m. The second actuator, MAP, is biased with permanent magnets placed outside the dynamic flux circuit and produces a field of about 90 kA/m bias. A 10 mm–thick coil permits using it against high loads, although because of the magnets and the coil, the diameter (excluding the masses) is about 70 mm. The third actuator, MAS, is biased with cylindrical permanent magnets placed in series between slices of Terfenol–D. The magnets' shape has been optimized [41] with FLUX2D [57], and produce a 90 kA/m bias field. MAS also has a 10 mm–thick coil, slighty longer than for the other types, but its diameter is only 50 mm. Some experimental properties of the MB, MAP, MAS drivers are compared in Table 6.4.

Fig. 6.21. MB, MAS and DCC actuators (respectively, from left to right)

Table 6.4. Experimental properties of the MB, MAP, MAS drivers

			MB	MAP	MAS
Bias	H_0	(kA/m)	100	90	90
Prestress	T_0	(MPa)	30	40	35
Coupling coefficient	k_{eff}	(%)	52	55	35
Max. magnetic energy density	$\varepsilon'_{\mathrm{m}}$	(kJ/m^3)	6.3	3.0	4.0
Max. elastic energy density	$\varepsilon'_{\mathrm{e}}$	(kJ/m^3)	15.3	30	48.5
Max. dynamic strains	S_{pp}	(ppm)	2020	3000	3500
Max. dynamic stroke	Δu	μm	202	300	350
Optimal mech. quality factor	$(Q_{\mathrm{m}})_{\mathrm{opt}}$		≈ 1.5	≈ 2.5	≈ 3.5
Max. dissipated energy density	$(\varepsilon")_{\mathrm{opt}}$	(kJ/m^3)	≈ 10	≈ 10	≈ 12

The MAS type of driver is an interesting example, both from the results obtained and the modeling point of view. It has the smallest coupling factor, due to the series magnets that introduce magnetic reluctances, uncoupled longitudinal compliances and radial stiffnesses [41]. These last mechanical effects cannot be correctly explained by simplified theory (see Sect. 6.3.1) but they are clearly predicted by ATILA software. In spite of these effects, the MAS presents high dynamic strains. The research of the absolute strain limits of linear drivers shows that the highest strains are obtained below resonance.

The curve of the absolute maximum strain against the frequency of the unloaded MAS (Fig. 6.22) has been calculated and tested taking into account both the field limit and the stress limit at each frequency. It defines a law of current that depends on frequency. This new strain curve is above the classical curve of strain at constant current, based on the maximum current acceptable at resonance. It possesses a large pass band, which might be used in several applications such as active damping, low frequency projectors, etc.

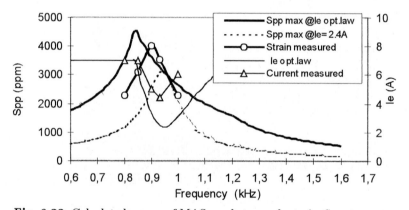

Fig. 6.22. Calculated curves of MAS: peak–to–peak strain S_{pp} at constant current $I=2.4$ A, strain with optimised current I_e law, and corresponding current Ie law; and measured values of strains of MAS and corresponding currents

The maximal dissipated energy density is the maximum energy per volume of Terfenol–D that can be dissipated in the load, which is achieved in the case of the optimal load. All the experimental values converge to 10–$12 \, \mathrm{kJ/m^3}$. This value is between five and ten times higher than that of PZT. It indicates that all these actuators can dissipate $0.4 \, \mathrm{J}$ providing, for instance, at $1 \, \mathrm{kHz}$, an output power of $2.5 \, \mathrm{kW}$ on optimal load. Linear actuators are studied for building micro–positioners [49, 50], fuel injectors [51], fast hydraulic drives [34], high pressure pumps [52], active damping applications [53, 54], helicopter blade control [55], etc. In all of these applications, the expected advantages over piezoelectric or conventional solutions are the large displacements and large forces associated with low voltages. Their main drawback is the rather high electric power requirement.

Transducers. The significance of the giant dynamic strains of Terfenol–D has been grasped rather quickly by transducer designers. Such strain levels, as well as high field limits, high coupling and high compliance, are well suited for high–power transducers both for acoustics (loudspeakers, sonars) [58]–[60] and for mechanics (welding, sealing, cleaning, machining, cutting, etc.) [30, 61].

The Tripode Tonpilz–type sonar transducer [45] (Fig. 6.23) is a good example for showing the high power capability of Terfenol–D. It is 31 cm long and 30 cm in diameter. It is based on three drivers, each of them including a 100 mm–long by 20 mm–diameter Terfenol–D rod. The maximum theoretical expectation was a head mass displacement of 110 μm, a Terfenol–D strain of 3250 ppm, an output power of 3.8 kW and a source level of 208.6 dB ref. 1 μPa at 1 m. Experimentation was performed to achieve about 90% of the theoretical performance. The head mass displacement was measured with an accelerometer, giving 98 μm at 1.2 kHz (Fig. 6.25). It corresponds to a 2900 ppm peak–to–peak strain in Terfenol–D, an output power of 3 kW and a sound level of 208 dB (Figure 6.24). This performance is achieved with an acceptable linearity. High power densities achieved now in Terfenol–D are ten times higher than those of PZT transducers. These results are interesting in the knowledge that the bias problem is now solved in different ways thanks to specific permanent magnet configurations (see earlier in this section) and are being applied [44, 62]. Such applications are seen to be promising candidates for development.

Motors. Magnetostrictive linear actuators are able to produce static displacements in the range 20 to 200 μm. These displacements being larger than mechanical tolerances, they render possible the successful building of inchworm motors [63, 64] offering high forces/torques and good resolution, at low speed. Such properties are very difficult to obtain with conventional electromagnetic motors. Inchworm motors need a gearbox to obtain high torques, which then introduce angular play: poor efficiency and wear are thus the weaknesses of inchworms, and these factors limit the number of applications.

J.M. Vranish [64] has constructed a rotating stepping motor with the highest torque (12.2 Nm) ever reported among all the piezoactive motors; its holding torque is also very high. Its speed limit (0.5 rpm) is small. Its angular resolution is better than 800 μrad. As typically with inchworms, its output power is low (<1 W) compared with the electric power required (600 W).

L. Kiesewetter [63] has built a linear inchworm motor which has been commercialised by Dynamotive in the paper industry. It uses both longitudinal and radial strains in the moving part of a Terfenol–D rod (Fig. 6.26). Modeling this type of motor [46] provides understanding of its mechanism (Fig. 6.27). According to Dynamotive, typical results for a motor based on

Fig. 6.23. Tripode sonar transducer

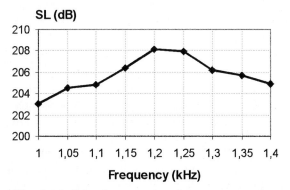

Fig. 6.24. Tripode sonar transducer: measured sound level versus frequency

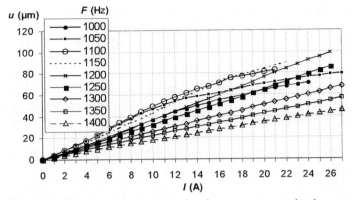

Fig. 6.25. Displacement n of the head mass versus excitation current I at different frequencies

a 120 mm–long by 10 mm–diameter rod are a maximum speed of 20 mm/s, a maximum force of 1000 N and a resolution of 2 μm.

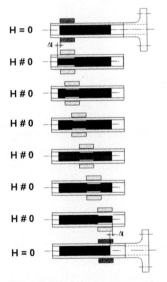

Fig. 6.26. Kiesewetter motor principle

Friction motors (also called ultrasonic motors) offer a new field of applications for magnetostriction. These motors use the vibrations of a stator to transmit a motion to a rotor or a driven member. Such motors based on piezoelectric ceramics already exist on a large commercial scale; they offer large dynamic and holding torques, along with low speeds and good efficiencies through resonance.

Following a principle used in piezoelectric ultrasonic motors [65], T. Akuta [66] has built the first magnetostrictive friction motor. This stator is made of pairs of orthogonal actuators excited with sinusoidal 90° phase–shift currents, which produce an elliptical vibration. The modeling of such magnetostrictive stators [47] has shown that in quasi–static operation a good elliptical motion is produced. It has also been shown that there are many coupled modes, but none of them provides a satisfactory elliptical motion. Therefore, unlike piezo-electric motors, this motor cannot operate at resonance. As a consequence and in relation to the previous analysis of power (Fig. 6.20), the efficiency is comparatively weak. Its other characteristics are a speed of 40°/s and a torque of 1.8 Nm [67].

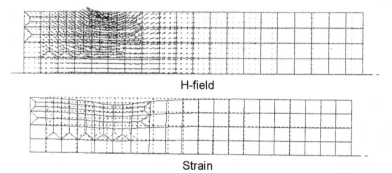

H-field

Strain

Fig. 6.27. ATILA computation of H–field and strain in the rod of a Kiesewetter motor at one step

It is difficult to convert existing piezo–motors to magnetostrictive versions; new designs have to be found. A first magnetostrictive motor using the mechanical resonance of two vibration modes has been built and tested by Cedrat Recherche [68] (Fig. 6.28). Its stator modules are made of a ring and two Terfenol–D linear actuators. The translation mode of the stator produces a vibration that is tangential to the contact zone (Fig. 6.31a). The flexure mode produces a vibration that is normal to the contact zone (Fig. 6.31b). These modes are coupled using a 90° phase shift, in order to produce ellip-tical vibrations (Fig. 6.29) that are used to transmit a motion to two rotors by friction. A low rotating speed of 100°/s, a torque of 2.1 Nm are achieved (Fig. 6.30). The goal was to show that Terfenol–D can be used for making high–torque motors and that resonance can be beneficial for that purpose and for improving efficiency.

The same principle can be applied to building various kinds of motor according to the number of modules and design choices: linear or rotating, stepping or ultrasonic, etc. It has been used recently to build an interesting ultrasonic piezomotor [69].

Fig. 6.28. Multi–mode magnetostrictive FLEX–M1 motor

Micro–motors and Actuators. T. Fukuda [70] has opened the field of miniature magnetostrictive actuators and motors taking advantage of wireless magnetic excitation. He has experimented with two small self–moving linear motors (some of cubic centimeter dimensions) based on a conversion–mode principle.

The first linear micro–motor, based on magnetostrictive thin films deposited on a $7\,\mu m$ polyamide film, was built in Japan in 1994 [71]. The 13 mm–long prototype used a 200 Hz vibration induced by magnetostriction to obtain one–way motion at 5 mm/s. This is a Mode Conversion Ultrasonic Motor (MCUM) according to the Japanese classification of piezoelectric motors.

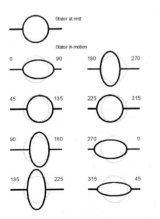

Fig. 6.29. Principle of FLEX–M1 stator Stator at rest and in motion versus the actuators phases

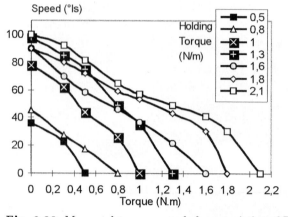

Fig. 6.30. Measured torque–speed characteristics of FLEX–M1 motor with different holding torques

The torsion–based, drift–free microactuator [72], invented by CNRS Grenoble, is basically a unimorph structure composed of a single magnetostrictive film deposited on a passive substrate. The new feature is a square shape maintained by hinges at three corners (Fig. 6.32).

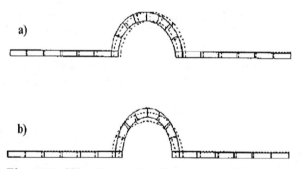

Fig. 6.31. Vibration modes of a stator module, computed with ATILA: **(a)** translation mode; **(b)** flexure mode

The useful displacement due to magnetostriction is obtained at the fourth (free) corner and without thermal displacement. The different deformed shapes are due to the anisotropy of magnetostrictive strains and the isotropy of thermal strains. Modeling with ATILA (Fig. 6.33) has permitted the design of appropriate micro–hinges. Prototypes have been realised by micro–machining a Silicon substrate and by depositing a magnetostrictive film by sputtering. Measurements using laser interferometry have confirmed the modeling expectations.

Fig. 6.32. Microactuator

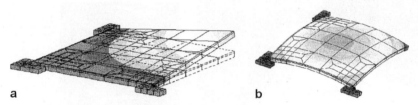

a   b

Fig. 6.33. Modeling of the microactuator shown in Fig. 6.32 **(a)** magnetostrictive deformation; **(b)** thermal deformation

Several standing–wave ultrasonic motors (SWUMs), have been designed at Cedrat [73] (Figs. 6.34 and 6.35). A linear motor is a self–moving Silicon plate including magnetostrictive film. It is submitted to a 10 mT dynamic field produced by an external coil, which may be placed at some centimeters' distance from the motor. At resonance, this field excites a flexure mode, producing vibrations in the plate, which in turn induces by friction a motor motion at 10–20 mm/s. A rotating version has been also created (Fig. 6.34b) that uses a slightly different principle [73]: the vibrating rotor is based a 100 µm–thick by 20 mm–diameter plate with 10 µm deposited magnetostrictive films, which are wireless and excited by a small coil. Typical performance is a rotating speed of 30 rpm and a torque of 1.6 µNm, with a 20 mT excitation field.

a   b

Fig. 6.34. Two wireless micro–motors: **(a)** standard form; **(b)** rotating form

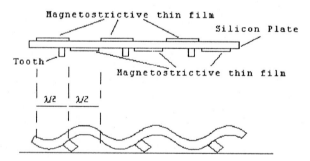

Fig. 6.35. Principle of the wireless linear micro–motor

These examples demonstrate some of the special advantages of magnetostriction, especially the fact that the moving parts are wireless. The disadvantage is the coil, which is difficult to make small because of the field requirements. These considerations are driving the development of films with magnetostriction at low field. Note that, as these devices are very small, the price of the material is not a problem, and so such actuators could find largescale applications — for instance in optics, in medicine, or in the automobile industry.

6.3.3 Summary

Several new Giant Magnetostrictive Materials have been developed in recent years for actuation applications. Bulk rare–earth–iron alloys Terfenol–D present typical static magnetostrains of 1000–2000 ppm, which permit the building of low–frequency actuators and transducers. New composite materials offer an interesting possibility for high–frequency ultrasonic applications. More recently, rare–earth–iron thin films have been also explored for actuation in microsystems.

Among all applications based on magnetostrictive materials, devices based on mechanical resonance, such as underwater transducers, ultrasonic transducers, resonant motors and micro–motors, are of special interest. Using resonance, giant dynamic strains (up to 4000 ppm) can be produced, which can lead to very large power/force densities and rather good efficiency. Although there is no large–volume application at the moment, some devices are already used for specific applications in domains such as pumps, micro–positioners, transducers, and research into other applications is growing. It is likely that we shall see magnetostricton finding application in the micro–actuator domain in the near future.

6.3.4 Acknowledgement

The authors would like to thank: D. Boucher (DCN) and A. Colin (DRET) for the financial support of Cedrat works on acoustic applications; C. Sol (French Ministry of Research) for the support on electrical engineering applications; the European Commission for the financial support on microsystems applications (BRE2–0536 MAGNIFIT); the partners of MAGNIFIT, especially Laboratory Louis Néel CNRS Grenoble, Kassel Universität and Forschungszentrum Karlsruhe, for their efforts in producing microsystems; R. Bossut (ISEN); and the team at ISEN acoustic laboratory for their continuous efforts in developing ATILA.

6.4 Shape Memory Actuators
J. Hesselbach

The shape memory effect was first discovered some sixty years ago in a gold–cadmium alloy. Since this extraordinary effect was recognized in the early 1950s as being caused by a so–called 'martensitic transformation', new and improved shape memory alloys have been found. As prices for shape memory alloys are dropping, more and more commercial applications – ranging from aviation to medicine – make use of the functional properties of those materials. In this contribution we will focus on the new and innovative field of actuator applications.

6.4.1 Properties of Shape Memory Alloys

Shape Memory Effect. The term shape memory (SM) refers to the ability of certain materials to annihilate a deformation and to recover a predefined or 'imprinted' shape. Even though the shape memory behavior is also attributed to some plastic materials, in this text only shape memory alloys are considered. The SM effect is based on a solid–solid phase transition of the shape memory alloy that takes place within a specific temperature interval.

The properties of the shape memory alloy vary with its temperature. Above the transition temperature, the alloy's crystallic structure takes on the so–called austenitic state. Its structure is symmetric and the alloy shows a high elastic modulus. The martensitic crystalline structure will be more stable for thermodynamical reasons if the material's temperature drops below the transformation temperature. Martensite can evolve from austenitic crystals in various crystallographic directions and will form a twinned structure. Boundaries of twinned martensite can easily be moved; for that reason SM elements can be deformed with quite low forces in the martensitic state.

When heated up, the austenitic structure will be established again. At the same time the SM element will return to its original shape because neither the phase transformation nor the de–twinning of the martensitic structure involves changes to the atomic lattice. The SM element may exert high forces when recovering its predefined shape; therefore the SM effect can be employed as a new actuator principle.

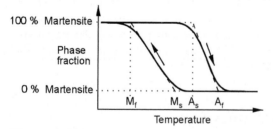

Fig. 6.36. Transformation temperatures and their hysteresis

Forward and reverse transformation occur at different temperatures, resulting in a hysteresis as can be seen in Fig. 6.36. The start and end of the transformation from martensite to austenite are given by A_s (austenite start temperature) and A_f (austenite finish temperature). The reverse transformation takes place in the temperature interval from M_s to M_f (martensite start and finish temperatures). The shape of the hysteresis curve in Fig. 6.36 strongly depends on the thermomechanical treatment of the shape memory alloy (see also Sec. 6.4.2).

The shape memory effect may be divided into three categories, each showing different functional behavior. They will be described briefly in the following sections. More detailed discussions can be found in [74] or [75].

One–way Effect. Stretching a shape memory coil spring when it is in a martensitic (i. e. cold) state will de–twin the differently oriented martensite. The result is an almost homogenously oriented martensitic structure. Similar to a common plastic deformation the SM coil spring will stay in the streched shape when unloaded (Fig. 6.37).

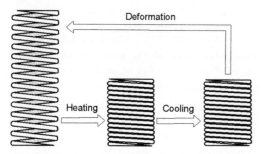

Fig. 6.37. One–way effect of a shape memory alloy spring

If the shape memory alloy spring is heated and the temperature surpasses the A_s temperature, the shape memory material starts transforming to austenite and the coil spring returns to the unstreched form. When reaching the A_f temperature, the transformation is completed. It is characteristic for the one–way effect that a shape recovery occurs only when the SM element is heated. There is no shape change when the element is cooled. The cold SM element must be deformed by an external force in order to achieve a movement when heated again.

The one–way effect is mainly utilized for fastening and clamping devices. Among the commercially most successful products are coupling sleeves made of shape memory alloy with the one–way effect, which guarantees very reliable connections of hydraulic pipes in airplanes.

Two–way Effect. Shape memory elements with a two–way effect will 're-member' a high–temperature shape as well as a low–temperature shape. The element flips between both shapes depending on the temperature.

If an SM coil spring with the two–way effect is heated, it will return to its predefined high–temperature shape — in Fig. 6.38 this is the compressed form. Upon cooling, the spring stretches to reach its low–temperature shape, ideally without the need of a supporting external force. SM elements with a two–way effect therefore represent thermally activated actuators.

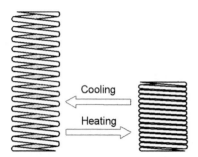

Fig. 6.38. Intrinsic two–way effect of a shape memory alloy spring

A shape memory element must be specially 'trained' to display the two–way effect. Training typically consists of several cycles of deformation to the desired low–temperature shape and subsequent shape recovery by heating [75]. Only very low forces can be exerted when the SM element changes from its high–temperature shape to its low–temperature shape; for this reason, steel springs or other elastic elements are added to the system to guarantee a full return to the low–temperature shape.

Superelasticity. Above the transition temperature, shape memory alloys show an extraordinary elasticity. Like typically elastic materials, a load will cause a shape deformation that disappears upon unloading. The difference is that shape memory alloys may be reversibly stretched or compressed five to ten times the amount of conventional materials. The restoring force is almost independent of the strain.

Because of its outstanding flexibility, superelastic shape memory wires are used as guide–wires for minimally invasive surgery or as orthodontic arch wires.

Actuator Activation. Shape memory alloys with the two–way effect can be utilized as actuators to generate repeated movements. To activate the shape memory effect and the corresponding movement, the SM element must be heated above the transition temperature A_f. Heating may be accomplished in different ways:

Thermal Activation. Most of the application examples of SM actuators presented so far rely on a thermal activation of the shape memory effect, i.e. the actuator element reacts according to the ambient temperature. Here is a short list of various application areas:

– automatic transmission (to shift points adjustment for cold start);
– ventilation flaps of greenhouses (temperature–dependent opening angle);
– fan clutches (control of engine temperature);

- headlamp concealment devices (open when light is switched on);
- fire protection (closes windows or opens sprinkler valves); and
- anti–scald safety valves (to cut off hot water).

As an example, Fig. 6.39 shows the principle of a control valve with a shape memory coil spring in the automatic transmission of a limousine [76]. Depending on the oil temperature, the SM spring lowers the pressure of the transmission oil, resulting in a smoother shifting of gears.

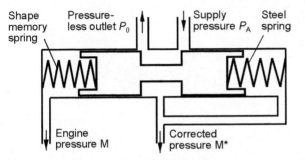

Fig. 6.39. Control valve with SM spring in an automatic transmission [76]

A major advantage of thermally activated SM actuators is the conversion of thermal energy of the surrounding medium to an actuator movement. These actuators do not need any additional (for instance electrical) power supplies, a property highly appreciated for all sorts of safety devices. Such an SM element makes up a complete system, consisting of temperature sensor and actuator.

Electrical Activation. The required heat energy may also be generated directly within the SM element by electrical current. Joule heating allows the construction of very small and compact electrically controlled SM actuators. Experimental applications for robotic devices and grippers (such as silicon wafer grippers) have proven the feasibility of such actuators [74]. The remaining part of this article will mostly cover electrically heated SM actuators because they can be utilized in automation devices.

Available Shape Memory Alloys. There are many alloys with a shape memory effect. They strongly differ amongst themselves with respect to transformation temperatures, effect amplitude and other material properties.

Shape memory alloys designated for actuator purposes (for example in automation systems) should meet the following requirements:

- large shape memory effect resulting in a long actuator stroke;
- high transformation temperatures: phase transformation should occur at high temperatures like 150–200 °C to avoid unwanted activation by warm ambient air, and high transition temperatures also guarantee a complete phase transformation to the martensitic state.

– a high number of activations and a stable SM effect; and
– a small hysteresis width between forward and reverse transformations.

Table 6.5 summarizes the most important properties of some commonly used shape memory alloys.

Table 6.5. Comparison of shape memory alloy systems

	NiTi	CuZnAl	CuAlNi	FeNiCoTi	Unit
Range of transformation temperature	−100 to +70	−200 to +100	−150 to 200	−150 to +550	°C
Hysteresis width	30	15	20		K
Max. one-way effect	8	4	6	1	%
Max. two-way effect	4	0.8	1	0.5	%
Fatigue strength	800–1000	400–700	700–800	600–900	N/mm^2
Admissible stress for actuator cycling	150	75	100	250	N/mm^2
Typ. number of cycles	>100 000	10 000	5 000	50	
Density	6450	7900	7150	8000	kg/m^3
El. Resistivity	80–100	7–12	10–14		10^{-8} Ω m
Young's modulus E_A	50	70–100	80–100	170–190	GPa
Corrosion resistance	very good	fair	good	bad	

A comparison of the required properties and the data of the shape memory alloys available (or under development) leads to the following conclusions:

– Nickel–titanium (NiTi) and nickel–titanium–copper (NiTiCu) have the best properties for actuator purposes. For that reason, industrial applications almost exclusively rely on nickel–titanium–based SM alloys. There is one single drawback to NiTi: its transformation temperature is limited so far to approximately 100°C.
– Copper–based shape memory alloys (CuZnAl, CuAlNi) can be designed for higher transformation temperatures and are less expensive than NiTi. Because of a lower lifespan and lower work output, they are not feasible for electrical actuator applications. Elements of CuZnAl are successfully implemented as thermal actuators in fire safety devices.

Table 6.6. Examples of actuator shapes

Actuator stroke	Material deformation	Actuator shape
Translation	Contraction	Tensile wire, bar, or tube 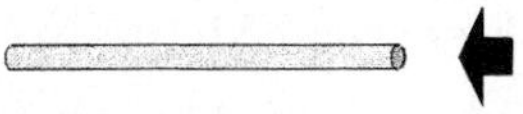
Translation	Extension	Compression bar or tube
Translation	Shear	Coil spring 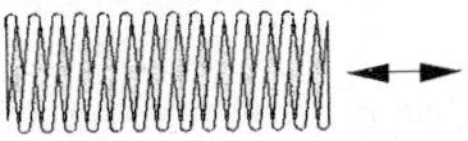
Rotation	Bending	Leaf spring 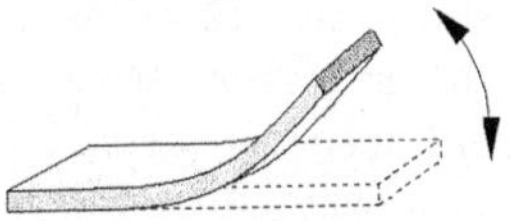
Rotation	Bending	Torsion helical spring 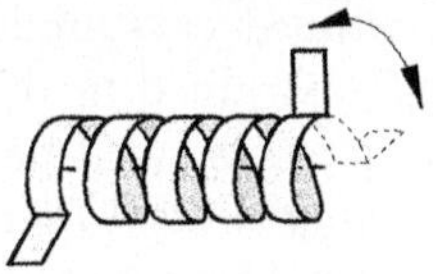
Rotation	Shear	Torsion wire, bar, or tube 

– Other shape memory alloys such as FeNiCoTi or NiTiHf, NiTiPd or NiAl are still in the state of early research and have not yet been perfected for commercial use. The properties being sought are high transition temperatures combined with good SM effects [77, 78].

Because of the superior actuator properties and the commercial impact of NiTi alloys, the following discussions will focus on these alloys. NiTi– and NiTiCu–based wires are commercially available with a range of transformation temperatures and in all diameters down to 25 μm. The same alloys are also supplied in the shape of flat–rolled wire or stripes in various sizes. Nickel–titanium is usually vacuum melted and then drawn or sheet–rolled. To reduce this time–consuming and expensive manufacturing process, new procedures suited especially for small–sized SM actuators are being investigated:

– rapid quenching: by pouring the melted alloy on fast–spinning cylinders, the alloy is cooled within milliseconds and forms thin (100 μm or thinner) films with the desired width [79]; and
– sputter–deposition: different sputtering techniques are available to deposit NiTi or NiTiCu on a substrate [80, 81]. The thin films have a thickness of

up to $10\,\mu$m. This technique opens up the possibility of using SM actuators in micromechanical systems.

6.4.2 Electrical Shape Memory Actuators

Actuator Shape and Stroke. The shape that the SM actuator recovers to when heated is 'imprinted' into the alloy by an annealing process. For instance, to fabricate a coil spring an SM wire is wound aroung a mandrel and annealed for 1–2 hours at 350–500°C. Annealing temperature and duration have a strong influence on the actuator's properties, such as the trainable two-way effect, the effect stability, and the hysteresis behavior.

The shape change between high–temperature and low–temperature shape defines the actuator stroke. Table 6.6 lists some commonly used actuator shapes and actuator strokes.

The two–way effect will be stabilized after 20–100 thermal and mechanical cycles. Because of the ability of the martensite (low–temperature phase) to form a twinned crystalline structure, different areas of the actuator element may be strained in different ways: Extension, compression, or shear are deformations that will be reverted to by heating. This variety offers the interesting opportunity to adapt the actuator's shape change to the special needs of the actuating task. By this means, transmission links or gears may be eliminated, which helps reduce the size and price of a system.

The actuator stroke is limited only by the reversible strain that the martensitic structure can accommodate by de–twinning — otherwise irreversible strain will occur. The admissible strain is determined by the type of shape memory alloy as well as the desired number of activation cycles. If the effect is to be employed only once (for example, for tube connectors), NiTi–based alloys may be strained up to 8%. For actuator use with more than 100 000 activations, only smaller strains are permitted, namely extensions $\varepsilon_{\text{adm}} < 3\%$ shear $\gamma_{\text{adm}} < 4\%$ and stresses up to $\sigma_{\text{adm}} < 150\,\text{N/mm}^2$ or $\tau_{\text{adm}} < 100\,\text{N/mm}^2$. Table 6.7 gives an overview of the design data of the most commonly utilized SM actuator geometries.

Dynamic Response. A central point of consideration when using shape memory alloys as electrically activated actuators is their response time between commanding signal and actuator movement. In theory, the phase transformation propagates with the speed of sound, but only if the necessary heat energy is supplied or dissipated fast enough.

Heating. Heating up the SM element is relatively simple. When conducting an electrical current, heat is generated because of Joule losses directly within the SM actuator. By controlling the current appropriately, very quick heating is possible. As an example, the response of an SM wire (diamter 0.22 mm) to different heating currents is shown in Fig. 6.40. With an additional short–time current pulse (line 'b') the actuator reacts much faster than with a constant heating current (line 'a'). The positioning time is faster than 0.5 s.

Table 6.7. Data of SM actuators

Symbols: F_{max}, M_{max}, ΔL_{max}, $\Delta\varphi_{max}$: maximum actuator force, torque, stroke, and angle respectively; σ_{adm}, τ_{adm}, ε_{adm}, γ_{adm}: admissible tensile stress, shear stress, extension, and shear respectively; D: SM wire diameter; L: SM wire length; D_m: coil diameter; i_f: number of turns; b, h: width and thickness of SM flat wires or bars.

Actuator shape	Max. force/torque	Max. stroke/angle
Tension wire or bar, compression bar (round cross section)	$F_{max} = \frac{\pi}{4}D^2\sigma_{adm}$	$\Delta L_{max} = \varepsilon_{adm}L$
Tension wire or bar, compression bar (rectangular cross section)	$F_{max} = bh\sigma_{adm}$	$\Delta L_{max} = \varepsilon_{adm}L$
Torsion wire or bar (round cross section)	$M_{max} = \frac{\pi}{16}D^3\tau_{adm}$	$\Delta\varphi_{max} = \frac{2L}{D}\gamma_{adm}$
Torsion helical spring (made of flat wire)	$M_{max} = \frac{1}{6}bh^2\sigma_{adm}$	$\Delta\varphi_{max} = 2\pi i_f\frac{D_m}{h}\varepsilon_{adm}$
Coil spring (tension or compression)	$F_{max} = \frac{\pi D^3}{8kD_m}\tau_{adm}$ $k \quad = \frac{2D_m+D}{2D_m-D}$	$\Delta L_{max} = \pi i_f\frac{D_m^2}{D}\gamma_{adm}$

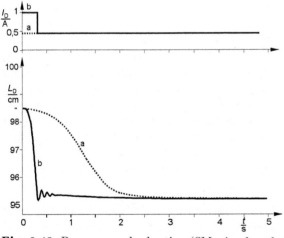

Fig. 6.40. Response under heating (SM wire, length L_D, diameter $0.22\,$mm) [82]

Cooling. The process of cooling down is strongly influenced by the medium surrounding the actuator. Therefore only external or constructural measures determine the actuator behavior at cooling time. Cooling can greatly be sped up by choosing a different surrounding medium, as shown by the plots in Fig. 6.41. It shows the cooling behavior of an SM wire in calm air, turbulent air, and water. As can be seen, an SM actuator in water will cool more than ten times faster than the same actuator in air at room temperature. Further possibilities to accelerate the cooling process are:

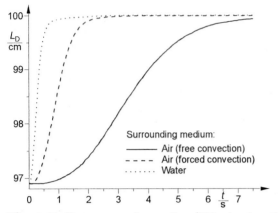

Fig. 6.41. Response under cooling (SM wire, length L_D, diameter 0.22 mm)

 – enlargement of the ratio between actuator surface and volume: one way to accomplish this is to make use of flat–rolled wire instead of round wire;
 – increasing the difference between the actuator temperature and the temperature of the surrounding fluid: for that reason SM alloys with high transformation temperatures should be preferred for actuators; and
 – active cooling by forced convection.

Position Control and Internal Sensoric Effect. If an SM actuator is employed only to switch between two different positions, a simple on/off–control of the heating current will be adequate. On the other hand, most SM actuator applications require fine positioning, which will be dealt with in this subsection.

To reach and hold a defined position cannot be accomplished by a feedforward control because the relationship between heating current and actuator stroke displays a hysteresis and is therefore ambigious. Considering the physical effects involved, a system with an electrically heated SM actuator may be described by a mathematical model consisting of three parts [83] (see Fig. 6.42):

 – The heat transfer model describes the heating of the actuator alloy by Joule energy as well as the heat losses to the surrounding air. The heat transferred from the actuator to the environment is a strongly nonlinear function of actuator temperature, ambient temperature and type of convection.
 – The model of the shape memory effect is based on thermodynamic laws of phase transitions in solids. Because of inner friction and losses of the phase transformation, the simulation of the hysteresis by means of the Preisach–model [83] must be modified and linked with the thermodynamic equations.
 – A kinematic model of the mechanical structure into which the SM actuator is integrated.

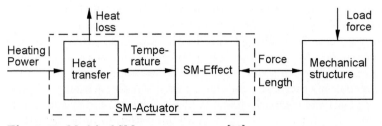

Fig. 6.42. Model of SM actuator system [83]

Based on this concept, the shape memory actuator system can be simulated by a nonlinear dynamical model.

Not only ambiguity due to hysteresis but also the influence of disturbances such as load force and heat loss on the actuator position make it clear that steady positioning of a SM actuator can only be achieved with a position sensor and feedback control.

The installation of an additional position sensor is not always possible. Reasons may be costs and/or unavailable space. In this case, the internal sensoric effect displayed by some NiTiCu–alloys may be employed for indirect position sensing [84]. This leads to the use of a smart actuator (cf. Sect. 6.1.6). In Fig. 6.43 the actuator length L_D of a NiTiCu shape memory wire is plotted against its electrical resistance R_D. The relation is free of hysteresis and is only slightly shifted by the actuator's load.

The almost–linear behavior between wire length and resistance can be explained by the fact that the actuator stroke is approximately proportional to the fraction of austenite and martensite in the alloy. Since the resistivity of martensite and austenite are different, the resistance of the SM actuator will vary according to the phase fractions: only in the fully martensitic or austenitic state does the relationship between actuator stroke and resistance

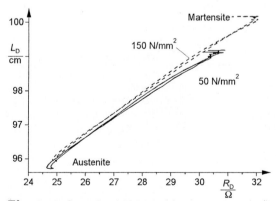

Fig. 6.43. Length of SM wire (diameter $0.22\,\mathrm{mm}$) with respect to electrical resistance

become non–linear. The stroke–resistance relation is independent of the ambient temperature (respectively, the type of surrounding medium or the type of convection) because it is affected only by the martensite fraction in the actuator material. On the other hand, a load force will induce a small amount of elastic strain, resulting in a shift of the stroke–resistance relation.

These explanations establish that the actuator's resistance may be used as an indirect positional feedback signal. A block diagram of such a feedback control circuit is displayed in Fig. 6.44. A PI–algorithm is implemented in order to calculate the electrical heating power P_{el}.

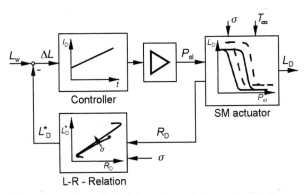

Fig. 6.44. Control circuit with resistance feedback

6.4.3 Perspectives for Shape Memory Actuators

The properties of electrically activated shape memory actuators described so far have indicated that these actuators are well–suited to drive mechanical mechanisms. The advantages and disadvantages of this kind of new actuator principle are summarized in Table 6.8. Shape memory actuators offer a

Table 6.8. Advantages and disadvantages of SM actuators

Advantages		Disadvantages	
+	Large energy density	−	Uncertainty about duration and
+	Small and compact		stability of the effect
+	Simple mechanisms	−	Relatively low velocity
+	Variable shapes	−	Very low efficiency
+	Linear or rotational motion	−	Limited range of transformation
+	Miniaturizable		temperatures
+	Usable in clean room environment		
+	Good corrosion resistance		
+	Low voltage ($< 40\,\text{V}$)		
+	Silent		
+	Intrinsic sensoric effect		

lot of advantages, but there are also some quite serious drawbacks. When comparing SM actuators with other actuator principles (such as piezoelectric stacks or solenoids), it should be taken into consideration that research for improved shape memory materials is relatively young (10–15 years). With shape memory alloys and actuators slowly gaining commercial importance it is expected that in the next several years new SM alloys will emerge that have higher transition temperatures and good effect stability [77, 78].

The disadvantage of low efficiency (below 2%) is determined by thermodynamics: most of the input energy is transformed to heat within the SM element. Furthermore, during the cooling process the heat is lost to the surrounding region and cannot be converted back to electrical or another reusable form of energy. Because of the low efficiency, limited effect duration, and low speed, it must be understood that shape memory actuators are not intended for applications where electrical motors or pneumatic cylinders are well established. Instead, electrical shape memory actuators offer a good choice for very special or new applications where conventional motors would either require expensive modifications or are not available.

The analysis of the advantages and disadvantages reveals good feasibility and opportunities for electrically heated shape memory actuators, especially in two fields of application:

– compact and light auxiliary actuating devices — as an example they may be used to increase the flexibility of automation devices, such as adjusting the range of grippers [85, 82]; and
– actuators for precision engineering and micromechanical systems.

The advantages of SM actuators listed in Table 6.8 gain importance where small mechanisms are concerned. The following properties recommend utilization of shape memory actuators in milli- or micrometer–sized mechanical mechanisms:

– Compared with SM actuators with large volumes, small SM actuators offer a much higher surface–to–volume ratio. Hence, heat transfer to the surrounding medium is strongly improved, resulting in faster response times of the actuators.
– Small NiTi–strips or thin–films may be fabricated by employing new methods, such as rapid quenching or sputter techniques. SM actuators fabricated in this way are less expensive than SM wires because less material is needed and the element is produced in the necessary size.
– Sputtering is a typical fabrication process also employed for micro parts. Therefore, sputtering of NiTi can be integrated in the manufacturing process more easily.
– The very high work–per–volume ratio of approx. $4\,\mathrm{J/cm^3}$ is highly valued if space is limited. SM actuators offer high forces and strokes. For example, a piezoelectric actuator with the same force and stroke would have to use up to ten times the space necessary for an SM actuator.

- Low efficiency is less important because overall energy consumption is low.
- As 'intelligent' actuators, the internal sensoric effect can be used for indirect position sensing. Again, this property is useful in small–sized applications where additional sensors cannot be accommodated for space and weight reasons.

There is a strong demand for miniature devices and sophisticated designs in many areas of technology. Examples for such fields of application are as follows:

- **Micro assembly:** While most of the technology used to fabricate parts of millimeter or micrometer size could be copied and modified from microelectronics fabrication processes, this is not true for micro assembly. There are scarcely any devices suitable for a small– or medium–scale automated assembly of micromechanical systems. Micro assembly opens a broad field of potential applications for new actuator principles. An example is the handling of millimeter–sized, lightweight parts under clean–room conditions and with small operational space available.
- **Inspection tasks:** It is often necessary to inspect inner cavities of machines or pipe systems without disassembly or destruction. Endoscopes are available for this task, but these (albeit flexible devices) are insufficient if the object under inspection has a very complex geometry or exact positioning is required. Here, SM actuators promise more degrees of freedom and more controllability, and are formed into smaller devices.
- **Medical devices:** Similar to the inspection of technical devices, more flexibility and controllability are desired for medical devices such as surgery instruments for minimally–invasive operations. Other medical applications may be drug–release systems implanted in the body.

6.4.4 Innovative Application Examples

In this section some examples of precision engineering prototypes are presented that apply electrically heated shape memory actuators as driving elements.

Mechanical Grippers for Micro Assembly.

General Aspects. Mechanical grippers have a variety of applications and for that reason will very likely be the most often used grippers also in micro assembly. There are some differences to common assembly procedures to be considered when assembling very small parts. For grippers, the required properties are briefly summarized:

- compact size;
- good controllability of gripping forces (in the range 1 to 100 mN);
- suited for a clean–room environment;
- control of adhesive forces;
- centering of gripping object.

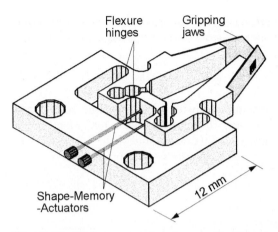

Fig. 6.45. Miniature toggle gripper with SM actuator [86]

Miniaturized copies of conventional grippers designed for 'macro' handling of systems cannot meet the special requirements posed by the small dimensions of micro parts. Small gripper size and clean–room suitability are achieved, for example, by observing the following rules.

– Conventional slide or roller bearings should be replaced by flexure hinges. Flexure hinges are created by specially formed notches in the material, causing a much lower flexural strength at that point. Miniature flexure hinges can be produced with little effort and are suited for clean–room usage.
– Use of solid state actuators. Small shape memory actuators can apply relatively high forces and strokes, can be well integrated into the gripper's mechanical structure, and do not emit particles into the clean–room environment.

Only small SM elements are required to actuate miniature grippers. Hence, fast opening and closing times of the gripping jaws can be expected. Two examples of micro grippers built according to these design principles are described in the next paragraphs.

Toggle Gripper. The toggle gripper mechanism in Fig. 6.45 consists of a single piece of plastic. For this prototype, flexure hinges were cut out of the material by milling, but for higher production quantities injection moulding is possible. The gripping jaws are closed by heating the SM wires (length 25 mm, diameter 0.15 mm). The toggle mechanism translates the actuators' strokes into a movement of the jaws of 0.5 mm. The gripping force is controlled by feeding back the signal of the strain gauges glued to the jaws.

Gripper with Antagonistic Actuator Design. Additionally to choosing actuators with small diameters, the opening time of the gripping jaws can further be sped up by using two SM actuators:

– to grip an object, the first actuator is activated and closes the jaws very quickly;
– to open the jaws, the first actuator is switched off and the second SM actuator is heated, and by this means, the jaws are pulled open even though the closing actuator is only slowly cooling down; and then
– both actuators cool down to get ready for the next gripping cycle.

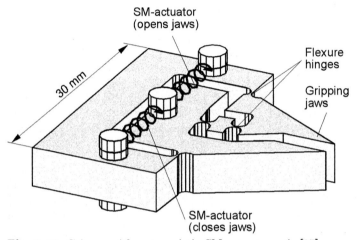

Fig. 6.46. Gripper with antagonistic SM actuator pair [87]

This design is called 'agonist–antagonist' because both actuators work in opposite directions [88]. Sect. 9.1 details the physiological aspects of the agonist–antagonist principle. Using this method, the work cycle of an assembly task can be optimized. Figure 6.46 shows a gripper with an antagonistic actuator design. As already pointed out in the previous example, the gripping mechanism is made of a single piece with flexure hinges milled in. Both gripper designs presented here can be manufactured very easily and at low cost. Also, further miniaturization is possible.

Miniature Inspection Robot. To inspect small cavities in machines or in medical examinations, endoscopes are used in most cases. The flexible endoscopes passively follow the contour of the cavities when pushed forward. For better control, the operator can bend the tip of the endoscope by wire pulls. Small and compact SM actuators could be integrated into the mechanical structure so as to obtain more sophisticated inspection devices. Placing the actuators close to the corresponding joints has several merits:

– more controlled axes due to integrated actuators;
– no kinematic coupling between adjacent axes; and
– no friction and stick–slip effects due to long transmission links or wires.

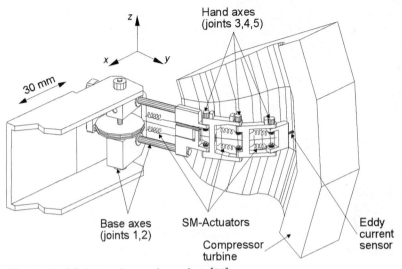

Fig. 6.47. Miniature inspection robot [89]

Figure 6.47 displays a miniature robot that was designed to inspect the inner cavities of a compressor turbine. At the tool center point, the robot is carrying an eddy current sensor that must be precisely positioned along the inner contour of the cavities. The aim is to detect small stress cracks and material wear.

The sensor must be oriented perpendicular to the turbine's surface. In order to reach the necessary movability the foremost links have a length of only 25 mm. SM coil springs in antagonistic design are integrated in each link close to the joints [89]. In spite of the compact size, the SM springs can move the links in the full operating range from $-110°C$ to $+110°C$.

Microvalve. Most of the applications published so far use SM wires or springs as actuators because this material is readily available. As an alternative, NiTi can also be sputtered. This fabrication technology allows for a better integration of the SM actuator into the mechanical structure. Thin–film actuators also have a very good dynamic behavior. For that reason, sputtering is being researched intensively.

A first product utilizing sputtered thin–film NiTi–actuators was presented by the TiNi Alloy Company (San Leandro, CA, USA) [90] (Fig. 6.48). It was used in a microvalve for gas chromatographic analysis. When the tiny NiTi–strips are powered, they lift the valve bonnet from the valve seat. If switched off, a small leaf spring presses the bonnet back and cuts off the gas flow. The

BeCu-spring

NiTi thin-film
actuator

Valve bonnet

Orifice die
5x8 mm²

Gas flow

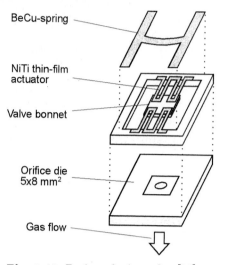

Fig. 6.48. Design of microvalve [90]

whole valve is assembled on a $5 \cdot 8 \, \text{mm}^2$ silicon substrate. Its technical data
are listed below:

- gas pressure: $< 3 \, \text{bar}$
- volume flow: $2 \, \text{l/min}$
- activation: $5 \, \text{V}, \, 300 \, \text{mW}$
- speed of operation: $10 \, \text{Hz}$

An advantage of using such a small valve in gas chromatographs is their
very small dead volume. Also, miniature valves are a first step towards
portable chromatographs and analysis devices (see Sect. 6.9).

6.4.5 Conclusion

The advantages offered by shape memory actuators become most obvious
where small–sized devices are concerned. Due to the very high work–per–
volume ratio, SM actuators of millimeter or micrometer dimensions have
large actuator strokes and forces. The response time strongly decreases with
shrinking actuator size.

SM actuators may have very many different shapes and offer a variety of
shape changes (i.e. actuator strokes). This property can be exploited so as to
adapt the SM element's shape to the actuating task. As application examples, miniature grippers and inspection robots with electrically heated SM
wires and springs integrated into their mechanical structures were presented.
The future opportunity for thin–film SM actuators to drive micromechanical systems and devices was demonstrated by a micro valve designed for gas
chromatography.

6.5 Electrorheological Fluid Actuators
W.A. Bullough

An electrorheological fluid, as defined for the present purpose, is a mixture of micron–sized, high–dielectric–strength particles carried in an insulating base oil. When an electric field is applied transverse to the direction of motion of the fluid, it causes an interaction between the particles, the field and the dispersant, and this results in an increase in the resistance to the flow of the mixture.

The first, and perhaps the most important thing to understand is that there is more than one so–called electroviscous effect [91]. Many researchers have reported these phenomena [92]. The fluids in this category have not all been of the slurrified electrorheological (ER) type that were first investigated in depth and developed by Winslow [93]. Indeed, high–speed recordings of the response of a currently 'most promising ER fluid' show at least two separate responses to the applied voltage excitation. However, so far as the engineer is concerned, this event may be considered to be part of a single response. The overall response, in this case the resistance to flow, which manifests itself as an increase in pressure or shear stress due to an applied voltage change, must not be confused with (say) the time response of a hydraulic servo valve that demonstrates basically a change of flow or piston displacement brought about by an electric input signal.

The mechanism(s) of an electroviscous effect is still not fully resolved and quantified. It is not strictly relevant to this work and is therefore not dealt with in detail. At this stage it can only be said that it is a very intricate and multidisciplinary event and, secondly, it should be understood that there is little change in the viscosity μ of the fluid as it is normally defined in its continuum context. The term electroviscous, which has often been used to describe the present class of fluids, is misleading in this sense. Rather, the field imposes a yield stress property on the fluid which is similar to, but not the same as, that which is a feature of the ideal Bingham plastic. This can readily be seen by referring to Figs. 6.49 to 6.52 inclusive. It is alternatively possible to claim that either the plastic viscosity changes with shear rate or the yield stress does — see later in this section.

Testing ER Fluids. The shear mode of operation is the term generally given to the simple shearing of the fluid, as in a Couette–type viscometer but with an electric field applied between the rotor and the stator interface with gap size h (Fig. 6.49). With zero voltage ($V = 0$) applied, most ER fluids exhibit near–Newtonian properties except at very low shear rates. When an electric field ($E = V/h$) is applied to the fluid, there is an increased resistance to its movement, which must be overcome before motion can take place (see Fig. 6.50 which is an idealised representation). Conventional constant temperature θ and speed ω in Couette laboratory techniques can normally only

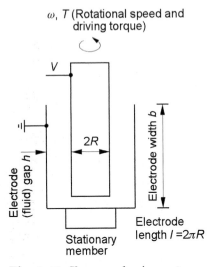

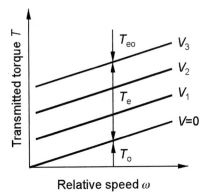

Fig. 6.49. Shear–mode viscometer

Fig. 6.50. Shear mode — diagrammatic test results

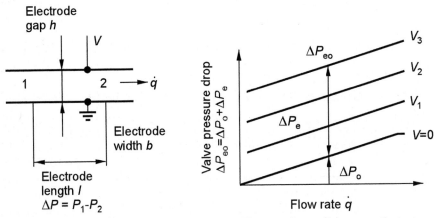

Fig. 6.51. Flow mode test electrodes

Fig. 6.52. Flow mode — diagrammatic test results

encompass shear rates ($\dot{\gamma} = \omega R/h$) up to a few thousand s^{-1} although cooled industrial clutch–type devices of similar geometry may reach $20\,000\,\mathrm{s}^{-1}$.

Because of the problems associated with the manufacture of sample slurries of the ER type and on account of difficulties brought about by high shear stress (τ) heating the small amount of fluid in a viscometer (at least, so far as scientific readings are concerned), much of the developmental testing of the fluids has been done in the 'static yield' situation, namely the point at which yield shear stress τ_e is overcome, at a given voltage, so that motion can commence. The shear–stress/shear–rate characteristic beyond this 'static yield' point has often been measured by rotating the viscometer with 'zero volts' applied, at a much lower shear stress than the yield point level. It will be noted that whilst this test procedure and configuration is obviously very useful for the small batch development of the fluids by chemists, physicists and rheologists, the data produced in this way needs to be treated with some caution. It is, after all, a 'system', comprising the same identifiable fluid matter throughout the tests, and in all a situation not often encountered in, say, a hydraulic power mechanism where, usually, a throughflow is required to procure the displacement of a linear or rotary piston, and a high level of compressive stress (pressure) is needed in order to keep the size of that piston down. Also, no allowance is made for the effect of $\dot{\gamma}$ on μ or τ_e.

The flow mode is thus perhaps of more interest to the engineer who is seeking to control and design high–force/torque, low–weight/volume, power transmission systems. Here it is necessary to have the facility to remove the working fluid from the source of heat generation to a convenient location, where it may be cooled and then recirculated, thus preserving the fluids lubrication properties. In this configuration the fluid is normally tested by pumping it between fixed parallel plates across which a voltage is applied (Fig. 6.51). Though this method clearly requires large amounts of fluid for the tests, it is nearer to the more familiar 'control volume' situation: the study of well defined and fixed geometrical configurations through which fluid is continually flowing. The high $C_v\rho$ (specific heat $\times$ density) product of the liquid serves to keep temperature excursions in check — a situation not always encountered in a closed system, where the negative temperature coefficient of resistance of its hydraulic semiconductor can cause a conductance–based thermal runaway.

Most of the tests performed in this (flow) mode have concentrated on keeping the flow rate $\dot{q}$ constant and measuring the pressure response ΔP to an applied voltage 'step'. The use of the pump over a period of time, with associated flow meters, pressure transducers, strainers, tanks, coolers, connectors etc. being in contact with the fluid, is a more convincing test of the serviceability and durability of the fluids, in a fluid power sense, than a closed system's shear mode or other low fluid volume tests. However, the simultaneous high–speed recording of pressure drop, flow rate, voltage, temperature and electric current I, given even the advanced instrumentation available nowadays, involves some problems. For example, it is very easy to erroneously measure some wave action in the hydraulic or electrical part of the test circuit or equipment, or indeed the drive motor regulation in taking the extra pressure load, unless great care is taken [94].

The separation of the true response time of the electroviscous, Winslow or electrorheological effect from unsteady pressure recordings of the step input type is a complex and tedious affair. Experimentation carried out in this domain is expensive and time–consuming, not least on account of the many variations of electrode separation and length, the number of variables and the different input frequencies involved. Again, because of these problems any data presented for appraisal should be treated with caution. It is necessary to ensure exactly what information has been put forward, and from what kind of test it was derived.

Flow visualisation tests are easier to carry out in the flow mode than the shear. A simple two–dimensional model, using a very dilute suspension of solid, shows clearly that the particles adopt a semi–regular matrix pattern in stationary fluid in a valve as the voltage is applied, and will then be distorted but held there until the forcing pressure reaches the yield value dictated by the voltage. At this juncture, particles appear to flow with the base liquid, though possibly not always at the same speed. Small electrode gap sizes prohibit true velocity–distribution studies across them; the usual technique of building oversized devices to facilitate such studies is prevented by the increased voltage demand. However, a Bingham–plastic–type velocity–profile/core–flow analysis on the problem [95] ties in quantitatively with each set of experimental flow–rate/pressure–drop results. Fairly recently [96], some photographic evidence of plug flow has become available. One interesting feature of an unyielded situation (when pressure is applied but the solid particles are still held by the field) is that small droplets of pure base oil are periodically observed leaving the outlet of the valve as if the matrix was behaving like a filter. The formation, distortion and breaking of the matrix is presumed to lead to complex rheological situations that have been observed as a hysteresis type of effect.

A further advantage of flow–mode testing is that the shear–rate magnitudes that would be encountered in a practical hydraulic device, often in excess of $40\,000\,s^{-1}$, can be achieved [97]. However, the definition of shear rate needs to be subjected to scrutiny: it is often derived from the Newtonian/Poiseuille formula, albeit when plug flow is present [98]. In a Couette viscometer care must be taken to avoid plug flow: the plug formed by radial field effects at low speeds distorts the $\tau, \dot{\gamma}$–plot [99], and electrostatic breakdown can occur at relatively low shear rates.

Characterizing the Fluid for Design Considerations. In many practical applications of hydraulic machine engineering it is not necessary to design to a high degree of precision. This is due to a number of factors such as, for example, the difficulty in predicting the load accurately over the cycle of operation; hence the working temperature of the fluid is not entirely assessable. Further, the advantages in capital–cost economy brought about by the mass production of pumps and motors etc. limits the number of truly purpose–made devices.

Within limits an electrorheological fluid (ERF) can have its properties fixed to suit a particular task [100]. This could be done by, say, adjusting the water content within the solid phase of a 'wet' fluid or its volume/mass fraction in relation to the base liquid. The method of production of the particles will have a bearing on their shape and size distribution, and thus the rheological behavior of the mixture will be affected. The same applies to 'dry' (no added water) fluids, which may be polymeric by nature.

There are many operating variables in an ER power system, not all of which can be controlled easily or simultaneously, and for this and for all of the above reasons it is probably not too productive at this stage of the development of ERF to spend an inordinate amount of time in perfecting precise steady–state and time–dependent analytical rheological models. These will no doubt be called for in due time when more 'standard' fluids are produced or as applications demand computer fluid mechanics packages, but they are not needed now.

The diagrammatic 'raw' steady–state test results of Figs 6.50 and 6.52 are typical approximations of those that would be achieved in shear and flow–mode tests, respectively. Notably, in this idealization the slope of the lines in Fig. 6.50 has the same constant value whatever the magnitude of the applied voltage or relative shear condition. In Fig. 6.52 much the same can be claimed but the slope, though constant, may not have the same value as in Fig. 6.50 — see later in this section. This infers that the field/yield effect can for, analytical reasons, be assumed to be independent of the fluid motion/shear rate/flow rate and that the field does not affect the flow–slope parameter — a very important approximation and often permissible simplification so far as a design procedure is concerned. Much the same applies for current flow. The overriding considerations outlined in this section have led to the choice of a simple Bingham plastic fluid model (Fig. 6.53) on which to base performance estimations for a conceptual electrorheological–shear–stress/electric–field device.

In order to do this, the viscous pressure drop ΔP_o of Fig. 6.52 is deducted from the total pressure drop ΔP_{eo} to give the electro or yield pressure ΔP_e for a particular voltage. Given the valve dimensions and using the dynamic viscosity as calculated from the zero–volts line and the valve dimensions, the yield stress at the wall may be isolated and $\dot{\gamma}$ calculated. Likewise, shear stress τ and $\dot{\gamma}(= \omega R/h)$ can be calculated from the shear–mode test data by neglecting radial effects. The well known relevant Poiseuille and Couette flow analyses are often used in these procedures. In both modes μ is derived from the zero–volts test. On this basis, flow– and shear–mode data will not correspond in the $\tau, \dot{\gamma}$–plane.

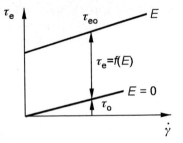

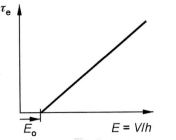

Fig. 6.53. Bingham–plastic–type model   **Fig. 6.54.** Electro yield property

Having produced results in the form of Fig. 6.53 for a given fluid, the yield stress dependence on voltage is produced at a nominal shear rate and is shown typically in Fig. 6.54. This property may be modeled in two ways: either as a deadband or barrier of field strength E_o of zero yield magnitude, followed by a linear relationship between τ_e and E (for $E > E_o$); or as τ_e proportional to E^2. In both cases the approximation will produce accurate enough designs over the full voltage range of operation. Only when the mechanism of the effect is fully understood will the correct presentation be clear — see section 6.5.

Design Formulas for Estimation Purposes. By adopting the approach explained above, calculations for a device may be approached in the following manner, the flow normally being laminar in fashion:

Clutch Type Controller. Torque T on a radial rotor element at a general radius r is given by

$$T_{eo} = 2\pi r^2 \tau_{eo} \delta r, \tag{6.20}$$

where

$$\tau_{eo} = \tau_e + \tau_o, \tag{6.21}$$

$$\tau_e = f(E), \text{and} \tag{6.22}$$

$$\tau_o = \mu \omega r / h. \tag{6.23}$$

For a cylindrical clutch this becomes

$$T = 2\pi R^2 l \tau, \tag{6.24}$$

where R is the mean radius and l the length of the cylinder.

In some applications there will be obvious limits to the use of the simple solution on account of heating, radial and centrifugal effects, and flow stability, as well as super slip — see later. The behavior of a device like this in 'pick up' and 'drop load' situations will depend to a major extent on the driver and load characteristics; only rarely will the speed of the ER effect *per se* come into question. Otherwise, a quasi–steady approach is adequate for this purpose.

Valve Controller. In this case

$$\dot{q} = bh^3 \Delta P_{\rm o}/12\mu l \tag{6.25}$$

and the wall shear–yield stress derived from a control volume placed around the electrode gap is

$$\Delta P_{\rm e} = 2\tau_{\rm e}bl/bh \tag{6.26}$$

To facilitate this simple design procedure, and for other reasons, the wall shear stress and shear rate in a valve should always be quoted in flow–mode–derived or imputed characteristics. Again, the pump and load characteristics and the inertia and compressibility of the fluid may have a significant effect on the time response of the system to a change of input voltage signal [101], especially under extreme conditions of operation.

Quasi–Steady Calculations and 'Super Slip'. Generally speaking, an ER fluid being operated at the correct temperature will respond rapidly to a voltage signal. The exact response time in a clutch situation is not easy to predict because of the difficulty of measuring torque in the shear mode (a stiff driveshaft gives an adequate natural frequency but insufficient deflection to measure with precision, and vice–versa). In the flow mode, compressibility and inertial effects [101] can mask the required signal. Fortunately, the implication is that so long as the delay between say a step voltage and the short–term steady–state shear–stress response to it ($t_{\rm m}^*$) is not great, then the ER fluid design can be treated the same as a normal hydraulic fluid for quasi–steady design purposes [102]. Typical values of $t_{\rm m}^*$ are 1 ms at the best operating condition (for $E, \dot{\gamma}, \theta$) and normally can be neglected for all except electrical supply–circuit and electronic–control purposes. For example, usually in the runup time for a clutch the load torque is essentially equal to the inertia of solid parts plus fluid weight times the angular acceleration, all at the correct operating temperature.

At very high acceleration rates, when the weight of fluid and solid load parts are competitive, the velocity profile during slip ceases to be linear and a kind of 'super slip' occurs. In this case the fluid will run up initially as if it were in normal linear–velocity profile mode and then encounter 'super slip' with a predicted final runup time of $t_{\rm o}$ very much greater than the normal (linear profile) acceleration time. This problem is so rarely encountered that the theoretical analysis has not yet been proved. Much the same considerations apply to valve flow, where fluid inertia and compressibility can often be neglected on account of the low frequency of operation ($< 10\,{\rm Hz}$ typically) and super slip is not encountered [103].

Electrical Quantifications. The resistance and capacitance of an ER device of electrode area A follow (approximately) the classic forms of $C \propto A/h$

and $R \propto h/A$, respectively, with a time constant RC fixed. Alas, both parameters depend on temperature, shear rate and voltage level/rate of slew. If the electrodes have too large a surface area, then peak current values are large, as well as the magnitude of the conductance. Rapid switching of electrostatic catches in the High Tension(Volts)/Direct Current (HT/DC) circuits can then be a limiting factor. Special drivedown facilities are required for these step–voltage–producing devices, and even then the controller may not discharge fully before charging is required again. Nevertheless, control of hysteresis in locking devices and proneness of the fluid to electrophoresis may require binary digital control. In general, modeling in an electrical sense is highly nonlinear [104] and its use is restricted more often than not to the design of control and excitation circuitry — see next subsection.

Typical Fluid Properties. There are many different types of ER Fluids: different base oils, solid material and solid fraction, different size and size distribution of particles and, if wet, different levels of water content therein can all be used. Dependent on the application, one characteristic is more important than another.

The comparison of ERF performance in fluids is made difficult through the lack of unsteady–state test data and an account of the differing effects of shear rate, temperature and form of field dependance from fluid to fluid. Table 6.9 shows a nominal comparison of kinematic yield–stress levels and corresponding conductances, drawn from the overlapping commercial fluid data available in the public domain at the time of writing.

Table 6.9. Comparison of kinematic yield–stress levels and corresponding conductances

Manufacturer	Viscosity mPas	Q'	A
ERFD	115	0.82	0.20
Bayer	65	2.1	0.19
Bridgestone	250	2.4	0.12
Nippon Shokubai	120	0.52	0.11

where

$$\tau_e \text{ kPa} = AE^2 \quad \text{and} \tag{6.27}$$
$$j \ \mu\text{A/cm}^2 = Q'E^2 \text{ with } E \text{ in kV/mm.}$$

A and Q are derived from the curve fitting of experimental data at $30°\text{C}$ and an arbitary but comparable range of values of j.

In general, static yield–stress data follows the threshold $k(E - E_o)$ law best, whilst kinematic yield stress is better fitted to the above square law expression (this portion of the text is supplied by courtesy of Dr. Linda Evans).

Although the effect of a change of temperature will be evidenced in the above properties (and may affect the settlement rate of the solid, which can only be matched perfectly with the base fluid at one temperature), none is more significant than its effect on current density j. Here, a small increase in temperature can raise the current consumption considerably, and vice versa. This is the main problem area of ERF. Certainly this factor alone is sufficient justification for the use of a simple approach to the initial characterization, fluid comparability problem and design procedures. Different temperatures can change slopes of the $\tau, \dot{\gamma}$–characteristic: increased temperature can often increase current to that for the optimum t_m^* and τ_e performance and give a level $\tau_e \neq f(\dot{\gamma})$ characteristic or thereabouts.

Work is still going on as to what the time domain response of the ER effect really depends on [94] or what is its meaning in terms of the fluid design. The initial and true ER effect in a valve, for example, is not the main concern: it is also not the true time of the full pressure rise, the valve geometry, flow rate and other factors such as super slip being involved. In the shear mode the position is similar yet the torque/input voltage response limit in response to small sine waves has been claimed to be as high as 1000 Hz and yet fluid temperature can be a critical factor in t_m^*, the electron–hydraulic time delay.

Design Variables and Controller Shape. From an inspection of Figs 6.50 and 6.52 and equations (6.24), (6.25) and (6.26), it can be seen that at any given relative speed or flow rate the ratio between the excited and unexcited torque or pressure drop, in the shear and flow mode, respectively, can be influenced by the choice of b and h for a particular fluid. The ratios τ_e/τ_{eo} and $\Delta P_e/\Delta P_{eo}$ are often very important considerations in a practical controller mechanism.

Equation (6.26) gives some indication of how to amplify the yield stress, i.e. by fixing the value of l/h in the valve. For example, if $l = 100\,\mathrm{mm}$ and $h = 0.5\,\mathrm{mm}$, then $\Delta P_e = 400\tau_e$. This type of manipulation is, of course, familiar to a hydraulics engineer, who often uses a small shear stress to effect a high pressure drop to drive a piston. It is, after all, why shock absorbers and actuators are usually of the piston type rather than of the shear plate variety, and why fluid–based damper devices are preferred to purely electrical types.

For the flow–mode situation, (6.25) and (6.26) show that the control ratio $\Delta P_e/\Delta P_o$ is independent of the valve width b and that the volumetric flow rate, for a given pressure ratio, is proportional to b. Similarly this pressure ratio for a given flow rate does not depend on l but the pressure drop does. If the surfaces of the electrodes are increased in area, the current demand will rise in proportion. If the gap h is increased, then ΔP_o falls dramatically — roughly as the cube of h since the flow is usually laminar and near–Newtonian. However, if (say) h were to be doubled, the voltage needs to be increased in line to maintain the yield stress, and this would in turn only give one half of the previous ΔP_e. This can be seen from the shear/pressure force balance around the electrode gap, as per (6.26). The designer has some choice — but at some cost. Also, consider the effect of these manipulations on, say, the capacitance and conductance of a controller.

Although this design situation poses many questions, the particular control duty that (say) a valve is required to perform will depend on the application and will not be further discussed here, since the flow rate may not always be constant as it is for the above example and, in addition, $\tau_{eo} = \tau_e + \tau_o$ and is therefore not a simple function of voltage.

There exist a few specialist combinations of the flow and shear modes. These comprise the 'squeeze' mode, which is basically the flow mode achieved by flow between plates that are approaching one another (as in a pedestal bearing), and the Rayleigh step mode in which pressure is generated hydrodynamically by a change in a flow section The former is usually associated with low–frequency operations of small displacement, e.g. in engine mounts. The latter has so far shown little promise in respect of rapidly centered bearings: the time constant t_m^* has proved too large for the intended operation. However, variable stiffness operation remains a possibility. Two dimensional flows are under investigation as a means of cooling slipping clutch drives.

6.5.1 Limitations to the Concept of Electrorheological Fluids

Electrorheological fluids are now considered in greater detail and with respect to their application in devices that are aimed at featuring electronically designated motion and flexible operation via adaptronics, or in third–wave machines. In effect, this also sets down the state–of–the–art position of research in the field and outlines the salient factors that determine research trends for fluid developers, whilst at the same time giving some idea of ERF machine performance limitations. The type of artifact under the spotlight is one wherein its function can be rapidly regulated without a change of geometry of the solid parts and at the behest of an electric signal alone.

ER Models and Characterization. Perhaps the most problematic area that obstructs the control of these changes by the use of electrorheological fluids is the particle mechanics or continuum conundrum for characterizing the flow of the dense semiconducting slurry (unexcited) or yielding plastic (excited). Whilst many years have been spent in computer analyses based on dielectric polarisation models of the multidielectric–bodied excited fluid structure, the yield–stress level remains underpredicted from the constituents by an order of magnitude. Only recently have single–sphere attraction models begun to relistically account for the yield–stress levels achieved in practical fluids; on the other hand it is not difficult to find texts that claim that the performance of the same fluid in Couette–shear flow and Poiseuille valve flow cannot be related i.e. the fluid cannot be treated as a continuum [98].

It is now evident that polarisation is not the only mechanism at work and that hydrodynamic effects [105] plus conductivity [94, 106, 107] at least need to be included in multiparticle models designed to illustrate the modus operandi of the effect and to link it quantitatively to fluid properties, flow conditions and excitation levels (see Fig.6.55).

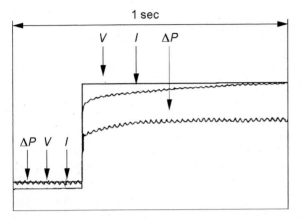

Fig. 6.55. Conductivity is seen to be important in the secondary response (at least). Steady flow in a valve with step voltage V applied; ΔP is valve pressure drop; and I is the electrical current. Similar behavior is seen in a clutch

Much insight has recently been gleaned into the relative importance of disparate fluid/particle conductivities and dielectric properties, where and when they are important, how they relate to the physical charge processes, and the dependence of the yield stress upon them. This has mainly been confirmed in steady–state–based investigations of the attractive forces between single spheres. There is, however, some way to go before any optimization procedure (for a given application) can be quantified in terms of materials make–up, especially in the time domain and where clustering of chains and particles are important. Meanwhile engineers need to use what empirical characterization data is available for commercial fluids, and this explains the layout of previous subsections.

The possibilities of characterizing ER fluids in flow as a continuum (otherwise the properties of viscosity and density have little significance and design techniques become entirely empirical) have been given a boost by work that indicates a link between such a fluid's performance in the above modes of flow and others (e.g. static shear) for different fluids. This is done by the use of nondimensional Hedström and Reynolds Numbers via use of Buckingham's relationships for a Bingham plastic [108] for steady flow, (Fig. 6.56). This is welcome yet perhaps surprising, since the thickness of the shearing fluid layer near a boundary can approximate to a particle (often of variable geometry) size.

The extended concentration by fluid developers on the details of slow steady flow belies the necessity to confront the intensely unsteady (and indeed

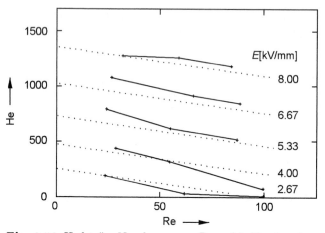

Fig. 6.56. Hedström Number versus Reynolds Number for a valve (full line experimental result) as predicted from clutch (Couette mode) experimental data (dotted) for different values of $V/h = E$

steady) high shear–rate motions that will be required in practical machine work cycles. Very often misleading appraisals of situations arise from the lack of fluid–performance details. Also, correspondence between shear and flow–mode time–domain performance details has not been shown across the plausible operating range.

Third Wave Machines. In a flexible machine capable of a high resolution of (smooth) force, velocity or displacement variation, there is little scope for the rapid generation of, say, large–scale motion by an inductive or relatively heavy rotor electromagnetic drive or the by generation of a shaped control current. In both cases, respectively, latching onto a high inertia source of steady motion (and the braking of it) and a high–capacity, high–tension supply line (and discharge by earth shorting) produces a digital event via engagement of the ERF [109]. Both AC and DC excitation components are present in step switching and dwell periods respectively, and yet there is a tendency to separate fluids research into AC or DC types. The realization is hence growing that potent AC fluids depend on particle polarisation in a poorly conducting fluid, whilst apparently good DC fluids need appreciable current flow — both conclusions being based on only steady–state strength and current–flow appraisal.

Returning to the Fig. 6.57 ultra–high–acceleration/low–inertia flexible–machine regime, it is predicted that the limiting change of (99%) speed response time t_0, in the digital mode of operation is determined by a factor $Kh^2\rho/\mu$, which is not a function of τ_e but which is heavily influenced by the inter–electrode gap size, and the fluid density and viscosity [103, 110]. This is important when the fluid self load, rather than the inertia of the solid part's power–transmission load, dominates the mechanics. This is somewhat contrary to the requirement for the control of the fluid viscous heating problem. K depends on the geometry of the controller.

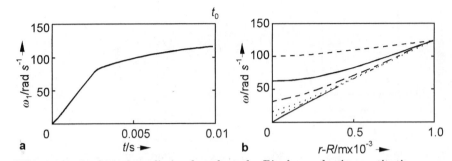

Fig. 6.57. Analytical prediction based on the Bingham–plastic constitutive equation of a lightly loaded cylindrical clutch performance: (**a**) limiting run up time t_o of output rotor speed; (**b**) corresponding super slip of velocity profile between electrodes. r is a general radius with R the radius of the inner driven member; ω_1 is the speed of the driven member at time t after switch–on of the step voltage between the constant–speed driving member and the driven member electrodes; ω is the speed of the fluid in gap

For a change–speed response time of less than, say, 20 ms a 4×10^5 V/s signal–rise/fall time rate is required. This has implications for the fluid capacitance, which is difficult to model as a function of shear rate [111]. When the voltage is rapidly applied, the yield shear stress follows at a time constant of approximately RC, the resistance and capacitance product of the inter–electrode space. There is little point in accelerating the load rapidly if the torque initiation lags much behind the step rates of change of excitation, although this lag can be difficult to measure [101, 112] Fortunately the lag seems to decrease the harder the fluid is being punished in terms of $E, \dot\gamma$, and θ (electric field, shear rate, and temperature respectively) (Fig. 6.58).

This factor becomes important if the generation of a motion profile in a third–wave machine is considered. Without getting involved with digital technology: if the x direction speed provided is constant, then the y penetration (driven by a bang–bang application of voltage and a yield stress of sufficient magnitude to give the relevant part high and instant acceleration) must be maintained over a very small time interval (fixed by the switching speed) if the resolution is not to be too crude. DC operation seems virtually mandatory, with any hysteretic and electrophoretic tendencies being arrested by a conjunction of binary switching and high $\dot\gamma$ (see Fig. 6.59).

Mechatronics and Testing. It does not seem possible to provide a figure of merit for a fluid that possesses these sundry needs, but the linear traverse mechanism will demonstrably test total capability in that respect [113]. In this device, two contra–rotating, high–inertia, constant–velocity rotors provide motion sources with HT and earth 'busbars', the excitation being con-

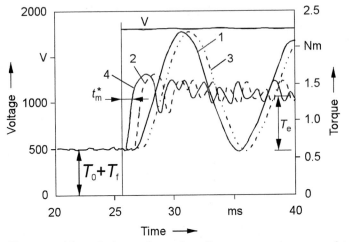

Fig. 6.58. Numerical transformation of step torque: 1. measured torque–transducer signal; 2. first estimate of ER clutch torque response; 3. predicted torque–transducer response for first estimate; 4. final estimate of ER clutch torque response. $T_O + T_f$ are viscous and real friction torques, with T_e due to application of step voltage V

trolled via switches. Two driven clutches, spaced from their driver co axially by the ERF, are each connected to a pulley, each of which is connected by a belt. The ERF, in opposing the clutch drives (Fig. 6.60), is excited alternatively to make the belt reciprocate, with typical steady speed of up to ±5 m/s separated by turnround times determined by the fluid properties: τ_e, μ, θ and t_m^*; high μ can distort the traverse profile, if excessive. A good–quality fluid should turn round in *circa* 20 ms. Thermal runaway should be avoided by as large a margin as possible; the heat transfer rate from the outer driving rotor is about the maximum per unit area that is achievable into the atmosphere. The full–speed centrifugal field on the particles is up to $100\,g$ and the belt acceleration around $50\,g$. Fluid degradation has been only generally described [114].

With an analysis of such performance data, the fundamental compatibility arises in relation to a low ERF time constant, the heating effect of the viscous shearing and conduction loads, and the level of voltage. Alas, a failure of fluid on this machine implies its separate analysis on each of several simple characteristics tests — in order to isolate the problem area. The test does, however, give a good example of the machine side of the overall electrical–chemical–rheological–thermal fluid/machine optimization.

It will be appreciated that the inertia, geometry and stress and strain in mechanical parts are linked to operating conditions, and particularly to acceleration in the unsteady mode. For example, the uniform speed of the traverse could be obtained (presumably) by having a long, small–radius clutch and having a large pulley and a low rotational speed. Likewise, fluid performance

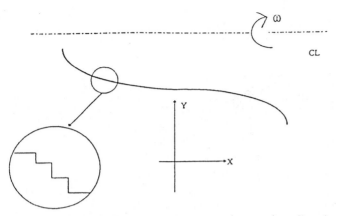

Fig. 6.59. Digital ER motion synthesizer (concept): y direction is shown ER controlled in switched steps of equal time elements, with steady speed traverse in (say) a lattice in the x direction

τ_e, t_m^*, and μ depends on the solids' content, the materials' properties, and the size and shape of particles, θ and $\dot{\gamma}$ [115]. With present lowish fluid yield–strength properties, the optimization process is made more hit–and–miss if full fluid data is not available [116].

In connection with (for example) the traverse mechanism, the need for a yield stress and a rising τ characteristic with $\dot{\gamma}$ is noted — Figs 6.50 and 6.52. This is necessary for any clutch drive where an overload may, for example, cause slippage if τ_e was to fall. However, the Bingham–plastic characteristic *per se* is not obligatory: in other types of flexible machines such as the vibration isolator, a linear force/velocity characteristic seems preferable [117]; and homogeneous liquid crystal–type fluids could become important here. In the flow mode of operation, much the same factors come into play as in the shear mode. The benefits of a high yield–stress magnitude is to reduce the amount of fluid volume for a given force/stroke requirement.

Hysteresis and Control. The valve–control application in general exemplifies an interesting control problem. Whilst good reasons have been given for digital control of a particulate ERF, the damper could be envisaged as a continuous ride member under analogue excitation/control [118]. Past studies have, however, shown a 'pseudo', or perhaps time–dependent, hysteresis [119, 120], which is better treated by bang–bang operation, otherwise more than a suggestion is apparent that voltage alone is insufficient as a control parameter [112]. These and the sometimes–experienced violent clutch (shuddering) and valve (choking) [121] may yet prove to be not separate phenomena but related characteristics linked to structure formation. These effects plus electrophoresis are to be avoided, save for their further investigation (Fig. 6.61). Shear modulus G', specific heat capacity C_v and bulk modulus β under field need to be known, since they can also determine the precision of any controlled positioning device.

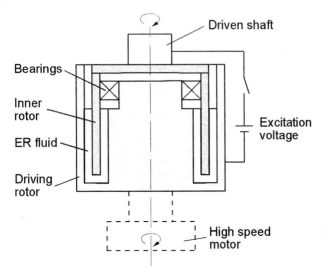

Fig. 6.60. Cylindrical clutch for ER traverse gear: on driven shafts (contra–rotating), pulleys are connected by belt; alternate excitation of clutches causes reciprocity with the belts, which carry the product to be wound on a bobbin (not shown)

All of the foreseen effects put a limit on the performance of a flexibly operated machine and set the requirements for τ_e and t_m^* $f(\dot\gamma, \theta, E)$ in the ER fluid. There may be competing limiting factors: fluid elasticity and volume, super slip, heating and cooling etc. Further limitations arise from lubrication (Fig. 6.62) — for example particles will only move through an elasto–hydrodynamic region at low speeds and anti–wear boundary lubricity is hence very important [122]. In film bearings of the hydrostatic Rayleigh type, the time delay t_m^* appears to be inordinately slow [123].

The self–weight/inertial–loading problem [124] can easily be avoided so far as solid material at its critical breaking length is concerned, but strain will limit the overall acceleration (on grounds of precision) — only a few materials will exhibit less than 0.01% strain at an acceleration of $100\,g$. Accelerations above $100\,g$ are regularly attained in flexibly operated machines and cause one to wonder at the rate of separation of particles and any possible cavitation effects in the fluid.

6.5.2 Future Aims and Present Problems

The whole aim behind present ERF developments is to provide a means of control (that is easy to apply and economical) to a mechanism that has to be by its nature flexible in force, displacement or speed and hydraulically operated. Electronic solid–state semiconductor devices and computers are powerful, inexpensive and adequate for many applications in control, and

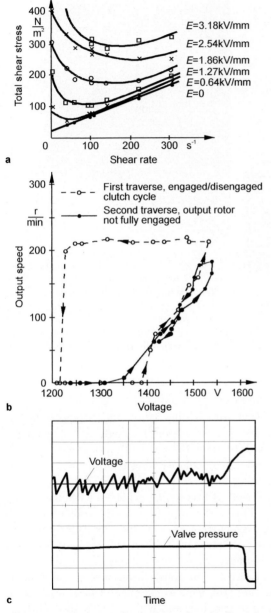

Fig. 6.61. Hysteresis/structure–related effects in **(a)** a Couette viscometer; **(b)** a clutch in on–off DC operation; and **(c)** a valve experiencing choking phenomena at nominally constant flow rate, where the uppermost trace is ongoing DC voltage and the lower is valve pressure drop *versus* time

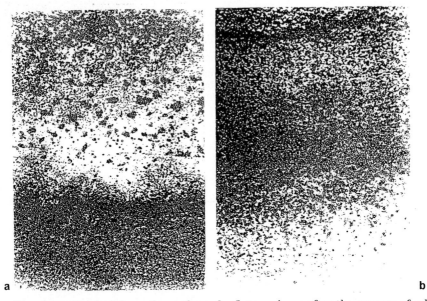

a   b

Fig. 6.62. Tracks left on the surface of a flat specimen after the passage of a ball rolling at (a) high speed (50 mm/s) and (b) low speed. At high speed, particles are unable to enter the contact between the rolling ball and plate surface, and a depleted track is observed; at low speeds, the particles enter the contact and have a lubricating effect. (Steel ball diameter 25 mm rolling on a glass disk; optical interferometic apparatus)

yet their interface with hydraulic machines usually involves a bulky solenoid or an expensive servo–valve, often with a pilot stage and a power supply. The aim of ER research is to be able to influence a hydraulic mechanism directly with a current low enough to allow the integration of a system made up of electronic transducing and signal–conditioning equipment, computer processors and feedback monitoring, solid–state controllable field–excitation supply and the hydraulic device itself. If this can be achieved without moving parts, then so much the better. This is the implementation problem of ERF.

At present, the problems of high current density and particularly its sensitivity to temperature, and low yield strength restrict the range of practical application of ER. An ER valve network is only competitive with (say) a servo–valve in certain passive control situations. Nevertheless, ER fluid is often preferable to a magnetic (i.e. long time–constant and large–excitation system) fluid, is under serious consideration, and will probably soon find application in such fields as: flexibly operated clutches, engine vibration isolation pads, [125], etc. Development is already nearing the level of performance where the inertial straining of solid components becomes a prior limitation (stresses due to acceleration under own weight).

The matching of the nonlinear yield characteristic, (or linearization) to a device has so far not proved to be a problem. The same cannot be said for high–pressure operation or heavy–duty position control. Here, very large and leaky valves would be required to give a locked position and load stiffness where large disturbing forces are involved: magnetic fluids or devices are better on heavy duty.

The subject of high–speed, flexibly operated electronically reconfigurable machines based on ER fluids is intensely multidisciplinary and highly non-linear in terms of analysis, furthermore the limits of operation cannot be graphically represented, such is the degree of compromise required between fluid design, motion and machine. This section cannot give comprehensive cover to the interface problems that exist; rather, it lists the more apparent and important factors. Having done this, it is hoped that the targets to suit both fluid developers and applications engineers can be set more effectively than hitherto.

Finally a word of caution: a $\tau, \dot{\gamma}$–characteristic for an ER fluid will not give exactly the same shape of torque, speed, pressure, or flowrate–curves for an ER device, and viscometers should be designed and operated so that the fluid rather than the device characteristic is measured [99, 115]. Other areas of ER fluid development requiring specialist attention include lubrication, hysteresis, stabilization and the exclusion of impurities.

6.5.3 Summary of Advantages of ER Fluids

The general comparative advantages of ER fluids in relation to other elec-trostructured fluids are:

– speed of action: the achievement of full yield stress occurs virtually when the capacitive charging process is completed, and usually, at optimum op-erating temperature, this takes place in 1–2 ms and is independent of con-troller size;
– the mass/inertia of any driven controller element can be low — ferrous components are not required;
– the powerpack that drives the electrodes can be remote from the controller, and its size is not usually a problem;
– steady–state currents can be low, albeit that they are provided at high voltage;
– the limiting run–up time t_o of the fluid can be short, and typically has the form (on a theoretical basis) of $t_o = 4\rho h^2/\mu$ in a Couette–type clutch; thus, all other things being equal, the self–weight problem of the fluid is greater as its density rises; although yield stress is not a factor in this t_o limitation; and
– there is low inherent susceptibility to abrasion/wear and centrifuging out.

6.6 Magnetorheological Fluid Actuators
J.D. Carlson

Magnetorheological as well as electrorheological fluids are materials that respond to an applied electric or magnetic field with a dramatic change in rheological behavior [126]. The essential characteristic of these fluids is their ability to reversibly change from free–flowing, linear, viscous liquids to semi–solids having a controllable yield strength in milliseconds when exposed to either an electric or a magnetic field. In the absence of an applied field, controllable fluids are generally well modeled as Newtonian liquids. A simple Bingham–plastic model is effective in describing the essential field–dependent fluid characteristics [127]. In this model, the total yield stress τ_y is given by

$$\tau_y = \tau_{y(\text{field})}\text{sgn}\dot{\gamma} + \eta_p \, \dot{\gamma} \qquad (6.28)$$

where $\tau_{y(\text{field})}$ is the yield stress caused by the applied field, $\dot{\gamma}$ is the shear rate and η_p is the field–independent plastic viscosity defined as the slope of the measured shear stress against the shear strain rate.

Magnetorheological (MR) fluids are less well known than their electrorheological (ER) fluid analogs, for which see Sect. 6.5. MR fluids are typically non–colloidal suspensions of micron–sized, paramagnetic or soft ferromagnetic particles. MR fluids should not be confused with colloidal ferrofluids in which the particles are about 1000 times smaller than those found in typical MR fluids. While both ER and MR fluids have an early history dating from the late 1940s, most research and development activities since then have concentrated on ER fluids. Only recently has a resurgence in interest in MR fluids been seen and a realization that MR fluids can provide the enabling technology for systems that provide practical, semi–active vibration control [130, 131, 132, 133].

The initial discovery and development of MR fluids and devices can be credited to Jacob Rabinow at the US National Bureau of Standards [134, 135, 136] in the 1940s. Interestingly, this work was almost concurrent with Winslow's early ER fluid work. In fact, the late 1940s and early 1950s probably saw more patents and publications relating to MR fluids than to ER fluids. While Rabinow's work is largely overlooked today, Winslow discussed his work on MR fluids in his 1949 paper on ER fluids [126].

MR fluids have recently been used to successfully enable both linear [137, 138, 139] and rotary dampers [138, 140, 141] in a number of specific semi–active control applications. Recent simulation and laboratory experiments using a small MR damper in a scale–model building indicate the efficacy of MR fluid dampers for semi–active control for seismic damage mitigation [142, 143, 144, 145, 146] MR fluid technology appears to be readily scalable to very large, high–force dampers appropriate for civil engineering applications [147]. Recent clinical studies at the Cleveland Clinic give preliminary indication of the efficacy of MR fluid devices for medical rehabilitation therapy [148].

6.6.1 Description of MR Fluids

A typical magnetorheological fluid consists of 20–40% by volume of relatively pure, soft iron particles, e.g carbonyl iron, suspended in an appropriate carrier liquid such as mineral oil, synthetic oil, water and/or glycol. A variety of proprietary additives, similar to those found in commercial lubricants, that inhibit gravitational settling and promote particle supension, enhance lubricity, modify viscosity, and inhibit wear are commonly added. Carbonyl iron is the common name given to iron particles formed from the thermal decomposition of iron pentacarbonyl.

The ultimate strength of an MR fluid depends on the square of the saturation magnetization of the suspended particle [149, 150, 151]. The key to a strong MR fluid is to choose a particle with a large saturation magnetization. Ideally, the best available particles are alloys of iron and cobalt known as Permendur, which have saturation magnetizations of about 2.4 Tesla. Unfortunately, such alloys are prohibitively expensive for most practical applications. The best practical particles are simply pure iron with a saturation magnetization of 2.15 Tesla. Virtually all other ferromagnetic metals, alloys and oxides have saturation magnetizations significantly lower than that of iron, resulting in substantially weaker MR fluids.

Particle diameter is typically 3–5 μm. Functional MR fluids may be made with larger–sized particles; however, in such cases particle suspension is generally more difficult. Smaller, easier–to–suspend iron particles could be used; however, manufacture of such particles is difficult. Commercial quantities of relatively inexpensive carbonyl iron are generally limited to sizes greater than 1 or 2 μm. Significantly smaller ferromagnetic particles are generally only available as oxides such as the pigments commonly found in magnetic recording media. Nano–MR fluids [152] made from such pigment particles are quite stable because the particles are typically only 30 nm in diameter; however, because of their lower saturation magnetization, fluids made from these particles are generally limited in strength to about 5 kPa and have a rather high plastic viscosity stemming from the very large surface area of the particles.

MR fluids made from iron particles exhibit a yield strength of 50–100 kPa for an applied magnetic field of 150–250 kA/m ($\approx$ 2–3 kOe). MR fluids are not highly sensitive to contaminants or impurities such as are commonly encountered during manufacture and usage; further, because the magnetic particle polarization mechanism is not affected by the surface chemistry of surfactants and additives, it is relatively straightforward to stabilize MR fluids against particle–liquid separation in spite of the large density mismatch. Antiwear and lubricity additives can also be included in the formulation without affecting strength and power requirements. A listing of typical MR fluid properties is given in Table 6.10.

Table 6.10. Typical magnetorheological fluid properties [153, 154]

Property	Typical Range
Max. yield strength, $\tau_{y(\text{field})}$	50–100 kPa
Max. magnetic field, H	$\approx 250\,\text{kA/m}$
Plastic viscosity, η_p	0.1–1.0 Pas
Operable temp. range	-40–150°C
Stability	unaffected by most impurities
Response time	$\sim$ milliseconds
Density	3–4 g/cm^3
$\eta_p/\tau_{y(\text{field})}^2$	10^{-10}–10^{-11} s/Pa
Power supply (typical)	2–25 V, 1–2 A (2–50 W)

6.6.2 Advantages and Concerns

Interest in MR fluids stems from the benefits they provide or enable in mechatronic systems. First and foremost is their inherent ability to provide a simple and robust interface between electronic controls and mechanical components. Much of the current interest in MR fluids can be traced directly to the need for a simple, robust, fast–acting valve necessary to enable semi–active vibration control systems [155, 156, 157]. Such a valve was the holy grail of semi–active vibration–control technology for nearly two decades. MR fluid technology provides the means for enabling such a valve to be made.

Rotary MR fluid brakes are finding one of the greatest advantages to be lower system cost. Typically, MR fluid brakes find their niche where eddy current brakes have traditionally been used. A key advantage of an MR fluid brake is its low electric power requirement. It is not unusual for an MR brake to replace an eddy current brake and to also allow the ancillary power supply to be replaced with a significantly smaller, lower–power and lower–cost unit [140]. The small MR brake described below that is used in exercise equipment requires only 1 A at 12 V compared with several amps at 90 V for the competing eddy–current brake.

The primary advantage of MR fluids stems from the large, controlled yield stress that they are able to achieve. Typically, the maximum yield stress of an MR fluid is an order of magnitude or more greater than the best ER fluids, while their viscosities are comparable. This has a very important ramification in the ultimate device size. As discussed in Sect. 6.6.4 below, the minimum amount of active fluid in a controllable fluid device is proportional to the plastic viscosity and inversely proportional to the square of the maximum field–induced yield stress. This means that for comparable mechanical performance the amount of active fluid needed in an MR fluid device will be about two orders of magnitude smaller than that for an ER device.

From a more fundamental physics perspective, the large strength of an MR fluid is related to the very high magnetic–energy density that can be established in the fluid before complete magnetic saturation of the particles occurs. For a typical iron–based MR fluid, this is of the order of $0.1 \, \mathrm{J/cm^3}$. ER fluids, on the other hand, are limited not by polarization saturation but by dielectric breakdown. This limits the maximum field strength and consequently the maximum energy density that can be established in an ER fluid to about $0.001 \, \mathrm{J/cm^3}$. For comparable device performance, MR and ER devices need to control the same magnitude of total field energy. Hence, the smaller amount of active fluid needed for MR.

From a more practical application perspective an advantage of MR fluids is the ancillary power supply needed to control the fluid. Total energy and total power for comparable performing MR and ER devices are approximately equal [130, 133]. The advantage of MR lies in the fact that they can be powered directly from common, low–voltage sources such as batteries, 12–volt automotive supplies, or inexpensive AC to DC converters. Further, standard electrical connectors, wires, and feedthroughs can be reliably used, even in mechanically agressive and dirty environments, without fear of dielectric breakdown. This is particularly important in cost–sensitive applications.

Another practical advantage of MR fluids is their relative insensitivity to temperature and contaminants. This arises from the fact that the magnetic polarization of the particles is not influenced by the presence or movement of ions or electric charges near or on the surface of the particles. Surfactants and additives that affect the electrochemistry of the fluid do not play a role in the magnetic polarizability of the particles. Bubbles or voids in the fluid can never cause a catastrophic dielectric breakdown in an MR fluid.

A unique advantage of MR fluids is that one is able to safely create a modest magnetic field throughout a large volume of fluid for special applications. As described in Sect. 6.6.7, it is possible to energize many liters of MR fluid. Creating an electric field in such a volume would be a frightening prospect.

MR fluids are not without disadvantages. Because of their high loading of dense iron, the fluids are heavy. In weight–sensitive applications this needs to be considered. The active volume of the MR fluid may be quite small, however, and the total volume may be significantly larger depending on the actual application (e.g. shock absorbers).

Of concern in many rotary applications are centrifugal effects. Because of the large density difference between particles and liquid, centrifugal separation can occur. For brakes, in which the housing is stationary, this is generally less of a concern because of the continual shear–induced remixing. The commercial MR brakes described below are routinely operated for extended periods of time at centrifugal accelerations of $300 \, \mathrm{m/s^2}$ and appear to be capable of over $6000 \, \mathrm{m/s^2}$.

The particle–fluid density mismatch is a concern for gravitational settling. However, this concern is generally mitigated by the great flexibility one has in choosing surfactants and additives. MR fluids exhibiting long–term stability with little or no sedimentation are achievable [154, 158].

6.6.3 Ways in Which MR Fluids are Used

Virtually all devices that use controllable MR (or ER) fluids can be classified as having (a) a fixed plate (flow mode), (b) a direct shear (clutch mode), (c) a squeeze film (compression mode), or any combination of these three. Diagrams of the basic modes of operation are shown in Fig. 6.63. Examples of flow–mode devices include servo–valves, dampers, shock absorbers and actuators. Examples of direct shear–mode devices include clutches, brakes, chucking and locking devices, dampers and structural composites. While much less understood and exploited than the other modes, the MR squeeze mode has been used in some small–amplitude vibration dampers [159].

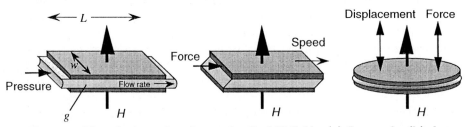

Fig. 6.63. Three basic modes of operation for MR fluids: **(a)** flow mode; **(b)** shear mode; **(c)** squeeze mode

Flow Mode. The pressure drop developed by a (valve)–flow–mode device can be divided into two components, the pressure ΔP_η due to the fluid viscosity and ΔP_τ due the field–induced yield. These pressures may be approximated by [127, 128]:

$$\Delta P_\eta = \frac{12\eta_\mathrm{p}\dot{q}L}{g^3 w} \qquad (6.29)$$

$$\Delta P_\tau = \frac{c\tau_{\mathrm{y(field)}}L}{g} \qquad (6.30)$$

The parameter c has a value that ranges from a minimum value of 2 for $\Delta P_\tau/\Delta P_\eta$ less than $\sim$1 to a maximum value of 3 for $\Delta P_\tau/\Delta P_\eta$ greater than $\sim$ 100. The total pressure drop in a valve–mode device is approximately equal to the sum of ΔP_τ and ΔP_η.

Direct Shear Mode. In a similar fashion, the force developed by a direct shear device can be divided into F_η the force due to the viscous drag of the fluid and F_τ the force due to field induced shear stress:

$$F_\eta = \frac{\eta_{\mathrm{p}} S L w}{g} \tag{6.31}$$

$$F_\tau = \tau_{\mathrm{y(field)}} L w \tag{6.32}$$

where S is the relative velocity. The total force developed by the sliding–plate device is the sum of F_η and F_τ.

Squeeze Mode. Squeeze–mode devices are much less well studied than those with valve and direct–shear modes. Some small–amplitude vibration dampers make use of squeeze mode. For small motions, this mode seems to offer the possibility of very large controllable forces, which may be particularly useful for end–stop of snubbing applications [160].

6.6.4 Basic MR Device Design Considerations

Minimum Volume of Active MR Fluid. While (6.29)–(6.31) are certainly useful and important in the design of controllable fluid devices, they often do not provide the clearest insight into the interplay and impact of the various parameters. For such comparisons it is often useful to algebraically manipulate the above equations so as to provide a different pair of derived equations [129, 161]. For valve–mode devices the following set of equations results:

$$V = \frac{12}{c^2} \left(\frac{\eta_{\mathrm{p}}}{\tau_{\mathrm{y(field)}}^2} \right) \left(\frac{\Delta P_\tau}{\Delta P_\eta} \right) \dot{q} \Delta P_\tau \tag{6.33}$$

$$w g^2 = \frac{12}{c} \left(\frac{\eta_{\mathrm{p}}}{\tau_{\mathrm{y(field)}}} \right) \left(\frac{\Delta P_\tau}{\Delta P_\eta} \right) \dot{q} \tag{6.34}$$

Equation 6.33 gives the minimum active fluid volume, $V = Lwg$, that is necessary in order to achieve the desired control ratio $\Delta P_\tau / \Delta P_\eta$ at a given flow rate $\dot{q}$ with a specified controlled pressure drop ΔP_τ. Equation 6.34 provides a geometric constraint in terms of a minimum value for wg^2. The key aspect to note in these equations is the dependence on the fluid viscosity divided by the square of the yield stress $\eta_{\mathrm{p}}/\tau_{\mathrm{y(field)}}^2$ that controls the fluid volume while the viscosity/yield stress ratio $\eta_{\mathrm{p}}/\tau_{\mathrm{y(field)}}$ exerts control over the shape or geometry of the valve.

Similarly, for sliding plate devices, equations that give the minimum necessary fluid volume and a geometric constraint result are easily derived:

$$V = \left(\frac{\eta_{\mathrm{p}}}{\tau_{\mathrm{y(field)}}^2} \right) \left(\frac{F_\tau}{F_\eta} \right) F_\tau S \tag{6.35}$$

$$g = \left(\frac{\eta_{\mathrm{p}}}{\tau_{\mathrm{y(field)}}} \right) \left(\frac{F_\tau}{F_\eta} \right) S \tag{6.36}$$

Once again minimum fluid volume depends on the ratio of viscosity to yield stress squared $\eta_p/\tau^2_{y(field)}$. In this case the geometric constraint is actually the minimum plate separation or gap g and is simply equal to the product of the control ratio, speed and $\eta_p/\tau_{y(field)}$.

Thus we see that for both fixed–plate and sliding–plate devices, the minimum volume of fluid that must be acted upon by the applied field is proportional to the viscosity and inversely proportional to the square of the induced yield stress $\eta_p/\tau^2_{y(field)}$. This factor can have a profound effect on the overall size of the device.

Speed and Electrical Power Considerations. The speed of an MR fluid device is largely determined by factors extrinsic to the MR fluid, such as the electrical characteristics of the electromagnet and the current source. No experimental time–response data are available on practical MR fluids. Ginder [151] estimates a lower bound for the fluid response time based on the flocculation time to form particle pairs as roughly $1\,\mu s$. Experimental transient response–time measurements on the linear damper described in Sect. 6.6.5 have shown that the damper can reach rheological equilibrium within approximately 6 ms after a step voltage input to the current driver [145]. For most practical MR fluid devices, the response time appears to be governed by the time it take for the current source to establish the magnetic field in the fluid, i.e. how fast the power supply can deliver energy. The key factors will thus be the resistance and inductance of the magnetic circuit and the output power of the supply.

Basic Application Feasibility. The above considerations may be condensed into a set of very simple expressions that apply to virtually any MR fluid application and are very useful for making a quick estimate of the feasibility of a given application. For a representative MR fluid having a maximum yield strength of 50 kPa and plastic viscosity of 0.25 Pa s, the expression for the minimum fluid volume reduces to that shown in (6.37). In this expression the fluid volume will be given in cm^3 if the mechanical power $(F_{on} \cdot s)$ is expressed in watts, i.e. force in N and speed in m/s. (For rotary applications, torque in Nm and angular speed in rad/s applies.)

$$V_{min} = \alpha \left(\frac{F_{on}}{F_{off}} \right) F_{on} \cdot s \cdot 10^{-4} \tag{6.37}$$

where $\alpha \cong 1$ for shear mode and $\alpha \approx 2$ for flow mode.

Also, since the maximum energy density that needs to be established in the fluid is approximately $0.1\,J/cm^3$, the maximum electrical power requirement (in watts) placed on the power supply is approximately 0.1 times the fluid volume divided by the time (in seconds) required to input the energy.

$$P_{electric} = \frac{0.1 \cdot V_{min}}{\Delta t} \tag{6.38}$$

Thus, for any application, the minimum information needed to determine active fluid volume and power is:

- F_{on}: minimum 'on–state' force or torque needed (N or Nm);
- F_{off}: maximum 'off–state' force or torque that may be tolerated (N or Nm);
- s: maximum speed or angular velocity for F_{off}(m/s or rad/s); and
- Δt: desired switching time (seconds).

6.6.5 Linear MR Fluid Dampers

Heavy–Duty Vehicle Seat Suspensions. The SD–1000 linear MR fluid damper shown in Fig. 6.64 is a small, monotube damper designed for use in a semi–active suspension system in large on– and off–highway vehicle seats. In this application the MR damper represents enabling technology for a variety of semi–active control schemes such as 'skyhook' damping or other displacement–, rate– and acceleration–based feedback strategies. The damper is capable of providing a wide dynamic range of force control for very modest input power levels, as shown in Fig. 6.65. Physically, this damper is a direct replacement for passive dampers currently used in seat suspensions.

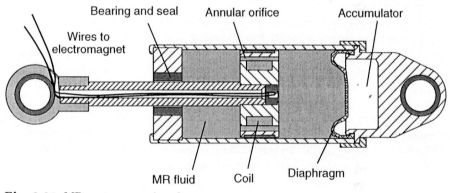

Fig. 6.64. MR seat–suspension damper

The damper is 3.8 cm in diameter, 21.5 cm long in the fully extended position and has a ±2.5 cm stroke. The MR fluid valve and associated magnetic circuit is fully contained within the piston. Current is to an electromagnetic coil via the leads through the hollow shaft. An input power of 4 W is required to operate the damper at its nominal maximum design current of 1 A. Although the damper contains about 70 cm³ of MR fluid, the actual amount of fluid that is activated in the magnetic valve at any given instant is only sabout 0.3 cm³. It is interesting to note that a comparably performing ER fluid damper based on an ER fluid with a maximum yield strength of 5 kPa would require about 30 cm³ of active fluid in the valve at any given instant.

The performance of a semi–active seat suspension equipped with an MR fluid damper is shown in Fig. 6.66. A suspended seat acts largely as a single degree of freedom system with a resonance at about 2.5 Hz. If the seat is

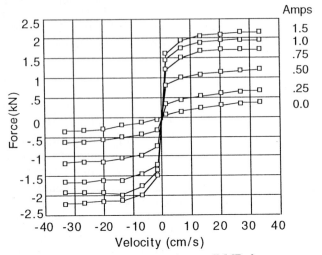

Fig. 6.65. Performance curves for small MR damper

lightly damped, it displays very desirable high–frequency isolation for good ride comfort. However, with low damping the seat displays an unacceptable amplitude magnification at low frequency. Adding a large amount of fixed damping can control the resonance; however, this degrades the high–frequency isolation and ride comfort. The power of the controllable MR fluid damper is shown by the curve labeled 'controlled damping' in Fig. 6.66. This is the transmissibility one obtains using a simple feedback controller that senses the real–time seat position. Proper tuning of the parameters of the feedback system enable the best of both worlds to be obtained — elimination of the resonance and retention of good high–frequency isolation.

Control of Seismic Vibrations in Structures. Because they combine the best features of passive and active control, semi–active control systems appear to have significant potential to advance the acceptance of structural control as a viable means for the dynamic mitigation of seismic damage. Both Spencer et al. and Dyke et al. [143, 147] have recently conducted pilot studies to assess the usefulness of MR dampers for semi–active seismic response control. Through simulations and laboratory model experiments using a Lord MR fluid seat damper of the type described above, Dyke et al. have shown that an MR damper used in conjunction with recently proposed acceleration feedback strategies, significantly outperforms comparable passive damping configurations.

Numerical examples have been used by Dyke et al. to compare the effectiveness of this MR fluid–based semi–active control approach with an ideal active control system. The performance of the semi–active control system

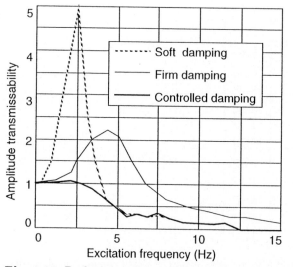

Fig. 6.66. Performance for an MR semi–active controlled suspended seat.

with the MR damper was found to be modestly better than that of the linear active controller. This indicates that a semi–active control system may not only be capable of approaching the performance of a linear active control but of actually surpassing its performance, while requiring only a fraction of the input power required by the active controller.

One way MR fluid dampers may be used for mitigating seismic–induced vibration in civil engineering structures is shown in Fig. 6.67. An array of MR dampers is distributed throughout a building in a chevron brace arrangement. Ground motion excites the low–order bending modes of such a structure, causing one floor to shear relative to the next. Without damping, vibration amplitudes grow until damage to the structure occurs. One approach to mitigating such damage is simply to make the structural components exceedingly stiff; this can work but results in added cost, added mass and loss of useful space. A semi–active damper system has the advantage of allowing a more flexible and lighter structure while providing a highly effective means for dissipating seismic energy before destructive motions occur.

In order to demonstrate that the technology for semi–active MR fluid dampers will ultimately be scalable to full–sized civil engineering structures, a full–scale, 200 000 N (20–ton) MR fluid damper has been designed and built [147]. A schematic of this full–sized, MR seismic damper is shown in Fig. 6.68. The damper utilizes a particularly simple geometry, in which the outer cylindrical housing is part of the magnetic circuit. The effective fluid orifice is the entire annular space between the piston's outside diameter and the inside of the tubular damper housing. Movement of the piston causes fluid to flow through this entire annular region. The damper is double–ended, i.e. the piston is supported by a shaft on both ends. This arrangement has

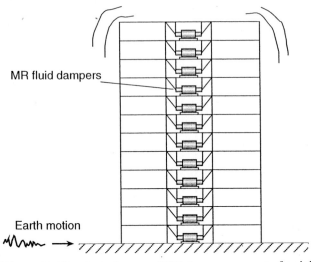

Fig. 6.67. Distributed array of MR dampers as part of a civil engineering structure. Seismic motion causes one floor to shear relative to next floor.

the advantage that a rod–volume compensator is not needed in the damper, although a small pressurized accumulator is required to accommodate the thermal expansion of the fluid. The damper has an inside diameter of 20.3 cm, a stroke of ±8 cm, and is designed to have dynamic force range F_τ/F_η greater than 10 at a speed of 10 cm/s.

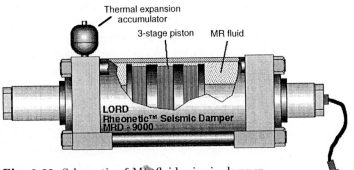

Fig. 6.68. Schematic of MR fluid seismic damper.

An electromagnetic coil is wound in three sections on the piston. This results in four effective valve regions as the fluid flows past the piston. The coils contain a total of $\sim$ 1.5 km of 16–gauge magnetic wire. An input power of 22 W is required to operate the damper at its nominal maximum design current of 1 A. The completed damper is approximately 1 m long and has a mass of about 250 kg. The damper contains approximately 5 l of MR fluid, but the amount of fluid energized by the magnetic field at any given instant is only approximately 90 cm³.

Expected performance of the MR seismic damper is shown in Fig. 6.69. A photograph of the completed damper is shown in Fig. 6.70.

6.6.6 Rotary MR Fluid Brakes and Clutches

An MR fluid brake (Lord MRB–2107–3) that is currently being manufactured and sold as a controllable rotary resistance element for programmable aerobic exercise equipment is shown in Fig. 6.71. This brake, which is shown in cross–section in Fig. 6.72, is 9.2 cm in diameter and provides a controlled dissipative torque of 0–6 Nm for speeds of up to about 1000 rpm.

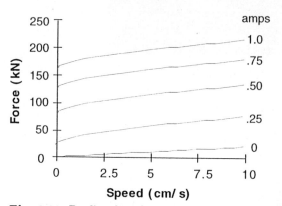

Fig. 6.69. Predicted performance for 20–ton MR damper

These MR brakes are at present being used as semi–active control elements in cycling and stair–climber types of aerobic exercise machines, as indicated in Fig. 6.73. They are typically used in conjunction with velocity feedback, wherein the torque is controlled in real time such that the user is forced to maintain a desired target speed profile. In this application MR fluid brakes require substantially less power to operate than the comparable eddy–current or magnetic–hysteresis brakes that have traditionally been used. Their simplicity and ease of control makes them a very cost–effective choice for a wide variety of applications ranging from controllable exercise equipment to precision active–tension control. While the exercise example is a rather subjective, non–quantitative type of semi–active control, it does, however, require the rotary damper to be extremely smooth, proportional and fast in order to achieve a satisfactorily subjective feel.

MR brakes are currently offered in the AS700 and AR300 machines sold by Nautilus.

Fig. 6.70. Completed 20–ton MR fluid damper.

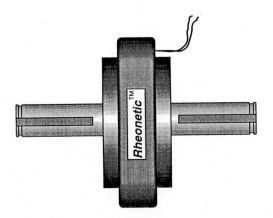

Fig. 6.71. Commercial MR fluid rotary brake.

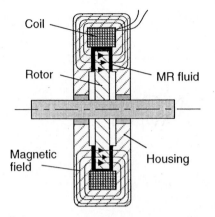

Fig. 6.72. Schematic of MR fluid rotary brake

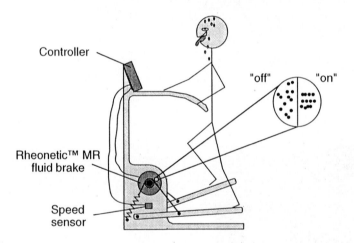

Fig. 6.73. Exercise machine with MR brake

6.6.7 Medical Applications of MR Fluids

A rather different and unique sort of MR fluid actuator is the universal, hand–and– wrist rehabilitation device shown in Figs 6.74 and 6.75.

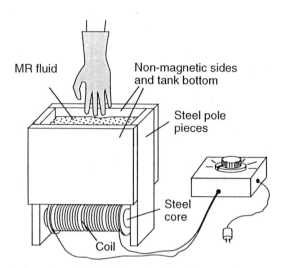

Fig. 6.74. MR fluid hand and wrist exerciser.

Fig. 6.75. MR fluid hand and wrist exerciser in use.

This actuator is essentially a large tank containing approximately 7 l of MR fluid. A large electromagnet provides a modest magnetic field (0–20 kA/m) throughout the entire fluid volume. The energy density in the MR fluid in this case is only on the order of $1 \, \text{mJ/cm}^3$, about two orders of magnitude smaller than that typical of MR valve and brake applications.

The MR fluid hand rehabilitator uses a thixotropic, high–volume fraction, non–toxic, water–based MR fluid having a consistency like a pudding mixture. The patient's hand is separated from the fluid by a glove, which may or may not be part of the vat. If the glove is separate and the MR fluid is exposed, a layer of immiscible mineral oil is floated on top of the MR fluid to prevent evaporation of the water from the MR fluid. The oil also serves to coat the glove prior to immersion so that the MR fluid does not stick when the glove is removed from the vat. A small heating element is normally included so that the fluid does not feel cold.

Unlike conventional hand rehabilitation therapy devices, e.g. rubber balls, the MR fluid tank is able to provide a controlled resistance for motions involving virtually all muscle groups in the hand and wrist. In particular, use is not limited to therapy involving a squeezing or gripping motion that affects only the finger flexor muscles. The MR tank is effective at providing execise of the opposing extensor muscles as well as the abductor and adductor muscles that spread and close the fingers laterally, and the various muscle groups involved in wrist motion. Such an MR device is particularly suitable for physical therapy programs where a progressively incremented level of resistance, based on a patient's progress, is desired.

6.6.8 Conclusion

Magnetorheological fluid actuators have provided technology that has enabled effective semi–active control in a number of real–world applications. Examples include controllable dampers for supension systems on vehicles and controllable rotary brakes for exercise and rehabilitation therapy equipment. Because of their simplicity, low power, and inherent robustness, MR fluid dampers appear to be capable of scaling to sizes suitable for civil engineering applications.

6.7 Electrochemical Actuators
D. zur Megede

Electrochemical actuators belong to the group of chemical actuators but they have their own additional properties. The function of chemical actuators is based on the change in volume that occurs during a chemical reaction. In this way it is possible to perform mechanical work. This can be an unwanted effect such as is seen as the force of expansion during the corrosion of iron used in concrete or engineering bricks in the construction industry. But the gas evolution in the course of a corrosion process can also be used for the slow dosing of lubricants. An example for a very fast actuator action can be found nowadays in the very widespread use of air bag systems in vehicles, which are an important element of security and which use chemical reaction that starts and takes place very rapidly.

The disadvantage of such pure chemical actuators is their irreversibility; therefore their usability is often limited to a single operation. In contrast, electrochemistry offers the possibility of performing chemical reactions in a controlled way that includes the reversal of the reaction path. This is possible because chemical reactions are determined by the transition of electrons, and this attribute is used in electrochemistry to control chemical reactions in a very specific and often reversible way. So electrochemical methods have been used successfully for a long time in the production of chemical substances and pure materials, as well as in the purification and the corrosion protection of metals or in energy technology. In addition, in the field of sensor technology, electrochemical sensors have shown their usability in the detection of harmful gases in air or other media, and such sensors are in commercial use.

To use electrochemical reactions for a propulsion system they must be associated with a change of volume. This can be either through a change of a substance or through a production of a new substance. In both cases, exactly determined oxidation or reduction processes take place, whereas the introduction of electrons or holes into a solid can change its geometry in a way that the solid expands during oxidation and shrinks again during the reduction process, and such structural changes are usually not entirely reversible and often end in complete destruction of the former structure of the solid.

A useful utilization of electrochemical reactions for actuator purposes is the reversible production of gases. The gas production is done by electrolysis while the gas consumption is performed in fuel cells or batteries, where much knowledge in the field of energy conversion exists.

Electrochemical actuators as reported in this article are comparable to pneumatic drives. In such systems the required pressure first has to be supplied by a costly pipe system and then has to be reduced again, which is often also not easily possible. In contrast, an electrochemical actuator (ECA) produces and consumes the pressure gas internally by a reversible electrochemical reaction. The only thing necessary is an external current supply. Such a drive is easy to control and therefore ideal suited for integration in microprocessor techniques [162].

6.7.1 Fundamentals

All actuators reported here are operated by a pneumatic principle, which means that the electrochemically controlled reaction is a reversible gas production in a system of two current–bearing electrodes immersed in an ionic electrolyte.

The gas reaction follows Faraday's law and so the amount of gas produced is directly proportional to the number of electrons per time brought into the system. With hydrogen as the reaction gas, the following equation is valid:

$$n(H_2) = (It)/(zF) \tag{6.39}$$

where $n(H_2)$ is the amount of hydrogen (in mol), I is the current (A), t is the time (s), z is the number of electron transitions, and F is the Faraday constant (As/mol).

In the case of hydrogen $z = 2$; in the case of oxygen $z = 4$. This means that with a given amount of current within one time unit, twice as much hydrogen can be produced as oxygen. Therefore hydrogen is usually used as the pressure medium despite known problems (e.g. lack of gas tightness).

Three reaction principles can be used for actuator purposes:

- Electrolysis/Fuel Cell;
- Gas Pump; and
- Solid Oxide Electrode/Fuel Cell.

The characteristics of electrochemical actuators can be optimised by the use of different electrolytes and electrodes [163].

Electrolysis/Fuel Cell. In the course of the electrolysis of an aqueous electrolyte, the gases hydrogen and oxygen are produced at the current leading electrodes. The following chemical reactions take place.

Positive electrode:

$$2\,H_2O \rightleftharpoons O_2 + 4\,H^+ + 4\,e^- \tag{6.40}$$

or

$$4\,OH^- \rightleftharpoons O_2 + 2\,H_2O + 4\,e^- \tag{6.41}$$

Negative electrode:

$$4\,H^+ + 4\,e^- \rightleftharpoons 2\,H_2 \tag{6.42}$$

or

$$4\,H_2O + 4\,e^- \rightleftharpoons 2\,H_2 + 4\,OH^- \tag{6.43}$$

The transport of the current within the electrolyte is performed by protons or hydroxyl ions, which have to be present in sufficient quantities in an acid or basic medium. In the event of an electrical short circuit between the electrodes, the electrolytic reaction breaks down. The reaction is influenced by physical parameters such as temperature and pressure. High temperature and low pressure have a positive effect on the gas production. More important, however, are the catalytic and geometric properties of the electrodes on which the reactions occur.

A special type of electrode is required for use in an actuator. It must possess a good electrical conductivity as well as sufficient mechanical stability. As an inexpensive material carbon can be used, and by the use of mixtures of different carbon types with individual particle sizes, large active surfaces, and good electrical conductivities, an optimal electrode can at least partially be tailored. Graphite can be added in order to get a higher inner conductivity.

The carbon particles are held together by a fiber structure that is formed by sintering a polymeric binder. The type of binder and its weight content allow the adaptation of the wetting properties of the electrode surface. An electrode produced in this way is highly porous and allows the formation of a three–phase boundary by simultaneous introduction of gases and electrolyte onto the electrode surface. In order to speed up the electrochemical reaction, a catalyst is added to the electrode, which should occupy the surfaces within the pores [164].

The reverse reaction of the gases to water, according to (6.40)–(6.43), is identical to the processes within fuel cells. Here, electrocatalysts are required, especially platinum and its alloys and palladium, all of which are well suited materials. The catalysts have to be distributed on the electrode surface in a very low concentration. Usually the catalyst is brought onto the electrode after completing the preparation process. For example this can be reached by soaking a porous electrode with a solution of a precious–metal salt. After a drying step, the metal salt is reduced to the base metal. A typical metal concentration is in the range of $1\,mg$ precious metal per cm^2 of electrode surface.

To complete the electrochemical cell an ion–conducting electrolyte between the electrodes is necessary. Its main tasks are to guarantee sufficient electrical conductivity within the cell and to deliver water for the gas–evolving reactions at the electrodes. It also has to store the water that is produced in the reversal reaction. Normally acid or alkaline aqueous solutions are used for this purpose. What kind of electrolyte is actually taken is determined by practical considerations, e.g. by the corrosion resistance of the materials used. Because the gases hydrogen and oxygen, which are produced during the electrolysis, can react directly to water when they come into contact, it is necessary to make sure that the gas spaces of the two electrodes are strictly separated from each other. This can be done by using ionic conducting membranes like those used in technical electrolysis plants or in fuel cells with polymer electrolyte membranes.

The principal arrangement of the electrochemical cell for a fuel cell reaction is shown in Fig. 6.76.

Gas Pump. Gas pumps also contain electrochemical cells with two differently polarised electrodes and an ion–conducting electrolyte. While at one electrode a gas (e.g. oxygen) is consumed, it is produced again at the other electrode. If the electrode areas are separated by an ion–conducting membrane, such an electrode arrangement will transport a gas from one place to another.

In this way it is possible, for example, to transport oxygen from the surrounding air into a closed housing where it can perform volume work. Again, such an actuator works reversibly, because consumption and production of oxygen can be performed on both electrodes on equal terms. Only the direction of the current flow through the cell determines the direction of reaction.

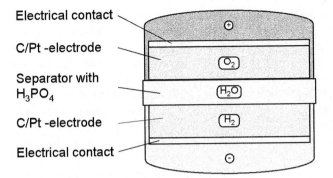

Fig. 6.76. Schematic electrode configuration for fuel cell reaction

However, for reasons mentioned in Sect. 6.7.1, such actuators are slower than actuators based on hydrogen.

Solid Oxide Electrode/Fuel Cell. A major disadvantage of the actuators so far described is the necessity of a gas–tight separation of the two electrode areas. This makes it very difficult to produce a cheap actuator. A practical alternative is the use of a solid–state reaction as a substitute for the oxygen evolution during the electrolysis. Such a reaction can be the oxidation or reduction of a metallic electrode.

In such an electrochemical cell, in the course of the electrolysis only one gas is produced. While at an active carbon electrode hydrogen is produced and builds up pressure (6.45), the unwanted production of oxygen can be overcome by oxidising a metallic electrode (6.44). We thus have:

Positive electrode:

$$Me + 2\,OH^- \rightleftharpoons MeO + H_2O + 2\,e^- \tag{6.44}$$

Negative electrode:

$$2\,e^- + 2\,H_2O \rightleftharpoons 2\,OH^- + H_2 \tag{6.45}$$

A possible candidate for the realization of the reaction in (6.44) is silver which is also used in battery technology. Nickel is also suitable and could be a practical alternative for cost reasons. During the electrochemical reaction it is only necessary to hold the cell voltage below the decomposition voltage of the electrolyte in order to avoid an unwanted production of oxygen.

To achieve a good corrosion stability of the above–mentioned metals, it is necessary to have an alkaline electrolyte. Very often an aqueous potassium hydroxide solution serves for this purpose, where silver, nickel and carbon can be used without significant corrosion. The highest specific conductivity ($0.5\,\mathrm{Ohm}^{-1}\,\mathrm{cm}^{-1}$) has a solution with 25 weight–% potassium hydroxide ($4.2\,\mathrm{mol/l}$). The reactions within the actuator are accompanied by strong changes in the water balance of the electrolyte: a solution with 20 weight–% KOH is normally used, which can get further concentrated during operation.

The overall amount of KOH needed is given by the uptake of fluid through the electrodes and the separator. Losses of hydrogen by diffusion during the lifetime of the actuator have to be considered too.

An actuator built up in the described way does not need a gas–tight membrane to separate the electrode areas any longer. The task of the separator then is to reduce the risk of electrode short circuits and for the storage of an electrolyte reservoir. Besides mechanical stability and high resistivity against alkalines, a good ability to store liquids is an important criterion for the choice of the separator. Well suited are mineral fibers, cellulose and polyolefinic web materials.

Both actuator systems based on the fuel cell reaction have in common that they are in principle similar to batteries: while the forward reaction during current input equals the charging procedure, the back reaction (which is accomplished by an external short circuit of the electrodes) can be compared to a discharge procedure. This behavior offers special advantages — for example, the energy of the reverse reaction can be stored in an external battery and used for the next actuator movement. This helps to reduce the energy consumption of the actuator and also gives the possibility of driving the actuator into a well defined position in case of a breakdown in the external energy supply.

6.7.2 Construction of Reversible Actuators

The electrochemical actuator fulfils its work of driving by a reversible gas reaction. This requires a gas–tight closed housing that can expand within given limits and transmit the necessary power by a plane surface. To enable a fast technical construction and an adaptation to existing driving systems, a solution based on stainless steel expansion bellows was chosen. The bellows allow the transmission of the required stroke and have sufficient pressure stability for power transmission.

Realization of a Fuel Cell Actuator. Up to the time of writing, there have only been some laboratory models of electrochemical actuators existing that work on the principle of electrolysis and its reversal in a fuel cell–type reaction [165]. It can be seen that this principle, with its very high reversibility, is suitable for actuators with over one million load cycles. A disadvantage is the necessity to compensate the different pressure levels between anode and cathode sides by additional measures so as to avoid damage of the ion–conducting membrane.

An additional difficulty arises from the low operating voltage of about 2 V DC, which is not easily supplied in stable form and without larger transmission losses. In summary, such an actuator can only be constructed at very high cost, which has hindered technical usability in spite of all of its advantages.

Realization of an Oxygen Pump. For reasons mentioned in Sect. 6.7.1, only continuously working dosages for lubricants that require a steady but slow flow have yet been made. Reversible working actuators based on the principle of an oxygen pump do not exist at the time of writing. However, as the electrical current flowing through a cell is limited by the amount of existing oxygen, it is possible to use the pump reaction for the quantitative detection of oxygen. Many oxygen sensors using this principle are already in commercial use. They normally possess an oxygen ion–conducting ceramic as the electrolyte and work at higher temperatures.

Realization of the Solid Oxide Electrode Concept. An important element of an ECA with a solid oxide electrode is the flexible housing that takes care of the power transmission and the integrated package of electrochemical cells located inside the housing. The cell package can be penetrated by hydrogen and has an electrical contact that is isolated against the housing. Additional measures for the gas–tight separation of the electrode rooms are in principle not necessary.

The cell package is similar to a battery built up as a staple of single cells that all have the same structure. Thus it is possible to combine the single cells in modules to form packages, and to construct serial as well as parallel circuits. Such arrangements can be tailored to almost any voltages and to an optimal current demand.

The structure of a single cell is shown in Fig. 6.77. The outer parts consist of compression–moulded and punched nickel. In an injection moulding form, they are coated with an alkaline–resistant thermoplastic material. The nickel sheets function as electrically conducting separation walls and close one cell against the other. The gas exchange takes place through holes in the polymer casing, which additionally serves for centring during assembly.

Furthermore, there can be seen on the figure from the upper left to the lower right

- a carbon electrode layered with platinum (dark);
- a polypropylene separator (light) for electrolyte storage;
- the solid oxide electrode (dark) consisting of a mixture of silver and silver oxide; and
- an additional separator as an additional reservoir for the electrolyte.

To build up an actuator, several single cells are combined to a staple and brought into a metal bowl. By a steel net lying on the bottom of the bowl the contact mass is obtained. The second electrical contact is made by a metallic tag that is welded to a nickel sheet. In this way the inner cells are also brought into electrical contact as a result of the bipolar structure of the package. For the gas exchange, the bowl has a drilled hole in its bottom plate.

The rigid bowl is built into a gas–tight housing that is flexible and able to expand. The housing consists of expansion bellows with a welded–in bottom

Fig. 6.77. Structure of a cell package with single electrodes

and a cover with a gas–tight contact lead–in. The bellows have to fulfill the following requirements:

– good corrosion stability against the electrolyte;
– very low permeability for hydrogen;
– a high number of load cycles with low fatigue of the material; and
– good weldability.

A material formed to fulfill these functions well is austenitic stainless steel. For the intended use as a drive for radiator valves, for example, a unit with an expansion of up to 5 mm with a force of 100 N is commercially available. The arrangement of the bowl with the cell package inside the housing can be seen in Fig. 6.78, which shows a cross–section through a completely assembled actuator.

Above the cell package there is a polypropylene ring to lower the deadband volume and thus reduce the gas demand. An additional O–ring holds the cell staple under elastic pressure and simultaneously takes care of the electrical isolation against the housing.

To the external surrounding medium, the actuator is gas–tight, closed by a lid that is simultaneously welded with the bellows and the bowl without

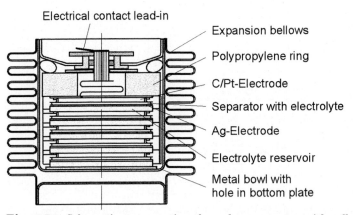

Electrical contact lead-in

Expansion bellows

Polypropylene ring

C/Pt-Electrode

Separator with electrolyte

Ag-Electrode

Electrolyte reservoir

Metal bowl with
hole in bottom plate

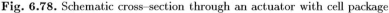

Fig. 6.78. Schematic cross–section through an actuator with cell package

any filler materials. Good results for this have been obtained when using a
laser welding technique. The electrical bonding is performed for the positive
pole by a glass lead–through with a stainless steel pin centred in the lid. The
actuator housing serves as the negative pole.

6.7.3 Possible Uses for ECA

Because in comparison with other arrangements it has a simple way of functioning, the ECA with a solid oxide electrode has so far come closest to
practical application. A version with five electrode packages in a serial arrangement is already at the stage of prototype assembly, and has been especially designed for driving of radiator valves. It needs a voltage supply of
12 V DC. Its operating characteristics are described below.

Characteristics of ECA. Fig. 6.79 shows a schematic of the measurement
equipment in which the functions of an ECA can be tested.

The equipment is designed in such a manner that the operational data are
similar to those of a radiator valve. The ECA has to produce an expansion
of 3 mm with a force of 100 N, with a spring simulating the valve. During
operation the developed actuator power as well as the current and voltage
are registered as a function of time. The testing equipment is automated,
and after the completion of the closure procedure an opening phase starts.
After closing the switches S1 and S2 with switch S3 opened, the hydrogen
evolution inside the actuator is started by a constant current of 200 mA and
an increasing force acts upon the spring. As soon as the desired force of 100 N
is reached, switches S1 and S2 are opened by a changeover relay and switch
S3 is closed. By this action the ECA is short–circuited across a load resistor
and drives back into its resting position. Then a new cycle can be started.
The changes of power, current and voltage as a function of time are shown
in Fig. 6.80.

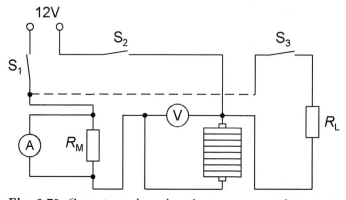

Fig. 6.79. Current supply and testing arrangement for actuator prototypes

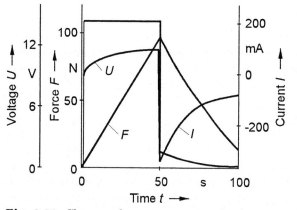

Fig. 6.80. Changes of power, current and voltage as a function of time

It can be seen from Fig. 6.80 that power is linearly increasing with time at a constant current feed. The measured actuator voltage only slowly nears the value of 12 V DC because the capacity of the electrodes is an obstacle to a rapid charging process. The current peak caused by the discharge in case of a short circuit is limited by the load resistor to below 500 mA.

So the ECA proves to be a real three–level controller, which only needs energy during the closure phase. During the static phase it remains without current, its closing energy is partially released while opening. These features are significant advantages over other systems.

Control Device of Heating Systems. At the moment the most interesting use of electrochemical actuators based on a solid oxide reaction is seen in heating systems that demand an individual control of radiators. In this case the technical requirements can already be fulfilled by the characteristics of the ECA. They are listed in Table 6.11.

Table 6.11. Requirements for an ECA used as a drive for radiator valves

Extension	5 mm
Driving force	100 N
Voltage supply	12 V DC
Operating current (during closing)	200 mA
Time for closing	100 s
Lifetime (dynamically)	100 000 cycles
Lifetime (statically)	10 years
Operating temperature	−5 to +60°C
Storage temperature	−30 to +80°C
Stability of extension	1–2 %/h
Power consumption	3 W
Weight	34 g

Whenever a building possesses a temperature–control device in each room and an ECA at every radiator, individual temperature regulation is possible. That means that in each room a controlled temperature can be adjusted in the way it is needed, which can be an advantage for big office blocks or hotel buildings. The ECA has here clear advantages over the electrothermal drives that are sometimes used. The comparisons are listed in Table 6.12.

Table 6.12. Comparison of features of electrothermal and electrochemical actuators

Electrothermal Drive	Electrochemical Actuator
High power demand in static operation.	No additional power needed in hold position.
Drives at the same controlled system influence each other.	No mutual interference, can be wired in groups.
Valve noise by two–level controller characteristic of drive.	No valve noise by stable three–level controller characteristic.
In case of energy shutdown, the valve is closed: the radiator cools.	In case of energy shutdown, the last valve position is retained.

Furthermore, it is possible to integrate the room temperature adjustment into a complete comfort system that is oriented to the individual wishes of its user. In Fig. 6.81 such an arrangement is shown schematically.

Other Uses. As the electrochemical actuator described in this section has a slow dynamic compared with other drives, it is advantageous to use it for the control of sluggish systems. This is the case for aeration and cooling systems, where the ECA can play out its advantages regarding reliability, low noise generation, compactness, low energy consumption, and so on. Besides the heating system already described, other examples are cooling systems for

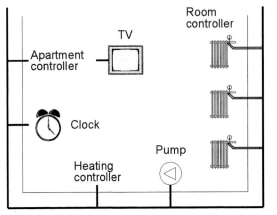

Fig. 6.81. Central control of single rooms for a residential building (schematic)

drives of all kinds (e.g. internal combustion engines) and air conditioning for mobile systems (vehicles, airplanes, etc.).

Further uses might be in areas where today irreversible drives are installed. The dosage of lubricants could be done in a way that meets the demand better than it is by a continuously working actuator. Also the possibility of an easy exchange of the lubricant housing without losing the drive could be an advantage.

Reliable drives are needed in space technology to open antennas or solar panels in space. An ECA could be a well–suited drive for this purpose because of its low energy demand and the high flexibility in power generation together with complete independence of position. The avoidance of sparking and the non–existence of moving parts are interesting features for safety–critical systems.

The adaptability of the ECA to the given task can be seen in the fact that it was possible, by simple enlargement of the cell package, to design an actuator that shows a three–times higher expansion together with a thirty–times higher driving force [166].

6.7.4 Summary

The electrochemical actuator shown int his section is in principle a pneumatic drive that uses an internal reversible gas evolution reaction in individual electrolytic cells. The generated gas pressure functions via expansion bellows as the driving force. So the actuator has several advantageous features:

– Its power can be adjusted to a given task by the amount of electrolyte in the cells and the electrical current flowing through the actuator. The same is valid for the elongation, which mainly depends on the bellows used.

- The operating voltage of the actuator is a multiple of the operating voltage of the individual electrochemical cells. So the voltage can be adopted to a given supply network. As the actuator in principle uses low voltage DC, it is compatible with most common bus structures used in computer–based systems.
- The actuator shows the behavior of a real three–level controller and has the additional advantage of needing energy only during the closure phase. The holding phase does not need any current, and during opening the spent energy can be partially recaptured. In addition, there is the possibility to hold a position in the event of energy shutdown, or to drive into a safety position.
- The drive's complete lack of noise is a very positive property for applications in comfort areas, as is the case in residential buildings. Examples are drives for radiator valves or for the opening and closing of windows and ventilation flaps.

An important objective of further developments is the extension of the operating temperature range to higher and lower values. This would allow use of the advantages described for the electrochemical actuator in technical applications in which today other much more expensive drives are used. Examples of this are chemical technology and the energy sector.

6.8 Chemomechanical Actuators
H. Wurmus and M. Kallenbach

To construct small but powerful drives is one of the most important aims in the field of drive technology. Simply making component parts smaller in scale, as for instance in microelectronics, is the wrong approach. Dimensioning problems and material problems (e.g. particle and crystal growth) prevent conventional drives from being scaled down; new drive principles, technologies, and materials have to be applied to achieve innovative solutions for this problem.

Materials that can transfer electromechanically generated energy, withstand high loadings, and have high electric and force densities are needed. Ideally, these materials should not have high electric or magnetic fields, nor large temperature gradients, because these fields can disturb the surrounding environment of the microsystem and its connected assemblies. Chemomechanical actuators are a promising alternative to conventional drives [167].

6.8.1 Chemomechanical Energy–Conversion Principles

Actuators are mainly characterized by the way they convert energy input into mechanical energy output. The energy–conversion process can occur in the simplest case directly in one step, as for example in electrostatic actuators, or in more complicated cases in more steps. Fewer steps result in higher efficiency.

The rarely used chemomechanical drives convert chemical energy into mechanical energy in one step. When a chemomechanical actuator has electrical energy applied to it, the actuator is an electro–chemo–mechanical energy converter. The energy conversion occurs in two steps: electrical energy into chemical energy and then chemical energy into mechanical energy. In this case, the electrical energy is the power–carrying energy. Electrical energy application and control are one of the most important basic demands, because electrical energy can be easy transported, transferred, and stored.

Unlike in technological systems, the energy conversion in natural muscle, which is the most important biological drive, occurs from chemical energy into mechanical energy only (see Sect. 9.1). In this case, the power–carrying energy is stored in reactive compounds, whereas the electrical energy (which is introduced via a nervefiber) triggers the power–switching module. That is why this kind of drive can be described as an electrically stimulated chemo–mechanical drive. Unlike in the electro–chemo–mechanical drives, the chemical energy is the power–carrying energy.

Different classes of reversible chemomechanical energy–converting materials have been investigated. Cross–linked organic materials that transfer chemical energy directly into mechanical energy include those below:

- **Synthetic rubber (in solvents).** When cross–linked rubber–like polymers are immersed in a solvent, the solvent fills the polymer network. The polymer changes its volume if the polymer segments attract the solvent molecules: the polymer net expands in order to achieve as much contact as possible between the polymer and the solvent. The polymer can swell by up to 300% of its original volume [168]. On the other hand, if the polymer segments repulse the solvent molecules, the rubber will contract. When the solvent evaporates, the rubber returns to its previous volume.
- **Net–linked collagens (contraction of collagen fibers).** When certain scleroproteins — so–called collagens — come into contact with formaldehyde, they become cross–linked, forming a collagen fiber network. After this, when they are immersed in salt solutions of varying concentration, they can change volume by up to 40% [169].
- **Polyelectrolyte gels.** A polyelectrolyte gel can be characterized as a cross–linked network of long polymer molecules that is filled with a water solution. When such a polymer net is immersed in water, the polymer stretches until the restoring force generated by the cross–links compensates for the hydrostatic pressure inside the gel. An applied electrical field is able to generate a pH gradient, and mechanical work results. Polymer gels can create large displacements and high force densities. Force density values are comparable to those of a human muscle [168].
- **Conducting polymers.** Conducting polymers are semiconductors in which the degree of doping and therefore the conductivity can be reversibly

changed [170]. When the degree of doping is changed by an applied potential, counterions move in or out of the polymer and the polymer changes its volume (in detail, see Sect. 6.8.2 below). The volume change that is generated can be a few percent of the original volume, which is large when compared with other materials like those used in piezo–ceramic or magnetostrictive drives. The principle of the change of volume is described later on polypyrrole.

– **Biological muscles.** In the muscle cells (fibers), the contraction is the result of an inward–sliding together of alternating arranged packages of macromolecules. The filament myosin forms cross–bridges to link a hexagonal framework made of actin filaments. Working like a stepping motor, these filaments encroach when chemical energy and nerve impact are supplied. Because this periodical arrangement is coupled in series and in parallel in space (a cascade), ways and forces can be added to macroscopically efficient work ($\Delta l/l$ max. 35%, $F = 20\,\mathrm{Ncm}^{-2}$). The muscle has to be restored by using strange forces (antagonists) after contraction, because the molecular reaction is an irreversible reaction (the one–way effect). The embedded connective tissue interlacing affects the transmission and causes passive mechanical properties (visco–elasticity).

In summary, rubber and cross–linked collagens are difficult to use as the drive element in microsystems because they cannot be voltage–controlled. Because the operation of all chemomechanical converters represented in this section are based on a diffusion process, these systems are not suited for highly dynamical processes: the time constants are in the range of some seconds.

6.8.2 Some Properties of Conducting Polymers

Polymers are often used as insulators because they are unable to conduct electricity. This picture has been changed in the last 20 years since a class of polymers have been researched that are able to conduct electrical current [174]. These materials belong to a larger class of organic materials, called synthetic materials.

Conducting polymers are chemically characterized by the so–called conjugation in which carbon double bonds alternate with carbon single bonds along a polymer backbone. Some examples of conducting polymers are shown in Fig. 6.82.

Conducting polymers are also characterized by their high conductivity when doped with ions (see Fig. 6.83). Their conductivity can be changed by orders of magnitude, and can be done controllably and reversibly by changing the doping level. But unlike in silicon, the dopants can be easily inserted and removed from the spaces they occupy between the polymer chains [172]. In comparison with other semiconducting materials, the doping level can be very high: approximately one dopant counterion per four monomers [173]. Because

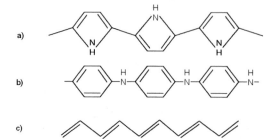

a)

b)

c)

Fig. 6.82. Different conducting polymers: **(a)** polypyrrole; **(b)** polyaniline; **(c)** polyacetylene. The materials are conjugated, which means that between carbon atoms there are alternating single bonds and double bonds

these materials are able to store a large amount of charge, the polymers are of interest for use in batteries [174].

Another interesting property is the band–gap that allows electron–hole recombination, which can also be found in conducting polymers. The band–gap allows electron–hole recombination and has made these materials interesting for use in light–emitting diodes (see [174] for example). The optical properties (especially light absorption) can be voltage–controlled, so these properties have also been investigated for use in electrochromic windows.

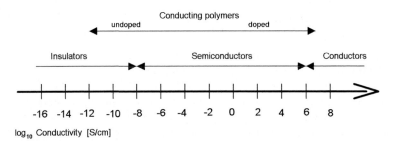

Fig. 6.83. Conductivity of conducting polymers at room temperature.

How can we use conducting polymers — an organic material — for converting energy? In order to dope conducting polymers, counterions must be physically inserted into the polymer. When one changes the doping degree (for example by changing the applied voltage), compensating ions move into or out of the polymer. This has the consequence that the polymer changes volume by several percent [167]. Using this effect, electrically controlled chemomechanical microactuators can be constructed. Some proposals for applications in micro–systems are published in [167, 175, 176, 169, 172, 177].

Polypyrrole (PPy) is one of the most stable conducting polymers, can be used in aqueous solutions, and has good mechanical properties (elastic, stable, cross–linked in a regular fashion). Likewise, the mechanical properties change with the anions. The actuator cannot be mechanically damaged (self–breaking) by a voltage overload; if too high a voltage is applied, the material will over–oxidize (and thus suffer chemical damage).

When a polypyrrole film is grown electrochemically onto a positively charged electrode, the film will be automatically p–doped or oxidized. Polypyrrole is stable in the p–doped state when no external potentials are applied. The fabrication is simple: it can be grown electrochemically onto an electrode or obtained commercially as a fiber from BASF (where it is called Lutamer).

How does the volume–changing process of polypyrrole work in detail? As the dopant, we used the salt sodium dodecylbenzene sulfonate (NaDBS) whose structure is illustrated in Fig. 6.84. When the salt is dissociated, the anion DBS$^-$ is built into the doped polypyrrole network to maintain charge neutrality. The anions are not bonded covalently but ionically. Because of its large size, DBS$^-$ is immobile and cannot diffuse out of the polymer.

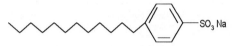

Fig. 6.84. The chemical structure of NaDBS. When Na$^+$ is split off during dissociation, DBS$^-$ can be incorporated into the polymer network to compensate missing electrons

If a negative potential is applied to the electrode, electrons flow back into the polymer and satisfy the missing charges; this is called 'reduction'. The anions are no longer necessary but cannot move out. To prevent the material from becoming negatively charged, cations simultaneously enter the polymer to compensate for the immobile anions, forming the salt NaDBS. The film is therefore neutral. Macroscopically the polymer changes volume. If a positive potential is applied, then the cations and electrons move outside the material again. This electrochemical switching process can be described with following electrochemical redox equation:

$$\text{PPy}^+(\text{bulky anion}^-) + \text{mobile cation}^+ + e^- \underset{\longrightarrow}{\overset{\text{Applied voltage}}{\longleftarrow}} \text{PPy}^0(\text{salt}) \qquad (6.46)$$

It has been found [167] that the following three effects are responsible for the volume change in conducting polymers.

- interactions between chains;
- conformations of chains (e.g. straight chains or twisted); and
- the insertion of counterions.

In the case of polypyrrole, the third effect is the most dominant [179]. Depending on the magnitude of the applied voltage, PPy can change volume by up to 2% [180].

Applying voltages between -1 V and $+0.35$ V to the reference electrode (where voltages depend slightly on the electrode positions within the cell due to voltage drops in the solution, and the reference electrode defines the zero potential in the electrochemical cell), the doping level can be sensitively controlled. Depending on the potential, there will be more or fewer cations in the polymer. Accordingly, the volume change can be sensitively stopped at any point (an analog system), although there is a small hysteresis effect. The facts of what happens in polypyrrole during electrochemical switching are summarized in Fig. 6.85.

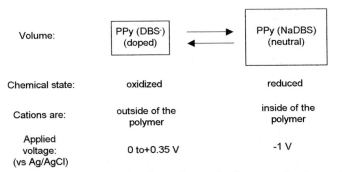

| Volume: | PPy (DBS⁻) (doped) | → ← | PPy (NaDBS) (neutral) |

| Chemical state: | oxidized | | reduced |

| Cations are: | outside of the polymer | | inside of the polymer |

| Applied voltage: (vs Ag/AgCl) | 0 to+0.35 V | | -1 V |

Fig. 6.85. Principles of volume change in the case of polypyrrole [182].

Influence of the Ions on the Movement. In the foregoing arrangement, the anions should be as large as possible in order to obstruct their diffusion out of the polymer. Some other anions of different sizes were [178, 179].

If $Li^+ClO_4^-$ was used for generating the volume change in polypyrrole (tosylate) [PPy(TsO)], one bad effect could be observed. When a negative voltage was applied to the electrode, Li^+ moved into the polymer and the volume started swelling. Over time, the volume shrank, presumably because LiTsO moved out of the polymer (see Fig. 6.86a). This effect damages the ability to change volume because the number of anions in the polymer decreases. This negative effect is called 'salt draining'. For the simple reason that the volume change reverses itself and the system is not stable in the longer term, this system cannot be used for actuation.

6.8.3 Polymer Microactuators

Polymers are attractive materials for microfabrication because one can create them to order, so that they have those specific properties that are required. By chemically synthesizing new monomer units and choosing from among different synthesis and processing methods, polymers can be tailor–made. In micromachining, polymers are already used every day in the cleanroom. The most important polymer is photoresist. Another frequently used polymer is

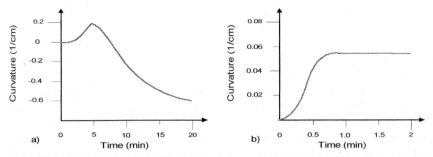

Fig. 6.86. Curvature of **(a)** PPy(TsO) and **(b)** PPy(DBS) [179]. The picture shows a difference in the contraction of both materials. The contraction of PPy(TsO) is not stable over a long time period, opposite to the volume change of PPy(DBS). The salt draining is responsible for the instability of the curvature of PPy(TsO)

polyimide, which serves as an insulating or structural layer [183]. Because polymers are made of twisted chains, conventional polymers are difficult to use when structures on the order of a few nanometers are demanded. But in micromachining, the structures are 1000 times larger, and these irregularities in the molecular positions are quite small in comparison. Thus, polymers are a class of materials that can be quite useful in making new micromachined devices.

The combination of polypyrrole and silicon is important because silicon is the main material for micromachining. Structures and devices made in silicon could be moved using the bilayer if integration were possible. In the following part of this section, the design steps and the fabrication of a simple device, which is a moving silicon plate actuated by a conducting polymer film, will be introduced.

The volume change of PPy amounts to around 2%. If we have a 1 μm–thick PPy layer, then we get a maximum displacement of around 20 nm. Admittedly this displacement is not sufficient for many technical applications. Because the displacements of volume–changing thin layers are so small, there is a problem using conducting polymer actuators as direct drives. Approaches to enlarge the displacement are needed.

It seems reasonable to suppose, first, that simply making layers thicker is one approach. That is not wrong, but because the actuator speed depends on the layer thickness (because of the diffusion process), then the thicker the material is, the longer is the reaction time. Because PPy is a poor ion conductor, it might in these circumstances be useful to add water–filled pores or tunnels in order to allow fast diffusion of the cations within the PPy.

Another approach to reach larger displacements may be to layer the polymer layers. This kind of microactuator is also known as a cascaded actuator and is normally very efficient because force (or torque) or displacements are increased when cascaded microactuators are used. It should be remembered that the most difficult thing to achieve is a small diffusion length that brings the ions into the polymer as fast as possible. For a liquid electrolyte, a second, fast ion–conducting layer might be alternated with the polypyrrole layers. Another idea is to store the ions instead of transporting them. This can be done using a rigid electrolyte or switching the ions between two different conducting polymer layers.

A third approach can be used to enlarge displacements, this time by a combination of an actuator and a movement–gearing element next in line. This transformer can be directly integrated into the actuator. If polypyrrole and gold are mechanically interlocked, then the swelling/shrinking polymer can be looked upon as the active element (the actuator) and the gold layer as a movement transformer. When the polymer contracts in the x and y directions (see Fig. 6.87) the passive gold layer cannot contract, and so it transforms the linear bilayer displacement into a curling movement around the z–axis. Depending on the bilayer shape, more or less complicated movement in free space can be achieved (Fig. 6.88). That means that a clever design of the bilayer can bring innovative solutions for movement generation in free space.

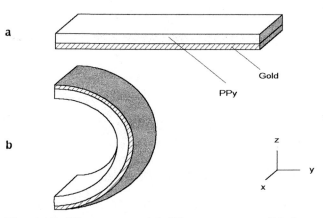

Fig. 6.87. PPy actuator: **(a)** Bilayer structure; **(b)** the contracting PPy causes the bilayer to bend. This principle is introduced in [172].

The displacement gearing in such a mechanism is high but the force gearing is poor. Most of the input mechanical energy is used to produce very high shear stress between the layers. By connecting a bilayer to a plate, the plate can be moved [184, 172, 170] (see Fig. 6.89 and 6.90). The first plates were made of the plastic photoprocessable benzocyclobutene (BCB) because its fabrication and processing are compatible with microsystem technologies.

An attempt was made to use electromechanical actuators to lift silicon plates of varying thickness (200–375 μm) and shapes (from 100×100 to $1000 \times 2000 \, \mu m^2$) driven by thin and flexible gold/polypyrrole bilayers. The gold layer was approximately 100 nm thick, the polypyrrole was roughly 1 mm

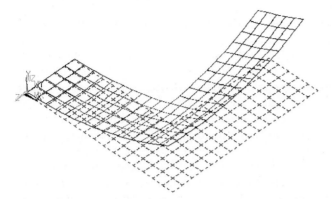

Fig. 6.88. An example for a more complex movement. This picture is a result of an FEM–modeling with ANSYS©.

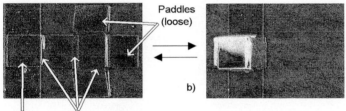

Fig. 6.89. Self–folding cubes were made using bilayers as hinges between rigid plates (paddles): **(a)** in the left part of the picture the unfolded structure of the cube is illustrated, the paddles being $300 \times 300\,\mu m^2$ and the active area of each plate $40 \times 300\,\mu m^2$; **(b)** the cube is folded on the right side and needs a closure time of less than one second parts (idea and source: [172]).

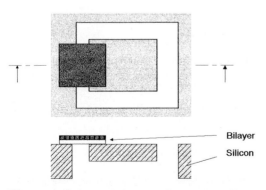

Fig. 6.90. Schematic layout of a moving silicon plate [185, 182]. The bilayer works as a hinge and links the silicon plate with the silicon wafer. The hinge can also be described as an active compliant body link [176].

thick. The construction operated in a salt solution of NaDBS; between $0.3\,V$ and $-1\,V$ were cyclically applied to the switch. It was possible to move the actuator in five–second–cycles.

With the help of this construction it was possible to calculate the energy–to–mass ratio and the energy–to–volume ratio for the actuator. Both factors can be used to compare this kind of actuator with others. Because the largest moved plate had a shape of $1000 \times 2000 \times 200\,\mu m^3$, the energy–to–mass ratio was given as $0.014\,J/g$, and the energy–to–volume ratio was given as $29\,550\,J/m^3$. Both ratios describe a minimum which can be obtained in practice, but larger and heavier plates have not been tested yet. Using a cycle time of 5 ms, the specific mechanical power (also called the power–to–mass ratio) can be given as $0.028\,W/g$. This is a very satisfactory result because these bilayers represent a breakthrough in drive–technology. Although smaller and lighter, the bilayers have a comparable value of specific power, such as that for electromagnets. Fig. 6.91 shows that smaller drives are not automatically weaker.

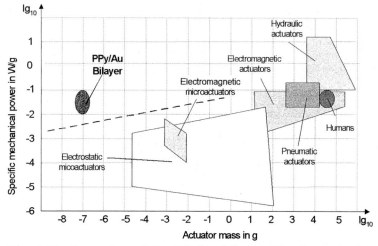

Fig. 6.91. Comparison of the PPy/Au bilayer with other actuators (source for the diagram without bilayer:[186]). In order to calculate the mechanical power, switching cycles of 5 s were considered. The figure shows that the bilayer is near the same power level as an electromagnet or a human muscle.

6.8.4 Synthetic Polymers as Muscle–Mimetic Actuator

The translation movement of a muscle is geared by a skeletal lever and is guided by hinge geometries of different complexities. In the simplest case, one degree of freedom is realized by one muscle loop. Hinges that have more degrees of freedom, serial hinge chains, and neuronal voltage–servos produce process–adapted force–way–functions and cause a typical fluently–lissome motion of organisms.

The working principle of a muscle is the superposition of nanosteps of macromolecules. The principle can be compared with a stepping motor in the smallest natural muscle unit called sarcomere [187]. The sarcomere and its model are shown in Fig. 6.92. In this model a serial and parallel cascade of bilayers is used to add the displacement and force generated by every single bilayer.

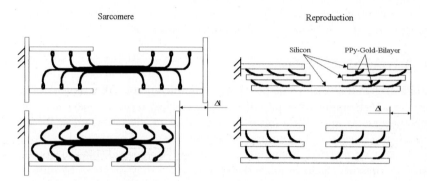

Fig. 6.92. The smallest natural muscle unit is the sarcomere (left). The muscle works like a stepping motor in the nanometer range. The movement of the muscle is the superposition of nanosteps generated by bending zones at the stick–like macro–molecules [188]. The model (right) bases on a serial and parallel cascade of bilayers.

The problem is that the mass of silicon in the system should be as small as possible because this mass must be used but has no energy share. On the other hand, the silicon is necessary to stabilize the system because the silicon is the base for the actuator.

6.9 Microactuators
H. Janocha

Microactuators are based on small–scale three–dimensional mechanical structures that are produced using lithographic and anisotropic etching techniques. The most diverse principles for generating forces such as the bimetallic effect, the piezoelectric effect, shape memory or electrostatics form the basis of actuation in microactuators. In a narrow sense, microactuators are characterized by monolithic integration of the force–generation mechanism; however, a broader interpretation includes also microstructures in which the force–generation mechanism is not monolithically integrated.

Various technologies are implemented in producing extremely small mechanical components with movable structures. The main technologies are classical micromechanics, based on single–crystal silicon, surface microma-chining for producing polysilicon structures, the LIGA technique for metals, plastics, and ceramics, and micromechanics based on quartz (more detail in [189, 190]).

By their very nature, microactuators allow only small displacements and forces, leading to a natural limitation of their application potential. However, the manufacturing techniques used in the production of microactuators are precisely those that permit the integration of sensory functions, including processing and perhaps control circuitry, onto a single chip. These so–called microsystems are characterized by multifunctional properties that, according to Chap. 1, are also characteristic of adaptronic systems.

6.9.1 Driving Mechanisms

Although electromagnetic principles play a significant role in actuators of a larger scale, their application provides rather modest results on the scale of microactuators. By applying a linear scaling factor m, electromagnetic forces generated in a mechanical structure are scaled down — i.e. reduced — by a factor as great as m^4 [191]. Pneumatic, hydraulic or biological (muscle) forces and surface pressures behave more reasonably since they decrease by only a factor of m to m^2. Electrostatic forces are of particular significance on the micro–scale: they may decrease by a factor of m^2, but since the breakdown field strength increases with decreasing dimensions (the Paschen effect), the electrical field strength can be increased, for example, by $m^{-0.5}$, leading to a net force scaling factor of m.

Piezoelectric driving mechanisms can be divided into three categories. Piezoceramic plates are typically attached to diaphragms. Application of a voltage generates a transverse contraction of the piezoceramic, leading to a vertical deformation of the layered composite. The displacements, however, are limited to a few micrometers at voltages of the order of 100–200 V.

Greater displacements of 10–30 μm (about 0.15% of the thickness), forces of several hundred newtons and surface pressures of 30 MPa can be achieved with piezo stacks. Cantilevered piezoelectric bimorph–type transducers are capable of generating displacements of several hundred micrometers — de-pending on the transducer dimensions — although at considerably lower forces. Piezoelectric drives exhibit hysteresis and are limited to operating tem-peratures much lower than their Curie temperature (typically 150–300 °C).

Relative strains of up to 0.2% are achievable with magnetostrictive ma-terials upon applying a magnetic field. The achievable displacements and surface pressures of up to 20 MPa are properties comparable with those of piezoelectric materials.

Electrostatic actuators are most often based on a parallel arrangement of two plates at a distance of just a few micrometers. One of the plates is flexible or flexibly suspended (Fig. 6.93). Application of a voltage causes the flexible electrode to be drawn toward the rigid electrode. Typical operating voltages lie in the range 100–200 V, and pressures in the range 0.1–100 kPa. The displacements are limited to a few micrometers because of the short range of electrostatic forces. Based on its operating principle, this form of actuation is temperature–independent. The response times of the mechanical components are of the order of less than one millisecond.

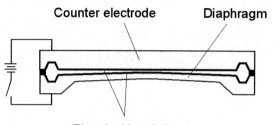

Fig. 6.93. Electrostatic diaphragm actuator

Either thermomechanical or thermopneumatic effects or phase changes (liquid–vapour) are exploited in thermal actuators. In the thermomechanical (bimetallic) principle, a metal layer is applied to the actuation diaphragm. Upon heating diffused resistances found in the actuator diaphragm, the metal experiences a greater strain than the base material. The resulting stress leads to a displacement of the diaphragm [192].

In the thermopneumatic effect, a closed chamber is heated, causing a change of pressure that displaces an elastic diaphragm. Through thermal losses, the electrical power consumption of actuators based on this effect is relatively high — typically 0.1–2 watt. Response times for heating are of the order of a few milliseconds, of cooling on the order of 100 milliseconds, and $100\,\mu m$ displacements are typical. The power consumption can be reduced and the achievable forces increased through the vaporization of a liquid in a partly filled chamber. The liquid phase can be converted to the gaseous phase with comparatively less power and without effecting a change in temperature.

In electrochemical actuators, the flow of an electrical current chemically converts a liquid to a gas in an electrochemical cell. Hydrogen is generated, for example, in a chamber filled with water, causing an increase of pressure in the cell. A reverse of the current decreases the pressure by oxidation of the hydrogen to form water. A diaphragm bounding the chamber can be brought into periodic motion as a result of the current (and thus pressure) changes. Typical response times of these actuator types are on the order of several seconds at displacements of several millimeters [193].

6.9.2 Ink–Jet Printer Heads

Ink–jet printer heads are an example of commercially available microactuators, which have been on the market for quite some time. Jets arranged
tightly in a row are driven either piezoelectrically or thermally using so–
called bubble principles. In the case of thermal activation, a short heating
pulse ($< 10\,\mu$s, $10\,$mW) generates a gas bubble in an ink chamber located behind a jet opening. A pressure wave is created in the liquid, causing a small
drop of liquid to be hurled from the jet. Typical drop sizes are on the order of
50 picoliters. After turning off the heating pulse, the bubble of gas collapses,
returning the printing head to its original condition. Actuation frequencies
in the kilohertz range are achievable.

6.9.3 Microvalves

A pneumatic valve, produced using conventional fine mechanics methods but
which can be regarded as a microvalve with respect to its power consumption,
is being produced and sold by the German company Hoerbiger [194]. The core
of the 3/2–way diverter valve is a piezoelectric bending transducer that closes
either the compressed air inlet or the air–bleed nozzle depending on the end
position, thereby controlling the pressure in the chamber connected to the
outlet (Fig. 6.94). The piezoactuator enables the valve to be switched within
2 ms without drawing power. The valve is available with working pressures up
to 200 kPa and has a nominal throughput of 1.5 l/min. It can be implemented
either as an on–off valve or as a proportional valve. Additional companies
(Bürkert, Joucomatic) are following the trend of powerless pneumatics and
offering similar valves based on piezoelectric bending actuators.

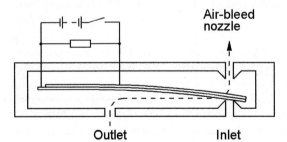

Fig. 6.94. Piezovalve according to [194] (dimensions: $30 \times 19 \times 8\,\mathrm{mm}^3$)

The American company IC–Sensors offers a thermomechanically driven
microvalve (with 2/2–way functionality) for gases (Fig. 6.95) [195]. The valve
consists of an elastically suspended valve reed and a rigid valve base with
a valve opening. The valve reed can be put into motion using the thermomechanical (bimetallic) effect. The temperature of the bimetallic structure

of aluminium and silicon is used to control the actuator force. The normally closed valves are designed for a maximum pressure of 100–200 kPa. Gas flows of up to 0.1 l/min are achieved with a driving power of 300 mW.

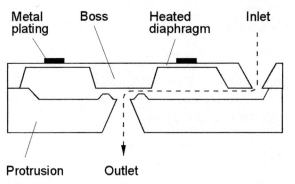

Fig. 6.95. Thermomechanical valve [195]

The American company Redwood MicroSystems offers thermopneumatically operated microvalves for gases. Both normally open and normally closed models are available. The basic structure consists of three components (Fig. 6.96). A drive chamber etched into silicon and filled with a liquid is bordered on one side by a thin silicon diaphragm and on the other side by a rigid glass cover plate (Pyrex). Heating of the chamber causes the liquid to vaporize and displace the diaphragm [196]. A valve opening located in a third component is opened or closed depending on the position of the diaphragm. The normally open valve variation can be converted to a normally closed variation with appropriate adaptation of the mechanics. The valves are designed for a maximum pressure of 700 kPa. A driving power of 1.5 W can control gas flows up to 1.5 l/min. A particle filter with a pore size of 10 μm must be used. The operating temperatures are limited to the range 0–55°C, and the switching times are typically of the order of one second.

In an approach pursued by the Bosch company, a valve reed in a normally closed design is opened by electrostatic forces (Fig. 6.97). Pressures of the order of 10 kPa and a throughput of 0.2 l/min are controlled with an operating voltage of 200 V [197]. In another approach realized by Hitachi, a thin conductive film is electrostatically set into motion over an opening (Fig. 6.98). The valve with 3/2–way functionality can be activated to control a maximum pressure of 30 kPa, and it requires nearly zero power at an operating voltage of 200 V [198].

A thermally driven bimetallic valve was developed at the Institute for Microtechnology and Information Engineering (IMIT) of the Hahn–Schickard Society, Villingen–Schwenningen, Germany [192]. The valve is suitable for use with both liquids and gases (Fig. 6.99). The valve consists of two components, a flexible valve reed and a valve seat. Placement of the valve seat on a flexible

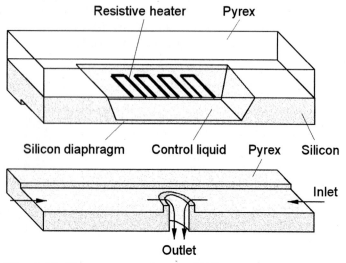

Fig. 6.96. Thermopneumatic valve [196]

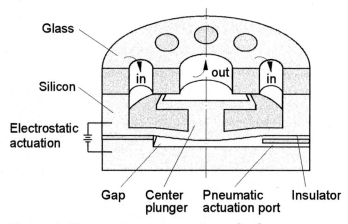

Fig. 6.97. Electrostatic valve according to [197]

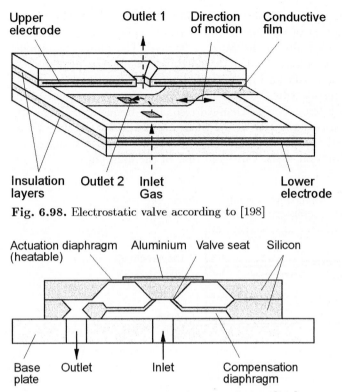

Fig. 6.98. Electrostatic valve according to [198]

Fig. 6.99. Thermomechanical valve according to [192]

compensating diaphragm decouples the required actuation force from the input pressure. The valve was designed for a maximum pressure of 100 kPa. The maximum throughput is 0.5 l/min for gases and 1 ml/min for liquids. The electrical power consumption is 1 W.

6.9.4 Micropumps

A second topic paralleling microvalves is the development of micro dosing elements and micropumps. The developments are concentrated primarily on the miniaturized diaphragm pump. These micropumps normally consist of a displacement diaphragm driven periodically using piezoelectric, thermal or electrostatic principles, and two passive check valves that direct the flow of liquid from the inlet to the outlet.

The electrostatically driven micropump displayed in Fig. 6.100 was developed at the Fraunhofer Institute for Solid–State Technology (IFT) in Munich. Maximum pumping rates of 1 ml/min and a maximum hydrostatic counterpressure of 30 kPa can be achieved with this device [199]. The external dimensions of the pump are $7 \times 7 \times 2\,\text{mm}^3$. The electrical drive signal is composed of a pulsed DC voltage with an amplitude of 200 V. The electrical power consumption of the pump unit depends upon the operating frequency, which lies typically in the range 1–20 mW. The pump is normally operated at frequencies between 1 and 1000 Hz. A volumetric displacement of approximately 0.01–0.05 mm³ is achieved in each cycle. Filters with a pore width of 5 μm are implemented to prevent contamination. An increase of the operating frequency above the mechanical resonance frequency of the valve (2000–6000 Hz) causes a reversal of the pumping direction; thus the pump can be implemented as a bi–directional unit. This effect results from a phase shift between the motion of the valve and that of the fluid [200].

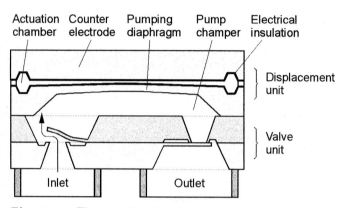

Fig. 6.100. Electrostatic micropump

Micropumps with flow nozzles fulfilling a rectifying function do not require a check valve [201]. A micropump driven by piezoelectrics was developed at the Technical University of Ilmenau. The privileged direction of flow is determined by two pyramid–shaped diffusers etched into silicon (Fig. 6.101). Due to their geometry, these diffusers exhibit different flow resistances for each direction at higher flow speeds (Reynolds number > 100). This characteristic enables alternating flows to be rectified through a 'two paces forward, one pace back' principle. With valve channel widths between 80 μm and 300 μm depending upon the type, these pumps are also less susceptible to contamination than those with check valves. Maximum pump rates of 400 μl/min and a maximum hydrostatic counter pressure of 7 kPa were achieved with a watery solution for a unit measuring 7 × 7 × 1 mm³. Pump rates of 1–10 μl/min are achievable with gases.

At Chalmers University in Stockholm, the nozzles are etched laterally into the silicon. A maximum hydrostatic counter pressure of 25 kPa (for a pump with larger external dimensions) was measured following an optimization of the flare angle [202]. The pumps are suitable for direct feed of fluids and gases. Care is to be taken when shutting off valveless micropumps be-

Vibration diaphragm with piezo-   Pump chamber
electric bimorph type actuator

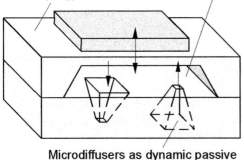

Microdiffusers as dynamic passive
valves at the inlet and outlet

Fig. 6.101. Valveless micropump. After [201]

cause the transported medium will flow back in the presence of a hydrostatic counterpressure.

An infusion pump for painkillers was developed at Trinity College in Dublin [203]. The pump will be worn on a wristwatch and is based on an electrochemical form of actuation. The flow through an electrochemical cell generates a gas and a corresponding pneumatic pressure. The pressure displaces the medication stored in a compressible reservoir (10 ml). The pump is currently undergoing clinical testing on patients receiving painkilling medication.

6.9.5 Microfluid Systems

In addition to valves and pumps, microtechnologies make available other modular fluid–flow components such as flow sensors, micro–mixers and reaction chambers. Customized fluid systems can now be produced solely on the basis of these modular components (microsystem≡adaptronic system, see Sect. 6.9). Typical applications are microanalysis systems and microdosing systems, for example for dosing medications, chemical reagents, lubricants and adhesives.

A microsystem for analysing water (Fig. 6.102) is being developed within the scope of a BMBF–funded joint project (VIMAS) under the leadership of the Fraunhofer Institute for Solid–State Technology. Using appropriate sensors, this system determines environmentally relevant parameters (concentrations of nitrates, oxygen and carbolic acid; pH values; opaqueness). The dimensions of the base plate are $31 \times 32\,\mathrm{mm}^2$.

Many modular designs of microanalytic systems are under study at the Research Centre of Karlsruhe. A system for detecting heavy metals in liquids consists of optochemical sensors, a microspectrometer and a fluid–handling system including a thermopneumatically operated micropump. As another example, an electrochemical microanalytic system (ELMAS) was developed on the basis of ISFET sensors for analysing body fluids.

Miniaturized lubricating systems are under development at the IMIT. The first application will be for improving the 'wick' lubricating process. In this process, a film of oil is carried by capillary action from a container to the part to be lubricated, typically a rotating part. Problems can sometimes arise when undesired excess lubrication and strongly varying oil consumption result from varying rotating speeds. A microsystem consisting of the dosing pump (as presented in Fig. 6.103), a microbuffer (volume $< 5\,\mathrm{mm}^3$) and an oil sensor offers a viable solution [204]. The oil sensor measures the level in the buffer and the pump provides the lubrication as needed.

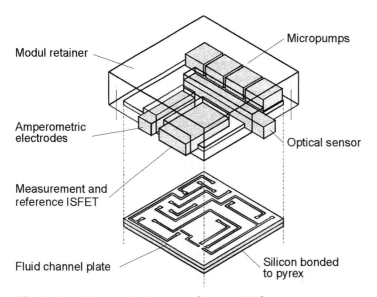

Fig. 6.102. Microanalysis system [source: IFT]

Dosing of the smallest quantities of liquid on the order of nanoliters and microliters was the goal in the cooperation between the Research Centre of Rossendorf and the GeSiM company of Dresden in the development of a microdrop injector [205]. The unit consists of a micro–injection pump (MEP) and a microsieve functioning as a diode for liquids (Fig. 6.104). The piezoelectrically driven injection pump functions similar to an ink–jet printer head, applying microdrops to the sieve. These droplets mix themselves with the liquid located below through surface tension. The microsieve makes use of surface tension effects to prevent the carrier liquid from soaking through into the injection chamber containing air. The disadvantage of half–opened systems is offset by the advantageous ideal liquid separation between the injection and carrier liquids by the air/sieve interface. Applications for this unit can be found in the fields of chemical sensing, pharmacy, medicine and biotechnology.

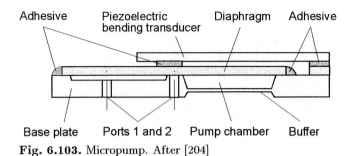

Fig. 6.103. Micropump. After [204]

6.9.6 Other Microactuators

Adjustment of Optical Waveguides. An integrated microsystem, designed at the Technical University of Ilmenau in Germany, is able to handle the tight mechanical tolerances of mono–mode waveguide couplings by a controlled adjustment. It basically contains a two–axis microactuator for moving a fibre or a microlens, an optical sensor for position detection and a control circuit. The piezoelectric drive has a bimorph cantilever movable normal to the wafer surface [206]. Its second direction, the in–plane movement, employs a compliant mechanism in order to enlarge the very small strains of a piezoelectric monomorph. It contains a set of elastic hinges arranged as two–stage gear. Fig. 6.105 shows the structure and the kinematic principle of the compliant gear.

The position sensor has the task of detecting any relative misalignment and providing a respective signal to the control circuitry. Several principles have been tested. The most promising one is the cladding–mode sensor that measures, by a photodetector, the intensity of the light that is not coupled into the core but into the cladding. Further development is especially aimed toward system miniaturization and integration. For instance, the fabrication of the position detector and the microactuator on a single substrate seems feasible. Also, the possibility is under consideration of analog control circuitry instead of the currently applied one–chip microprocessor in order to further minimize the volume occupied.

Micromotor with Integrated Micro Gear Box. The Institut für Mikrotechnik, Mainz in Germany, has developed a highly reliable micromotor [207]. A synchronous motor scheme with a rotating permanent magnet has been employed (Fig. 6.106 a). The micromotor featuring an outer diameter of 1.9 mm exhibits a maximum measured torque of 5 μNm in continuous operation. The lifetime is considerably longer than 6 months at 10 000 rpm. A micro gear

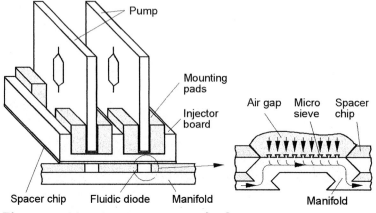

Fig. 6.104. Microdrop injector. After [205]

box of the Wolfrom type has been developed using individually modified in-
volute tooth profiles, whose components are fabricated in metal and polymer
materials by means of the LIGA process (Fig. 6.106 b). The motor, with integrated gear box, increases the available torque and helps to open new fields
of application — for example, communication and information technology as
well as consumer electronics.

Hybrid concepts make use of the most suitable material and the most
appropriate process in the fabrication of each component. Such a 'hetero-
morphic' construction is typical of many microsystems and also demonstrates
a broad need for efficient construction, connection and microassembly techniques, and standardized electrical and mechanical interfaces.

Electrostatic Linear Actuator. In the linear actuator depicted in Fig.
6.107, the slide moves over the stator supported by air. Electrostatic forces are
generated between comb–like or striped electrodes located on the opposing
surfaces of the stator and slide, causing motion of the slide along the x–
axis, binding in the y direction and attraction in the z direction. Additional
electrodes act as sensors for determining the position in the x direction and
the distance of separation in the z direction. All structures are sputtered onto
a glass substrate using conventional methods.

The actuator is a 3–phase stepping motor represented by an open control
loop. This actuator has a range of displacement in the x direction of 25 mm
with a positioning uncertainty of 5 μm. It can also perform small rotations
φ_x, φ_y. The holding force can reach 50 mN and the maximum displacement
speed 50 mm/s. The electrostatic actuator offers a graphic representation
of the trend toward 'milliactuators': this species of actuator is constructed
using microtechnologies but generates forces and displacements on a more
macroscopic scale.

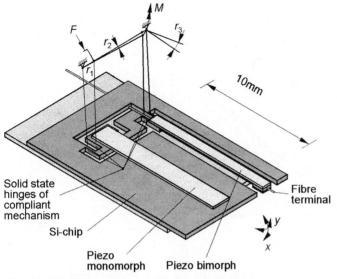

Fig. 6.105. Piezoactuator with compliant gear

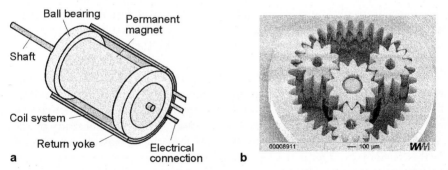

Fig. 6.106. Micromotor with integrated gear box: **(a)** schematic of the micromotor; **(b)** assembled 'planetary' gear system [207]

6.9.7 Conclusion and Outlook

The market outlook appears to be particularly favorable for microvalves. Microvalves driven by piezoelectric bending transducers or thermal principles are already being produced and sold by companies. The possible applications of these valves will increase when they can be mass–produced inexpensively and operated powerlessly. Stimulus can be expected particularly from valves with 3/2–way functionality, to be introduced as pilot valves in many areas of automation.

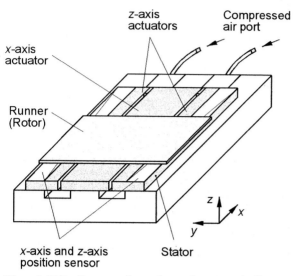

Fig. 6.107. Construction of an electrostatic linear actuator based on micro-technologies. (Source: PASIM Mikrosystemtechnik, Suhl in Germany)

Micropumps for transporting and dosing small liquid quantities represent another main group of microfluid actuators. The fact that these pumps are not capable of producing suction currently impedes their use in industry, but should be solvable in the future.

A group exhibiting particular prominence among microactuators currently under development is the group of microactuators used for controlling fluids. Based on the current state of development, this branch of microactuator technology is closest to industrial application. A 1994 market study by the American System Planning Corporation stresses the worldwide significance of this technology. The market potential for elements used to control fluids was studied according to the sectors: infusion pumps for use in medicine; industrial valves; micromechanical valves; flow sensors; and ink–jet printer heads. A market volume of US $2.6 billion for individual components and US $ 6.5 billion for related systems is predicted for the year 2000.

References

1. Janocha, H. (Hrsg.): *Aktoren — Grundlagen und Anwendungen* . Springer–Verlag Berlin Heidelberg 1992
2. Schäfer, J.: Design magnetostriktiver Aktoren. Dissertation, Universität des Saarlandes, 1994
3. Janocha, H.; Schäfer, J.: Compensation of hysteresis in solid state actuators. Sensors and Actuators. **A 49**, 1995, pp. 97–102

4. Mücklich, F.; Janocha, H.: Smart Materials — The 'IQ' of Materials. Systems. Z. Metallkd. **87**, (1996) 5, pp. 357–364

5. Queensgate Instruments: *Nano Positioning Book.* Berkshire, UK (1997)

6. Koch, J.: *Piezoxide.* Valvo (1988)

7. Sundar, V.; Newnham, R.E.: Electrostriction and Polarization. Ferroelectrics **135** (1992), pp.431–446

8. Zickgraf, B.: Ermüdungsverhalten von Multilayer–Aktoren aus Piezokeramik. VDI Fortschrittsberichte, VDI–Verlag Düsseldorf (1996)

9. Stettner GmbH & Co; Techn. Prosp. Piezoceramic Components. Lauf, Germany (1996)

10. Nelson, J.G.; Neurgaonkar, R.R.; Oliver J.R.; Dewing, D.C.; Larson, C.; Dobbi, S.K. and Rosenthal, J.S.: Piezoelectric Technology: Research and Applications. Proc. SPIE **3324**, San Diego, USA (1998)

11. Gebhardt, W.: Ultrasonic Measurement of the Exhaust Gas Mass Flow. Proc. ISATA, Florence, Italy, (1997)

12. Physik Instrumente: Techn. Prosp. Products for Micropositioning. Waldbronn, Germany (1998)

13. Dyberg, J.: Magnetostrictive Rods in Mechanical Applications. Proc. First Int. Conf. on Giant Magnetostrictive Alloys, Marbella, Spain (1986), 192–214

14. Tichý, J.; Gautschi, G.: *Piezoelektrische Meßtechnik.* Springer, Berlin (1980)

15. Vishnewsky, W.; Glöß, R.: Piezoelectric Rotary Motors. Proc. Actuator 96, Bremen, Germany (1996), pp.245–248

16. Uchino, K.: *Piezoelectric Actuators and Ultrasonic Motors.* Kluwer, Boston (1997)

17. Wehrsdorfer, E.; Borchardt, G.; Pertsch, P. and Karthe, W.: Vibration Drive. Proc. Actuator 96, Bremen, Germany (1996), pp.209–212

18. Burleigh Instruments: Techn. Prosp. Micropositioning Systems. NY, USA (1989).

19. Clephas, B.; Janocha, H.: Extended Performance of Hybrid Actuators. Proc. 4th European Conference on Smart Structures and Materials, Harrogate, UK (1998), 81–88

20. Spanner, K.: Piezo Actuators Move in the Nanometer Regime. Laser Focus World **32**, (1996), 161–168

21. Janocha, H.; Clephas, B.; Kuhnen, K.: Inherent Sensory Capabilities of Solid State Actuators. Proc. Euromech., Magdeburg, Germany (1998)

22. Chen, P.C.; Chopra, I.: Induced Strain Actuation of Composite Beams and Rotor Blades with Embedded Piezoceramic Elements. Smart Mater. Struct. **5** (1996), pp.35–48

23. Tzou, H.S.; Bao, Y.: Modeling of Thick Anisotropic Composite Triclinic Piezoelectric Shell Transducer Laminates. Smart Mater. Struct. **3** (1994), pp.285–292

24. Takigami, T.; Oshima, K. and Yoshikazu, H.: Application of Self–Sensing Actuator to Control of a Cantilever Beam. Proc. American Control Conference, Albuquerque, USA (1997), pp.1867–1872

25. Jones L.D.; Garcia E.: Novel Approach to Self–Sensing Actuation. Proc. SPIE **3041**, San Diego, USA, (1997), pp.305–314

26. Bozorth, R.M.: *Ferromagnetism.* 2nd ed., Ed. Van Nostrand, NY, 1951, p. 980.

27. De Lacheisserie E.: *Magnetostriction: Theory and applications.* Ed. CRC Press, USA, 1993, p. 410

28. Clark A.E.: *Magnetostrictive rare earth–Fe_2 compounds, Ferromagnetic materials.* Ed.E.P. Wohlfarth, US ,Tome 1, 1980, pp. 531–588

29. Verhoeven J.D.: The effect of composition and magnetic heat treatment on the magnetostriction of TbDyFe twinned single crystals. J. Appl. Phys. **66** (2), 1989. pp. 772–779

30. Sandlung L.: Magnetostrictive powder composite with high frequency performance. Proc.Third int. Workshop on power transducers for sonics and ultrasonics, Ed.Springer, Fl, May 6–8, 1992, pp. 113–120
31. Quandt E.: Magnetostrictive thin film actuators. Proc. Actuator 94, Ed.Axon (Bremen, G.), pp. 229–232
32. Claeyssen F.: Giant dynamic magnetostrain in rare earth–iron magnetostrictive materials, IEEE Trans. MAG.27, N 6, Nov.1991, pp. 5343–5345
33. Moffett M.B.: Characterization of Terfenol–D for magnetostrictive transducers. JASA, **89** (3),1991, pp. 1448–1455
34. Kvarnsjö L.: *On characterisation, modelling and application of highly magnetostrictive materials.* Doct. Thesis TRITA–EEA–9301, ISSN 1100–1593, 1993, p. 173
35. Body C.: Non linear finite element modelling of magneto–mechanical phenomenon in giant magnetostrictive thin films, Proc.CEFC 1996
36. Engdhal G.: Loss simulations in magnetostriction actuators. J.Appl.Phys. **79** (8), 1996
37. ATILA – A 3D CAD software for piezoelectric and magnetostrictive structures. ISEN, Lille (F), Distr. CEDRAT, Meylan (F) & MAGSOFT, Troy NY (US)
38. Debus, J.C.: Finite element modeling of PMN electrostrictive materials. Proc. Int. Conf. on Intelligent Materials, ICIM96–ECSSM96, SPIE vol. 2779, Lyon (F), 1996, pp. 913–916
39. Claeyssen, F.: Design and building of low–frequency sonar transducers based on Rare Earth Iron magnetostrictive alloys. Doct. Thesis, Ed. Defence Research Inform.Cent. Ed. HSMO, MoD, London, also Conception et réalisation de transducteurs sonar basse fréquence à base d'alliages magnétostrictifs Terres Rares–Fer, Thèse INSA Lyon: 1989, p. 414
40. Claeyssen, F.: Modeling and characterization of the magnetostrictive coupling, Proc.2nd int. Workshop on power transducers for sonics and ultrasonics, Ed.Springer, 1990, pp. 132–151
41. Claeyssen, F.: Giant Magnetostrictive Alloys Actuators. Proc. Magnetoelastic Effects and Applications Conf. Ed. L.Lanotte, Pub. Elsevier, Holland, 1993, pp. 153–159, or Journal of Applied Electromagnetics in Materials **5** (1994) pp. 67–73
42. Bouchilloux, P.: Dynamic Shear Characterization in a Magnetostrictive Rare Earth–Iron Alloy, M.R.S. Symp. Proc., **360**, 1994, pp. 265–272
43. Bossut, R.: Finite element modeling of magnetostrictive transducers using Atila. Proc. ATILA conf joint with 2nd int. Workshop on power transducers for sonics and ultrasonics, Ed.B.F.Hamonic, Isen, Lille(F), 1990, pp. 19–26
44. Claeyssen, F.: Progress in magnetostrictive sonar transducers. Proc. UDT93, Ed.Reed Exhib. UK, 1993, pp. 246–250
45. Le Letty, R.: Combined finite element — normal mode expansion methods for ultrasonic motor modeling. IEEE Ultrasonic Symp. Proc, 1994, pp. 531–534
46. Claeyssen, F.: Analysis of magnetostrictive Inchworm motors using f.e.m.. Proc. Magnetoelastic Effects and Applications Conf. Ed. L.Lanotte, Pub. Elsevier, Holland, 1993, pp. 161–167
47. Claeyssen, F.: State of the art in the field of magnetostrictive actuators. Proc. Actuator 94 conf., Ed. Axon, Bremen (G), 1994, pp. 203–209
48. ETREMA Terfenol–D Magnetostrictive Actuators Information. Etrema Products, USA, 1993, p. 6
49. Eda, H.: Ultra precise machine tool with GMA. Annals of the CIRP Vol.41/1, 1992, pp. 421–424
50. Wang, W.: A high precision micropositioner based on magnetostriction principle. Rev.Sci.Inst. **63** (1), Jan. 1992, pp. 249–254

51. Cedell, T.: New magnetostrictive alloy for rapid conversion of electric energy to mechanical motion. Proc. Actuators 90, Ed.Axon (Bremen, G.), 1990, pp. 156–161

52. Suzuki, K.: Magnetostrictive plunger pump. Int. symp. on GMA&A, Japan, Poster 5, 1992, p. 6

53. Hiller, M.W.: Attenuation and transformation of vibration through active control of magnetostrictive Terfenol. J.Sound Vib., **133** (3),Pap. 364/1,1989, p. 13

54. Janocha, H.: Design criteria for the application of solid state actuators. Proc. Actuator 94 conf., Ed. Axon, Bremen (G), 1994, pp. 246–250

55. Giurgiutiu, V.: Solid–state actuation of Rotor Blade Servo–Flap for Active vibration control. J. of Int. Mat. Systems and structures, **7**, 1996, pp. 192–202

56. Lhermet, N.: Actuators based on biased magnetostrictive Rare Earth–Iron Alloys, Proc. Actuator 92, Ed.Axon (Bremen, G.), 1992, pp. 133–137

57. FLUX2D — A 2D CAD software for electric engineering. LEG, Grenoble (F), Distr. CEDRAT, Meylan (F) & MAGSOFT, Troy NY (US)

58. Claeyssen, F.: Design of Lanthanide magnetostrictive sonar projectors. Proc. UDT91, Microwave Exh.&Pub.Ltd, 1991, pp. 1059–1065

59. Moffett, M.B.: Comparison of Terfenol–D and PZT4 power limitations. J. Acoust. Soc. Am. **90** (2), Lettters to editor, August 1991, pp. 1184–1185

60. Jones, D.F.: Recent transduction developments in Canada and US. Sonar Transducers'95 Conf., Proc. Inst. Of Acoustics, Ed. Univ of Bath, UK, **17**, Pt 3, 1995, pp. 100–106

61. Wise, R.J.: Ultrasonic welding of plastics by a Terfenol driven magnetostrictive transducer. Proc of 3rd Conf. on Welding and Adhesive Bonding of Plastic, Dusseldorf, Nov.1992, pp. 10–12

62. Dubus, B.: Low frequency Magnetostrictive Projectors for Oceanography and Sonar. Proc. 3rd Europ. conf. on underwater acoustics, 1996, pp.1019–1024

63. Kiesewetter, L.: Terfenol in linear motor. Proc. of 2nd. int. conf. on GMA, Ed. C.Tyren, Amter (Fr),1988. Ch.7, p.15

64. Vranish, J.M.: Magnetostrictive direct drive rotary motor development. IEEE Trans. MAG.27, **6**, Nov.1991, pp.5355–5357

65. Mori, K.: European Patent, N 0155694A2, 1985

66. Akuta, T.: Rotational–type actuators with Terfenol–D rods. Proc. Actuators 92, Ed. Axon (Bremen, G.), 1992, pp.244–248

67. Akuta, T.: Improved Rotational–type actuators with Terfenol–D rods. Proc. Actuator 94, Ed. Axon (Bremen, G.), 1994, pp.272–274

68. Claeyssen, F.: Design and construction of a new resonant Magnetostrictive Motor. Proc. Intermag 96, Seattle, WA, June 1996, IEEE Trans. MAG also Proc. Actuators 96, Ed. Axon (Bremen, G.), 1996, pp.172–274

69. Claeyssen, F.: A new multi–mode Piezo–electric Motor. Proc. Int. Conf. on Intelligent Materials, ICIM96–ECSSM96, SPIE vol. 2779 , Lyon (F), 1996, pp.634–637 also Proc. Actuators 96, Ed. Axon (Bremen, G.), 1996, pp.152–155

70. Fukuda, T.: GMA applications to micromobile robot as microactuator without power supply cables. IEEE Micro–electro–mechanical Systems proc.1991, pp.210–215

71. Honda, T.: Fabrication of Magnetostrictive Actuators Using Rare–Earth (Tb,Sm) — Fe Thin Films. J. Appl. Phys. **76** (10), 15 Nov 1994, pp.6994–6999

72. Betz, J.: Torsion based, drift–free magnetostrictive microactuator. Proc. Actuators 96, Ed. Axon (Bremen, G.), 1996, pp.283–286

73. Claeyssen, F.: Micromotors Using Magnetostrictive Thin Films. Proc. of SPIE Conf. Smart structures and Materials, San Diego, US, March 98

234   References

74. Duerig, T.W.; Melton, K.N.; Stöckel, D.; Wayman, C.M.: *Engineering Aspects of Shape Memory Alloys*. Butterworth–Heinemann Ltd., London, 1990
75. Funakubo, H. (Ed.): *Shape Memory Alloys*. Gordon and Breach Science Publishers, New York, 1984
76. Krämer, J.: Formgedächtnislegierungen in der Automobiltechnik und im Maschinenbau. Lecture notes of the course *Konstruieren mit Formgedächtnislegierungen*, Technische Akademie Esslingen, course no. 17964/50.051, Ostfildern, Feb. 21–22, 1994, pp.79–111 (in German)
77. Russel, S.; Sczerzenie, F; Clapp, P.: Engineering Considerations in the Application of NiTiHf and NiAl as Practical High–Temperature Shape Memory Alloys. Proc. Int. Conf. on Shape Memory and Superelastic Technologies, March 7–10, 1994, Pacific Grove, CA, USA, pp.43–48
78. Touminen, S. M.: High Transformation Temperature Ni–Ti–Hf Alloys. Proc. Int. Conf. on Shape Memory and Superelastic Technologies, March 7–10, 1994, Pacific Grove, CA, USA, pp.49–54
79. Furuya, Y.; Matsumoto, M.; Matsumoto, T.: Mechanical Properties and Microstructure of Rapidly Solidified TiNiCu–Alloy. Proc. Int. Conf. on Shape Memory and Superelasticity, March 7–10, 1994, Pacific Grove, CA, USA, pp.905–910
80. Holleck, H.; Kirchner, S.; Quandt, E.; Schlossmacher, P.: Preparation and Characterization of TiNi SMA Thin Films. Proc. 4th Int. Conf. on New Actuators, June 15–17, 1994, Bremen, Germany, pp.361–364
81. Ishida, A.; Takei, A.; Miyazaki, S.: Formation of Ti–Ni Shape Memory Films by Sputtering Method. Proc. Int. Conf. Martensitic Transformations ICOMAT–92, July 20–24, 1992, Monterey, CA, USA, pp.987–992
82. Kristen, M.: *Untersuchungen zur elektrischen Ansteuerung von Formgedächtnis-Antrieben in der Handhabungstechnik*. Braunschweiger Schriften zur Mechanik, No. 15–1994, Mechanik–Zentrum der TU Braunschweig, 1994 (in German)
83. Hesselbach, J.; Stork, H.: Simulation and Control of Shape Memory Actuators. Proc. 5th Int. Conf. on New Actuators, June 26–28, 1996, Bremen, Germany, pp.396–399
84. Hesselbach, J.; Hornbogen, E.; Mertmann, M.; Pittschellis, R.; Stork, H.: Optimization and Control of Electrically Heated Shape Memory Actuators. Proc. 4th Int. Conf. on New Actuators, June 15–17, 1994, Bremen, Germany, pp.337–340
85. Hesselbach, J.; Kristen, M.: Shape Memory Actuators as Electrically Controlled Positioning Elements. Proc. 3rd Int. Conf. on New Actuators, June 24–26, 1992, Bremen, Germany, pp.85–91
86. Hesselbach, J.; Pittschellis, R.: Greifer für die Mikromontage. Werkstattstechnik — Production und Management, wt 85 (1995), pp.595–600 (in German)
87. Hesselbach, J.; Pittschellis, R.; Thoben, R.; Oh, H. S.: Handhabungsgeräte für die Mikromontage. Zeitschrift für wirtschaftliche Fertigung, ZWF 91 (1996) 9, pp.437–440 (in German)
88. Ikuta, K.; Beard, D.C.; Moiin, H.: Direct Stiffness and Force Control of a Shape Memory Alloy Actuator and Application to Miniature Clean Gripper. Robotics Research 1989, The Winter Annual Meeting of the ASME, San Francisco, CA, USA, Dec. 10–15, 1989, Vol. **DSC**–14 (1989), pp.241–246
89. Pritschow, G.; Kehl, G.; Hesselbach, J.; Stork, H.: Integration of Shape Memory Actuators in Miniature Robots. Proc. 5th Int. Conf. on New Actuators, June 26–28, 1996, Bremen, Germany, pp.405–408
90. Johnson, D.; Busch, J.: Recent Progress in the Application of Thin Film Shape Memory Alloys. Proc. Int. Conf. on Shape Memory and Superelastic Technologies, March 7–10, 1994, Pacific Grove, CA, USA, pp.299–304

91. Pickard, W.F.: Electrical Force Effects in Dielectric Liquids. Progress in Dielectrics. Vol. **6**, Temple Press, London, (1965) p.3

92. Bullough, W.A. and Stringer, J.D.: The Incorporation of the Electroviscous Effect in a Fluid Power System. Proceedings of the 3rd Int. Fluid Power Symposium, Turin, B.H.R.A. pp.F3–37 (May 1973)

93. Winslow, W.M.: Induced Fibration of Suspensions. Journal of Applied Physics. Vol. **20**, p.1137 (December 1949)

94. Whittle, M.; Peel, D.J.; Firoozian, R. and Bullough, W.A.: Dependence of E R. Response Time on Conductivity and Polarisation Time. Phys. Rev. E. Pt6A, Vol. **49**, pp.5249–5259 (1994)

95. Philips, R.W. and Auslander, D.M.: The Electro Plastic Flow Modulator. source unknown (May 1971)

96. Tsukiji, T. and Utashiro, T.: Flow Characterisation of E R. Fluids Between Two Parallel Plate Electrodes. Proceedings ASME Int. Congress and Expo, San Francisco, Developments in Electrorheological flows FED Vol. **235**, MD Vol. **71**, pp.37–42 (Nov. 1996)

97. Bullough, W.A. and Foxon, M.B.: The Application of an Electroviscous Damper to a Vehicle Suspension System. Proceedings of the Third International Conference on Vehicle System Dynamics, Blacksburg, Virginia, Swets and Zeitlinger, Amsterdam, p.144 (August 1984)

98. Janocha, H.; Rech, B. and Bölter, R.: Practice–Relevant Aspects of Constructing ER. Actuators. Proc. 5th Int. Conf. on ERF/MRS held SMMART Sheffield. World Scientific Publ. pp.435–447. (July 1995)

99. Atkin, R.; Xiao, S. and Bullough, W.A.: Solutions of the Constitutive Equations for the Flow of an Electro–Rheological Fluid in Radial Configurations. Jnl. of Rheology, Vol. **35**, pp.1441–1461 (1991)

100. Winslow, W.M.: Field Responsive Force Transmitting Compositions. United States Patent Specification No. 3, 047, 507, (July 31st 1962)

101. Whittle, M.; Bullough, W.A.; Peel, D.J. and Firoozian, R.: Electrorheological Dynamics Derived from Pressure Response Experiments in the Flow Mode. Jnl. Non Newtonian Fluid Mech., Vol. **57**, No. 1, pp.1–25 (April 1995)

102. Bullough, W.A.; et al.: The Electro–rheological Clutch: Design, Performance, Characterisation and Operation. Proc. I Mech E. Vol. **207**, pp.82–95

103. Whittle, M.; Atkin, R. and Bullough, W.A.: Dynamics of an Electro Rheological Valve. Proc. 5th Int. Conf. on ERF/MRS held SMMART Sheffield. World Scientific Publ. pp.100–117 (July 1995)

104. Whittle, M.; Firoozian, R.; Peel, D.J. and Bullough, W.A.: A Model for the Electrical Characterisation of an ER Valve. Intl. Jnl. Mod. Phys. B., Vol. **6**, No 15+16, pp.109–140. World Scientific Publishers

105. Melrose, J.; Itoh, S.I. and Ball, R.C.: Simulations of ERF with Hydrodynamic Lubrication. Proc 5th Int. Conf. ERF/MRS, held SMMART. World Scientific Publ. pp.404–410 (July 1995)

106. Wu, C.W.; Chen, Y.; Tang, X. and Conrad, H.: Conductivity and Force between particles in a model ERF. I–Conductivity II – Force. ibid pp.525–536

107. Boissy, C.; Atten, P. and Foulc, J.N.: The Conduction Model of Electro Rheological Effect Revisited. ibid pp.156–165 — see also pp.710–726 and pp.756–763 by same authors

108. Peel, D.J.; Stanway, R. and Bullough, W.A.: The Generalised Presentation of Valve and Clutch Data for an ER Fluid and Practical Performance Prediction Methodology. ibid pp.279–290

109. Tozer, R.; Orrell, C.T. and Bullough, W.A.: On–Off Excitation Switch for E.R. Devices. Int. Jnl. Mod. Phys. B. Vol **8**, No. 20 and 21. pp.3005–3014 (1994)

110. Whittle, M.; Atkin, R. and Bullough; W.A.: Fluid Dynamic Limitations on the Performance of an ER Clutch. Jnl. Non Newtonian Fluid Mech. Vol **57** No. 1 April 1995, pp.61–81

111. Sianaki, A.H.; Bullough, W.A.; Tozer, R. and Whittle, M.: Experimental Investigation into Electrical Modelling of Electro–rheological Fluid Shear Mode. Proc. I.E.E, Sci. Meast. & Tech. Vol. **141**, No. 6, pp.531–537

112. Bullough, W.A.; Makin, J.; Johnson, A.R.; Firoozian, R. and Sianaki, A.H.: ERF Shear Mode Characteristics: Volume Fraction, Shear Rate and Time Effects. Trans. ASME, Jnl. Dyn. Syst. Meast. and Control pp. 221–225 (June 1996)

113. Johnson, A.R.; Makin, J. and Bullough, W.A.: E. R. Catch/Clutch Simulations. Int. Jnl. Mod. Phys. B. Vol. **8**, No. 20 and 21, pp.2935–2954

114. Wolfe, C. and Wendt, E.: Application of ERF in Hydraulic Systems. Proc. AXON–VDI/E Actuator'94 Conference, Bremen, pp.284–287

115. Bloodworth, R. and Wendt, E.: Materials for ER Fluids. Proc. 5th Int. Conf. ERF/MRS, held SMMART Sheffield 1995. World Scientific Publ. pp.118–131 (July 1995)

116. Bullough, W.A.; Makin, J. and Johnson, A.R.: Requirements and Targets for ER Fluids in Electrically Flexible High Speed Power Transmission. Am. Chem. Soc. Fall Meeting Washington DC. Plenum Pub p.295–302 (1995)

117. Naem, A.S.; Stanway, R.; Sproston, J.L. and Bullough, W.A.: A Strategy for Adaptive Damping in Vehicle Primary Suspension Systems. 1994 Proc. ASME. Winter Annual Meeting Chicago 1994, pp.395–399

118. Peel, D.J.; Stanway, R. and Bullough, W.A.: A Design Methodology based upon Generalised Fluid Data. Int. Conf. Intel Materials, Lyon, Engineering with ERF. pp.310–316 (1996)

119. Hosseini–Sianaki, A.; Makin, J.; Xiao, S.; Johnson, A.R.; Firoozian, R. and Bullough, W.A.: Proc. Soc. Fluid Power Transmission and Control, 1st Fluid Power Trans. & Control Symp., Beijing, pp.591–595, Operational Considerations in the Use of an Electro–Rheological Catch Device. Beijing Inst. Tech. Press. (1991)

120. Block, H. and Kelly, J.: Electro–rheology. J Phys D **21** (1988) p. 1661.

121. Peel, D.J. and Bullough, W.A.: The Effect of Flowrate, Excitation Level and Solids Content on the Time Response of an Electro–rheological Valve. Jnl. Intel. Matl. Systems and Structures, Vol. **4**, No.1, pp.54–64

122. Dwyer–Joyce, R.; Bullough, W.A. and Lingard, S.: Elastohydrodynamic Performance of Unexcited Electro–Rheological Fluids. Proc 5th Int. Conf. ERF/MRS, held SMMART Sheffield. World Scientific Publ. pp.376–384 (July 1995)

123. Leek, T.H.; Lingard, S.; Bullough, W.A. and Atkin, R.J.: The Time Response of an Electro–rheological Fluid in a Hydrodynamic Film. ibid. pp.551–563

124. Yates, J. R.; Lau, D.S. and Bullough, W.A.: Inertial Materials: Perspective, Review and Future Requirements. Proc. AXON–VDI/E Actuator'94 Conference, Bremen, p.275–278 (1994)

125. Stanway, R.; Sproston, J.L.; Predergast, M.J.; Case, J.C. and Wilne, C.E.: ER fluids in the squeeze–flow mode an application to vibration isolation. Journal of Electrostatics, 1992 **28** p.89–94

126. Winslow, W.M.: Induced Fibration of Suspensions. J. Appl. Phys., **20** pp.1137–1140 (1949)

127. Phillips, R.W.: *Engineering Applications of Fluids with a Variable Yield Stress.* Ph.D. Thesis, University of California, Berkeley (1969)

128. Duclos, T.G.: An Externally Tunable Hydraulic Mount Which Uses ER Fluid. Soc. of Automotive Engineers, SAE Paper # 870963 (1987)

129. Duclos, T.G.: Design of Devices Using Electro–rheological Fluids. Soc. of Automotive Engineers, SAE Paper # 881134 (1988)

130. Carlson, J.D.; Catanzarite, D.M. and St. Clair, K.A.: Commercial Magneto–Rheological Fluid Devices. Proc. 5th Int. Conf. on ER Fluids, MR Fluids and Assoc. Tech., pp.20–28, July 1995

131. Kordonsky, W.: Magnetorheological Effect as a Base of New Devices and Technologies. J. Mag. and Mag. Mat., **122** pp.395–398 (1993)

132. Carlson, J.D. and Weiss, K.D.: A Growing Attraction to Magentic Machine Design, pp.61–66, Aug. 8 (1994)

133. Carlson, J.D.: The Promise of Controllable Fluids. Proc. of Actuator 94, pp.266–270

134. Rabinow, J.: The Magnetic Fluid Clutch. AIEE Transactions, **67** pp.1308–1315 (1948).

135. Magnetic Fluid Clutch. National Bureau of Standards Technical News Bulletin, **32** pp.54–60(#4) (1948)

136. Rabinow, J.: Magnetic Fluid Torque and Force Transmitting Device. U.S. Patent 2, 575, 360 (1951)

137. Vogt, W.: Semi–Active Dampers Innovate Shock Control. OEM Off–Highway, Sept. (1995)

138. Lord Corporation: Rheonetic Magnetic Fluid Systems, Lord Corp. Pub. #PB8003, pp.1–10 (1996)

139. Vogt, W.: Creating A Smoother Ride. Equipment Today, Sept. (1995)

140. Design News Editorial: Brake Cuts Exercise–Equipment Cost. Design News, p.28 Dec. 4 (1995)

141. Lord Corporation: Product Specification Sheet–Rheonetic Rotary Resistance System MRB–2107–3, Lord. Corp. Pub. #DS8030 (1996)

142. Dyke, S.J.; Spencer Jr., B.F.; Sain, M.K. and Carlson, J.D.: Seismic Response Reduction Using Magnetorheological Dampers. Proc. IFAC World Cong., San Francisco, California (1996)

143. Dyke, S.J.; Spencer Jr., B.F.; Sain, M.K. and Carlson, J.D.: Experimental Verification of Semi–Active Structural Control Strategies Using Acceleration Feedback. Proc. 3rd Int. Conf. on Motion and Vibr. Control, Chiba, Japan, **3**, pp.291–296 (1996)

144. Dyke, S.J.; Spencer Jr., B.F.; Sain, M.K. and Carlson, J.D.: Modeling and Control of Magnetorheological Dampers for Seismic Response Reduction. Smart Mat. and Struct., in press

145. Dyke, S.J.; Spencer Jr., B.F.; Sain, M.K. and Carlson, J.D.: Phenomonological Model of a Magnetorheological Damper. J. Engineering Mech., ASCE, in press.

146. Spencer Jr.,B.F.: Recent Trends in Vibration Control in USA. Proc. I3rd Int. Conf. on Motion and Vibr. Control, Chiba, Japan (1996)

147. Carlson, J.D. and Spencer Jr., B.F.: Magneto–Rheological Fluid Dampers for Semi–Active Seismic Control. Proc. I3rd Int. Conf. on Motion and Vibr. Control, Chiba, Japan (1996)

148. Grabiner, M.A.; Cleveland Clinic Foundation, Dept. of Biomed. Engr., Cleveland, Ohio, private communication (1996)

149. Carlson, J.D. and Weiss, K.D.: Magnetorheological Materials Based on Alloy Particles. U.S. Patent No. 5, 382, 373 (1995)

150. Ginder, J. M.; Davis, L.C. and Elie, L.D.: Rheology of Magnetorheological Fluids: Models and Measurements. Proc. 5th Int. Conf. on ER Fluids, MR Suspensions and Assoc. Tech., pp.504–514, July 1995

151. Ginder, J.M.: Rheology Controlled By Magnetic Fields. Encyclopedia of Appl. Phys., **16**, pp.487–503 (1996)

152. Kormann, C.; Laun, L. and Klett, G.: Magnetorheological Fluids with Nano–Sized Particles for Fast Damping Systems. Proc. of Actuator 94 (H. Borgmann and K. Lenz, Eds., AXON Technologie 1994) pp.271–274

153. Weiss, K.D.; Duclos, T.G.; Carlson, J.D.; Chrzan, M.J. and Margida, A.J.: High Strength Magneto– and Electrorheological Fluids. Soc. of Automotive Engineers, SAE Paper #932451 (1993)

154. Lord Corporation, VersaFlo Product Information Sheet, (Lord. Corp. Pub. #PI02–MRX–135CD, 1995)

155. Crosby, M.J.; Harwood, R.A. and Karnopp, D.: Vibration Control Using Semi–Active force Generators. Trans. of ASME, No. 73–DET–122 (1973)

156. Karnopp, D. and Crosby, M.J.: System for Controlling the Transmission of Energy Between spaced Members. U.S.Patent No. 3,807,678 (1974)

157. Ivers, D.E. and Miller, L.R.: Semi–Active Suspension Technology: An Evolutionary View, Lord Corp. Pub. #LL–6005 (1994)

158. Lord Corporation, VersaFlo Controllable Fluids, Lord Corp. Pub. #PL01–2000A, pp.1–8 (1993)

159. Carlson, J.D.: Multi–Degree of Freedom Magnetorheological Devices and Systems Using Same. U.S. Patent No. 5,492,312 (1996)

160. Jolly, M.R. and Carlson, J.D.: Controllable Squeeze Film Damping Using Magnetorheological Fluid. Proc. of Actuator 96, pp.333–336

161. Carlson, J.D. and Duclos, T.G.: ER Fluid Clutches and Brakes — Fluid Property and Mechanical Design Considerations, Electrorheological Fluids, pp.353–367 (1990)

162. Janocha, H.(Ed.): *Aktoren, Grundlagen und Anwendungen.* Springer–Verlag, Berlin, Heidelberg, New York; 1992, pp. 236–242

163. Hamann, Vielstich: *Elektrochemie II; (Taschentext Band 42).* Verlag Chemie, Weinheim 1981, pp. 205–206

164. Vetter, K.–J.: *Elektrochemische Kinetik.* Springer–Verlag, Berlin, Heidelberg, New York; 1961, p. 37

165. Kempe, W.; Schaper, W.: Elektrochemischer Aktor. Forschungsbericht BMFT AS–00402; 1991

166. Kuhnen, K.; Jendritza, D.J.; Kickel, H.: Aktuelle Bauformen und Anwendungspotentiale elektrochemischer Aktoren. In: *D.J. Jendritza: Technischer Einsatz Neuer Aktoren.* Expert Verlag 1995 (Kontakt & Studium; Bd. 484) ISBN–3–8169–1235–4

167. Baughman, R.H.; L.W. Shacklette; R.L. Elsenbaumer; E.J. Plichta and C. Brecht: Microelectromechanical applications based on conducting polymers. *P. L. Lazarev (ed.), Molecular Electronics* (Kluwer Academic Publishers, The Netherlands, 1991)

168. Hunter, I.W. and S. Lafontaine: A Comparisation of Muscle with Artifical Actuators. IEEE 0–7803–0456–X/92, 1992

169. Schünemann, M.; H. Wurmus: Chemomechanische Aktoren. D.J. Jendritza (ed.), *Technischer Einsatz Neuer Aktoren* (expert–Verlag, Renningen–Malmsheim, 1995)

170. Smela, E.; O. Inganäs and I. Lundström: New devices made from combining silicon microfabrication and conducting polymers. *Molecular Manufacturing,* edited by C. Nicolini, Plenum Press, New York, 1996

171. DeRossi, D. and P. Chiarelli: Biometic macromolecular actuators. K.S. Schmitz (editor): *Macro–ion characterization.* American Chemistry Society, Washington, 1994

172. Smela, E.; O. Inganäs and I. Lundström: Science **286**, p.1735 (1995)

173. Skotheim, T.A.: *Handbook of conducting polymers*: Vols 1 and 2 (Marcel Dekker, New York, 1986)

174. Berggreen, M.: Organic Light Emitting Diodes. Linköping Studies in Science and Technology. Dissertation, Linköping University, Sweden, (1996)

175. Baughman, R.H.: Synthetic Metals, **78**, p.399 (1996)

176. Bögelsack, G.: Nachgiebige Mechanismen in miniaturisierten Bewegungs-systemen. Proceed. of the Ninth World Congress IFToMN, Politecnico di Milano, p.3101, (1995)

177. Smela, E.; O. Inganäs and I. Lundström: Differential adhesion method for microstructure release: an alternative to the sacrificial layer. Transducers '95, Stockholm, Sweden, June 25–29, p.218, (1995)

178. Pei, Q. and O. Inganäs: Journal of Physical Chemistry, **96**, p.10507 (1992)

179. Pei, Q. and O. Inganäs: Solid State Ionics, **60**, p.161 (1993)

180. Pei, Q. and O. Inganäs: Synth. Met., **55**, p.3718 (1993)

181. Schoch, K.F. and H.E. Howard: IEEE Spectrum, June, p.52 (1993)

182. Kallenbach, M.: polymer micro–drives. diploma work, TU Ilmenau, Germany, 1996

183. Suzuki, K.; I. Shimoyama; H. Miura and Y. Ezura: Creation of an insect–based microrobot with an external skeleton and elastic joints. Micro Electro Mechanical Systems '92, Travemünde, Germany, February 4–7, p.190 (1992)

184. Smela, E.; O. Inganäs and I. Lundström: J. Micromech. Microeng. **3**, p.203 (1993)

185. Kallenbach, M. and E. Smela: Moving silicon plates actuated by conducting polymer films. Microsystemstechnologies, Potsdam, Germany, September 17–19, p.469 (1996)

186. Kallenbach, E.; A. Albrecht; O. Birli; M. Eccarius; K. Feindt and V. Zöppig: Magnetische Mikroaktuatoren — Entwicklungsstand und Perspektiven. 15. Internationales Kolloquim Feinwerktechnik, Mainz, Germany, September pp.25–29, (1995)

187. Schilling, C.; Wurmus, H.; Bögelsack, G.: Klein aber komplex, Der Betrag der Mikrosystemtechnik für die Bionik. edited by W. Nachtigall, Technische Biologie und Bionik — was ist das?, Biona–Report 9, 2. Bionik–Kongress, Saarbrücken, (1994)

188. Holdenried, J: Polymere für Mikroantriebe. diploma work, TU Ilmenau; Germany, (1997)

189. Heuberger, A. (Hrsg.): *Mikromechanik.* Berlin: Springer–Verlag 1989

190. Büttgenbach, S.: *Mikromechanik.* Stuttgart: Teubner 1991

191. Trimmer, W.S.N.: Microrobots and Micromechanical Systems. Sensors and Actuators, **19**(1989), pp. 267–287

192. Freygang, M.; Haffner, H.; Messner, S.; Schmidt, B.: A New Concept Of A Bimetallically Actuated, Normally–Closed Microvalve. Proc. Transducers '95, Stockholm, 1995, pp.73–74

193. Hamberg, M.W.; Neagu, C.; Gardeniers, J.G.E.; Ijntema, D.J.; Elwenspoek, M.: An Electrochemical Micro Actuator. Proc. MEMS '95, Amsterdam, The Netherlands, 1995

194. Wirtl, J.: Die piezokeramische Pille öffnet den Weg zur leistungslosen Ventil-ansteuerung. Firmenschrift der Hoerbiger Fluidtechnik GmbH, D–86956 Schongau

195. Jerman, H.: Electrically–Activated, Normally–Closed Diaphragm Valves. Proc. MEMS '91, Nara, Japan

196. Zdeblick, M.J.; Anderson, R.; Jankowski, J.; Kline–Schoder, B.; Christel, L.; Miles, R.; Weber, W.: Thermopneumatically Actuated Microvalves and Integrated Electro–Fluidic Circuits. Proc. Actuator 94, Bremen, 1994 pp. 56–60

197. Mettner, M.; Huff, M.; Lober, T.; Schmidt, M.: How to design a microvalve for High pressure Application. Robert Bosch GmbH, 70469 Stuttgart, 1990

240   References

198. Shikida, M.; Sato, K.: Characteristics of an Electrostatically driven Gas Valve under High Pressure Conditions. Proc. MEMS '94, Osio, Japan, 1994
199. Zengerle, R.: *Mikro–Membranpumpen als Komponenten für Mikro–Fluidsysteme.* Verlag Shaker, Aachen, 1994, ISBN 3–8265–0216–7
200. Zengerle, R.; Ulrich, J.; Kluge, S.; Richter, M.; Richter, A.: A Bidirectional Silicon Micropump. Proc. MEMS '95, Amsterdam, 1995, pp.19–24
201. Gerlach, T.; Wurmus, H.: Working principle and performance of the dynamic micropump. Proc. MEMS '95, (1995), Amsterdam, The Netherlands, pp. 221–226
202. Olson, A.; Enoksson, P.; Stemme, G.: Proc. Transducers '95, Stockholm, 1995, pp.291–294
203. Keefe, D.O.; Herlihy, C.O.; Gross, Y. and Kelly, J.G.: Patient–controlled analgesia using a miniature electrochemically driven infusion pump. British Journal of Anaesthesia 1994, pp.843–846
204. Stehr, M.; Messner, S.; Sandmaier, H.; Zengerle, R.: A New Micropump With Bidirectional Fluid Transport and Selfblocking Effect. Proc. MEMS '96, San Diego, in press
205. Howitz, S.; Wegener, T.; Fiehn, H.: Mikrotropfeninjektor (1996). FZ Rossendorf e.V., GeSiM mbH Dresden
206. Gerlach, T.; Enke, D.; Frank, Th.; Hutschenreuther, L.; Schacht, H.–J.; Schüler, R.: Towards an integrated microsystem for the automatic adjustment of mono–mode optical waveguides. Workshop MicroMechanics Europe MME '96, Barcelona, 21–22 Oct. 1996
207. Kämper, K.–P.; Ehrfeld, W.; Hagemann, B.; Lehr, H.; Michel, F.; Schirling, A.; Thürigen, Ch.; Wittig, Th.: Electromagnetic permanent magnet micromotor with integrated micro gear box. Proc. Actuator 96, Bremen 1996, pp.429–436

7. Sensors in Adaptronics

7.1 Advances in Intelligent Sensors
J.E. Brignell and N.M. White

The emergence of intelligent sensors arises from the fortunate conjunction of technological demands and technological feasibility. There was a time when engineers made do with a few basic measurements of physical quantities that they knew they could measure, rather than seek sensors that could accurately convey the information they really needed. As society and industry have become more complex this option has become less and less realistic. There is an increasing need to determine precise values of physical and chemical measurands independently of any other variables present. Large–scale integration has appeared just in time to provide a solution to the major problems posed by such needs.

In the days of linear continuous electronics, the available sensors were limited by stringent requirements on linearity, cross–sensitivity, freedom from drift, etc. This meant that most of the vast panoply of possible sensor mechanisms had to be rejected out of hand. The magnitude of change wrought by the appearance of digital electronics would be difficult to overstate. The existence of a drift–free storage mechanism alone provides a solution to many problems, but coupled with an increasing availability of processing power it diminishes once insurmountable barriers almost to the point of negligibility.

Of equal importance with the steadily increasing power of devices is the remarkable decrease in cost. Not only has the density of transistors been doubling every three years, but the cost of a logic gate has been halving every two years [1] and there is no sign of this trend abating. We have related elsewhere [2] how the first suggestions for intelligent sensors were ridiculed on the grounds of the high cost of microprocessors; nowadays a microprocessor is simply a library element that can be incorporated in an ASIC design and manufactured on a large scale for a few pence.

7.1.1 Primary Sensor Defects

Before undertaking a brief review of some fundamental principles of sensing, we need to define terms. Searching through the literature in the area of measurement, the reader is faced with many different and sometimes conflicting definitions of 'transducer', 'sensor' and 'actuator'. Some authorities contend that a transducer should only be applied to energy–conversion devices and that sensors are something different. We have chosen to define the terms with reference to the measurement (or control) system. Thus transducers divide into two subsets: sensors that input information to the system from the external world, and actuators that output actions into the external world.

The intelligent sensor approach means that sensors that were initially thought to be unusable because of fundamental flaws (such as nonlinearity, cross–sensitivity, etc.) are now realizable. Before proceeding, it is worth noting the five major defects found in primary sensors [3]. They are:

- nonlinearity;
- cross–sensitivity;
- time (or frequency) response;
- noise; and
- parameter drift.

In the days of linear, continuous electronics, nonlinearity was a major problem. Such compensation techniques as were available were based on diode networks having reciprocal characteristics, but by their nature these were relatively crude. As a result, all nonlinear primary sensor mechanisms tended to be ignored. Now, linearization processes such as look–up tables or polynomials are easily realizable with digital electronics.

No primary sensor is sensitive to a single physical variable, and this fact gives rise to the important defect known as cross–sensitivity, the dominant form of which is in relation to temperature. Virtually all physical and chemical processes are temperature–dependent and hence so are all our uncompensated primary sensors.

The materials and structures associated with primary sensors contain dissipative, storage and inertial elements. These translate into the time derivatives appearing in the differential equation that models the sensor system. Hence another major defect is represented by the time (or frequency) response. Unless response is ideal (i.e. an impulse in the time domain or a constant in the frequency domain) the output waveform of the sensor will not be a faithful reproduction of the waveform of the sensed variable. The means to neutralize this imperfection involve filtering, which may be thought of in terms of pole–zero cancellation. If the device has a frequency response $H(s)$, then a cascaded filter of response $G(s) = 1/H(s)$ will compensate for the non–ideal time response. The realization of such a filter in analog form presents a major obstacle that has greatly diminished the digital case.

Noise is generally any unwanted signal, but the term is often used to imply random signals. Random noise will always be present, if only because the universe is in a state of continuous agitation. The almost ubiquitous low–frequency $(1/f)$ noise can cause great difficulties with primary sensors. The nature of $1/f$ noise is not well understood but, by definition, its amplitude per bandwidth is inversely proportional to frequency. Hence measurements of signals down to zero frequency are particularly difficult.

Having looked at the various defects in sensors, we will now address four fundamental techniques of compensation for the defects [4].

− Structural compensation;
− tailored compensation;
− monitored compensation; and
− deductive compensation.

Structural compensation refers to the most traditional form. It concerns the way in which the material forming the sensor is physically organised to maximise the sensitivity of the device to the target variable and to minimise the response to all other physical variables. A good example here is the load cell (see Sect. 7.1.2). Not only is the mechanical structure of the device symmetrical, but so is the electrical structure (i.e. the Wheatstone bridge). This illustrates the fundamental manifestation of structural compensation, which is design symmetry. The target variable is thus arranged to produce a difference signal, while all other physical variables produce a common mode signal.

Inevitably, there will be a residual effect after applying structural compensation techniques for which it cannot cater, and this residual effect will vary between nominally identical sensors. Further techniques of minor adjustment are thus needed to minimize the residue. The term 'tailored compensation' refers to trimming techniques that require action, as determined by the individual sensor and not the overall design — a major cost item in the traditional industry.

The third class, monitored compensation, relies upon taking a measurement of the cross–sensitive variable and compensating computationally, either by reference to a model of the sensor, or by making use of data obtained from a calibration cycle. The tool for monitored compensation is the 'sensor–within–a–sensor'. In the extreme case, such as chemical sensors, where the cross–sensitivity is so severe that it becomes one of lack of specificity, the sensor array approach is preferred.

The final class of compensation is deductive compensation. This is resorted to in special circumstances when, for one reason or another, the test object is not physically accessible. Examples of such objects would include a nuclear reactor, the human brain, or the cylinder chamber in an internal combustion engine. Deductive compensation requires reference to a model, and because all models are imperfect, it is only used as a last resort.

Illustrative Examples. In this section, our main objective is to provide examples of sensors that can benefit directly from the intelligent sensor approach. It is not feasible to cover even a significant fraction of the range, so we have chosen two illustrative examples in the forms of a load cell and chemiresistor.

Perhaps the most common element found in mechanical sensors such as load cells is the strain gauge. This may take a variety of forms, such as semiconductor, thick–film or thin–film, but the most readily available is the metal foil gauge. This is attached to the structure by means of an adhesive. The positioning of the gauges is often critical, and great care must therefore be taken to ensure correct positioning. This labour–intensive task is one factor that accounts for the relatively high cost of sensors based on foil gauges. The metal foil strain gauge typically has a resistance of either 120 W or 350 W, thereby limiting the excitation voltage to about 10 volts in order to prevent self–heating effects. Thick–film strain gauges, in contrast, do not suffer from this problem because they exhibit a high resistance, typically greater than 10 kW. The use of microelectronic fabrication techniques also permits such sensors to be deposited quickly and accurately [5]–[7].

Fig. 7.1 shows a representation of a precision load cell together with the electrical configuration of the strain gauges in the form of a Wheatstone bridge arrangement. Metal foil strain gauges are normally used with these devices. The mechanical structure offers a considerable degree of immunity from errors due to eccentric loads. The residual effects still need to be removed, however, and traditionally this is accomplished by tailored compensation in the form of trimming. An eccentric load is applied by attaching a beam to the load cell with a fixed mass at the free end. This is then rotated and small areas are filed off the structure to optimize the immunity to eccentricity. First–degree temperature compensation of the device is traditionally achieved by adding a length of resistance wire, of known temperature coefficient, to one arm of the bridge. These trimming activities represent major cost items.

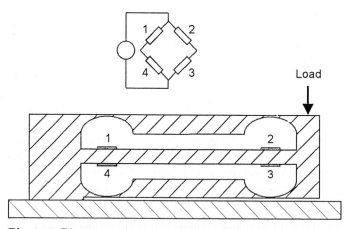

Fig. 7.1. Diagrammatic representation of the electrical and mechanical structure of a precision load cell, illustrating structural compensation by design symmetry

Chemically sensitive resistors (chemiresistors) are devices comprising a planar electrode pattern deposited onto an insulating substrate. The electrodes are then coated with a suitable chemically sensitive layer. The basic idea is that the conductivity or permittivity of the layer changes in the presence of a chemical measurand, and this is measured by monitoring the impedance change between the electrodes. Unfortunately, a single sensing element will not only respond to the desired quantity, but will also exhibit a marked cross–sensitivity to other variables, including temperature, humidity and different chemical species within the local environment.

Fig. 7.2 shows a gas sensor array fabricated using thick–film technology [5]. The chemically sensitive layer is an organic semiconductor. The device also has a platinum heating element situated underneath the electrode pattern. By supplying current to the heating element, the localized sensor site can be heated in order to promote the chemical reaction between the organic layer and the sample gas. The resistance of the platinum heater can be monitored to infer the temperature of each sensing site. The cross–sensitivity to other gases is significant, and a sensor array is needed to increase the specificity of the device. Within the array, each element is coated with a different reactive organic layer. Elaborate pattern–recognition techniques, implemented in software, are required to establish quantitative analysis of a mixture of gases flowing over the sensor array.

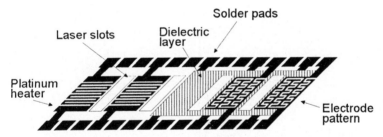

Fig. 7.2. An example of an array of chemiresistors fabricated using thick–film techniques. The slots are cut by a laser and help to isolate each sensor site from its neighbors

Current research at the universities of Southampton and Warwick is aimed at the production of arrays of gas–sensing elements on silicon substrates. The operation of the devices is similar to that described above. Polymeric materials are being used as the gas sensitive layers, and the front–end electronics are being fabricated as an ASIC.

The devices described above are examples of intelligent sensor systems that would not have been realizable in the early days of analog electronic systems. The response of each individual sensor element exhibits a nonlinear characteristic and is also cross–sensitive to a variety of other variables.

7.1.2 A Case in Point: The Load Cell

We have already referred to the precision load cell, and this has been a useful test–bed for many of the ideas for the implementation of intelligent sensor processes. It is a good example of a class of devices that pose a particular difficulty, in that the measurand is also an important parameter of the physical sensor system. One of the problems posed by mechanical sensors is that they tend to exhibit the oscillatory frequency–response characteristic of second–order systems. In the load cell the load being measured contributes significantly to the inertial parameter of the system. The old–fashioned way of dealing with the response was to provide massive damping, either mechanically or electrically, but this sledgehammer to crack a nut did nothing to improve the response time; indeed, it only made it worse. By means of digital filtering we could remove the response precisely, but there is an interesting paradox: as the load increases, the resonant frequency and the damping of the system both decrease. So, in order to measure a given load rapidly, we have to know the load before we can produce the correcting filter that corresponds to it. The way this chicken–and–egg conundrum can be solved provides a powerful illustration of the capabilities of the intelligent sensor approach [2].

The locus taken by the roots of the characteristic differential equation of the load cell as the applied mass changes can be determined by automatic system–identification techniques.

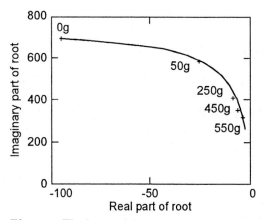

Fig. 7.3. The locus of the roots varies as a function of mass

Such a locus is illustrated in Fig. 7.3, and the roots of the compensating filter need to follow it. For each value of mass there is a corresponding final output of the compensating filter once oscillation has ceased. The trick is to make the parameters of the filter vary with its own output, as dictated by the locus. When a load is first applied, the compensating filter is set to the parameters for zero load, and as the signal begins to rise the parameters follow it. As the output signal crosses the correct value, the compensating filter is exactly right and the system locks in at the steady value. A typical output is shown in Fig. 7.4. The precision of such an approach compared with the sledgehammer of massive damping means that response times can be improved by at least an order of magnitude.

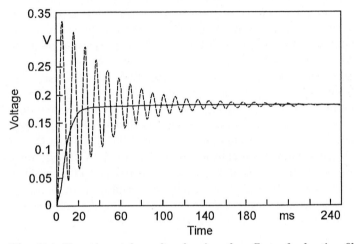

Fig. 7.4. Experimental results showing the effect of adaptive filtering of the response of a load cell. The broken line is the uncompensated output and the solid line shows the output from the adaptive filter

7.1.3 The Impact of ASICs

Techniques such as those discussed above were first developed on large computers and ultimately implemented on microprocessors. These were still comparatively cumbersome, requiring a circuit board to be associated with each sensor. At this stage it is worth emphasizing why the compensation needs to be done at the sensor site. In a large industrial instrumentation system, the central computer could be overburdened with sensor compensation processing, while the communication system could be overloaded by raw uncompensated sensor data. Ideally the compensation and communication electronics should be contained in the sensor housing and should be functionally invisible to the user.

Now a substantial analog subsystem can be accommodated on the same chip as an embedded microprocessor, and so it is conceivable that the entire compensation and communications system can be constructed in single–chip form. It is important, however, not to understate the scale of the problem of developing and debugging such a system, and unless resources are very substantial it is preferable at the present stage of technology to keep the processor as a separate programmable device. Not least of the problems is the fact that analog simulators are not nearly as easy to construct as digital ones.

Also at this point we come up against one of the major problems and a source of cogent criticism of the very concept of intelligent sensors. It has always been a truism that the more complex a system is, the less reliable it is. Fortunately this principle can be reversed by the introduction of two concepts — self–test and autocalibration — and there is one simple component of ASICs that make these realizable: the digitally controlled analog switch. By means of such switches an analog subsystem can be made to reconfigure itself to perform various checks (gain, offset, linearity, etc.) as well as to monitor possible disturbing variables, such as temperature. In a typical design [8] there are 16 such switches.

The development process on such systems could be fraught with complexity, and so it is important to establish methods that give the designer maximum support. A very useful technique is to embed the ASIC in a PC as shown in Fig. 7.5. Data acquisition (DAQ) boards are used to provide intimate access to the functions of the chip. Software is developed in a portable language, such as C, which allows it to be ported onto a suitable microprocessor once it has been developed and tested. This leaves the question of support software, which is discussed in the following section.

7.1.4 Reconfigurable Systems

From the above, the importance of advances in electronic hardware — in particular ASICs — on the development of intelligent sensors is clearly evident. The role of software drivers is equally essential as these control and perform the necessary tasks in test, calibration and operating modes [8]. Additionally, the software is responsible for ensuring correct communication between the sensor and the host system, and can also be used to ensure that hazard conditions are eliminated during hardware development.

Returning to the ASIC chip mentioned above, it can be seen that with the 16 analog switches there are 65 536 potential configurations. Many of these are forbidden conditions that would cause catastrophic failures if they were to arise. The problem can be solved by only allowing the use of a predefined set of standard combinations. This approach however, is extremely inefficient in terms of storage and operating speed, and also restricts the user to the preset list that may not be desirable for futures applications of the ASIC. The solution is to provide a software driver in the form of a filter that prevents any destructive configuration being set up, but allows all other combinations.

The switch configuration is stored as a vector of noughts and ones in two bytes. Each configuration is therefore represented by a unique 16–bit word that is stored in the memory space of the controlling digital processor. The subsystem can be switched into a specific self–check or auto–calibration

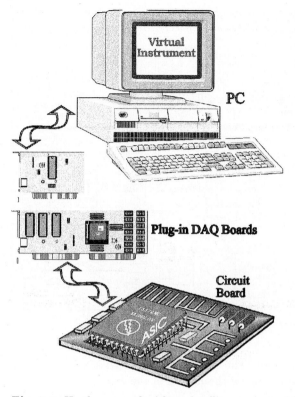

Fig. 7.5. Hardware used with an intelligent sensor ASIC [9]

mode with only a few control instructions. For the purpose of the initial prototyping, a PC is used as the controller. The code is written in C, the portability of which means that it can be downloaded into an embedded processor for the final implementation.

Fig. 7.6 shows the virtual instrumentation panel for controlling the ASIC. Note that the layout of the ASIC is an important part of the display. The switches can be activated on screen using a pointing device such as a mouse. The values are then passed to the software filter, which initially searches for forbidden settings. A process of binary masking is used to detect the forbidden conditions. The control word from the switch settings is logically ANDed with a mask. The number of bits in the result is counted and this is used to detect an invalid condition.

As an example, switches 1 to 5 in the analog front–end ASIC are used to provide a signal source to a differential amplifier. If more than one switch is on, a hazardous condition will occur. The status of these switches is represented by the lower five bits of the 16–bit control word ($1 = $ on, $0 = $ off). The process is explained in Fig. 7.7. The number of logic 1s is counted and if the total is greater than one, then more than one switch is set and therefore a

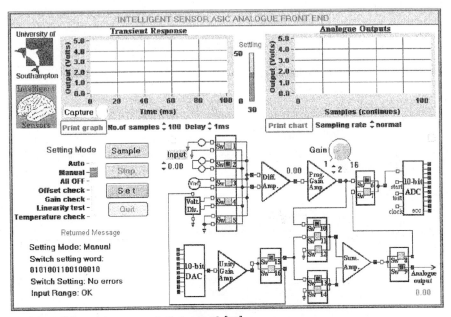

Fig. 7.6. Virtual instrumentation panel [10].

forbidden condition has occurred. The user is presented with a message on screen that provides further details of the nature of the error.

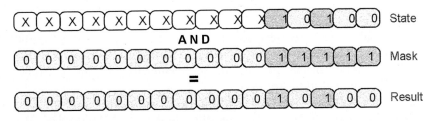

Summation of result bits = 2

Fig. 7.7. Filtering of forbidden switch settings

Forbidden connection paths through the subsystem can also be detected. Each path can be assigned a unique digital word, which can again be filtered to detect errors.

As the complexity of the intelligent sensor hardware increases, there is a requirement for more sophisticated software for testing, calibration and modes of operation. For example, quantitative analysis of gas mixtures can be performed using the so–called electronic nose [53] utilizing an array of chemiresistors, as in Fig. 7.2. The pattern–recognition techniques needed for such systems are becoming increasingly complex. Approaches such as neural networks and fuzzy logic mean that there will be additional emphasis placed on the importance of the associated software for sensor applications.

7.1.5 Communications

When we begin to consider multisensor systems, one of the most important factors is the nature of the topology of the network connecting the sensors to the central processor. Fig. 7.8 shows four possibilities of methods of networking sensor systems together. Fig. 7.8(a) gives the classical arrangement of the star topology. Here each sensor is connected to the centre by at least one pair of wires. There are a number of disadvantages associated with this approach. Firstly, a great deal of cabling is required and this could easily become the dominating cost for a large industrial system. Secondly, as more sensors are added, a bottleneck occurs at the centre where all the cables arrive.

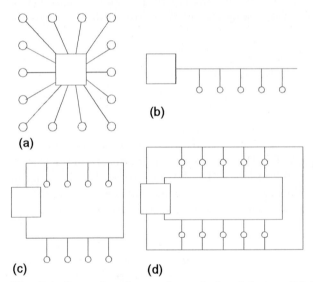

Fig. 7.8. Examples of network topologies: **(a)** star; **(b)** bus; **(c)** ring; **(d)** double ring

A more attractive idea is based on the bus topology shown in Fig. 7.8(b). The transducers share a common pair of wires. We now have the requirement that each device must have a unique address to distinguish it from its neighbors. Another potential problem is that if the shared data highway is severed at any point, all devices beyond that point are disconnected from the system. A third problem is that, as the number of sensors increases, their share of the bus, under time–division multiplexing, decreases, though this is not a new consideration as input/output (I/O) resources have always had to be shared.

The vulnerability of the simple bus to cable severance can be overcome by the modification illustrated in Fig. 7.8(c). Here the bus is arranged in a complete loop and provision is made for it to be driven from either end. This means that if the cable is severed at one point, the system can carry on by addressing both ends separately. This arrangement also allows the position of the fault to be determined by observing which devices fail to respond from either end. In cases where there is a particular danger of disruption, for example where there is an explosion hazard, the configuration in Fig. 7.8(d) can be adopted: four separate drivers address the double looped bus. The bus separation, and hence the length of the stubs connecting the transducers, is made large enough to minimize the probability of both buses being disrupted by a catastrophic event.

General Requirements for a Low–Level Protocol. There is an obvious requirement for a procedure that maintains and initiates communication throughout an overall system. Consider a continuous stream of bits being received by a station on a bus. In the absence of a protocol, a number of questions need to be asked concerning the nature of the digits:

– Where does a message begin or end?
– Is the message for me or another station?
– What is the actual information contained in the message?
– How is the message formatted?
– Has the message been transmitted correctly?

It is clear from these questions that a number of fields are required to establish a working protocol. Fig. 7.9 illustrates a well–known protocol HDLC. The first field is the opening flag that is a unique signal that cannot occur by accident anywhere else in the message. The bit pattern is 01111110, and in order to preserve the uniqueness, bit–stuffing is used in the non–flag section of the message. A logic zero is added whenever a sequence of five logic ones occurs.

Flag	Address field	Control field	Information field	Frame check sequence	Flag
01111110	8 BITS	8 BITS	VARIABLE	16 BITS	01111110

Fig. 7.9. The HDLC protocol

The next field is the address field, 8 bits in length and thereby allowing up to 256 devices to be uniquely addressed. This is an essential requirement for a shared bus system. The control field makes a statement about the purpose and nature of the message. For example, it could convey a series of instructions such as:

– carry out a self–test;
– set the amplifier gain to 10;
– transmit values of temperature; and/or
– the following field comprises 32 bits divided into four octal sub–fields.

The all–important information field is of variable length in the HDLC protocol, although other protocols use a fixed length. The condition of this field is highly conditioned by the fields that have already gone before it. The length of the field is contained in the control field and may vary from data packet to data packet.

The penultimate field is the frame check sequence, which is a number derived from a process in the preceding field. The process is repeated in the receiver and checked for a match. If no match occurs, a request is issued for re–transmission of the message. Finally, the end–of–packet flag 01111110 is transmitted.

Instrumentation Communication Systems. One of the most common systems is the 4–20 mA current loop. Typically, this is a two–wire system where one wire supplies the excitation for the transducer, and the other wire supplies the output current. The minimal use of cabling is one of the advantages of this technique. The output current is constrained between 4 mA and 20 mA, with 4 mA representing zero and 20 mA corresponding to full scale. The technique of using a live zero allows failures to be detected — for example, an open circuit condition would result in no current flowing (a 4 mA reading) while a short–circuit fault would be seen as an output current exceeding the 20 mA limit. In practice, an external reference resistor is connected between the output and the ground so that the recipient measures a voltage signal in the range 0.4–2 volts.

A popular method of connecting laboratory devices in a network is the General Purpose Instrumentation Bus (GPIB, or IEEE488). This consists of sixteen parallel data lines and eight control lines. It is widely used for inter-connecting oscilloscopes, multimeters, PCs, etc., but is limited to distances of about 20 m or so. It has a relatively high data–transfer rate of around 1 Mbyte/s, with full 'handshaking' among up to fifteen devices. In general, GPIB is regarded as being too expensive to implement for interfacing trans-ducers outside the laboratory owing to the associated wiring costs.

The system that is expected to become the standard for the networking of transducers is called 'fieldbus'. There is currently no defined international standard, but the International Electrotechnical Commission is responsible for defining one. The term FIELDBUS (in upper case letters) is reserved for the emerging international standard. The lowest layer of the protocol, the physical layer, is in a high state of maturity, but the remaining definitions for other layers are yet to emerge. This is largely as a result of political, rather than technological, uncertainties.

HART is a communication protocol developed by Rosemount Inc. and is an acronym for the Highway Addressable Remote Transducer protocol. It is a digital system but preserves the integrity of the 4–20 mA current–loop signal. The physical layer utilizes the same hardware as the 4–20 mA convention, but a frequency–shifted keyed current signal is effectively superimposed on the analog signal. The mean value of the digital signal is zero and therefore no additional DC level is added to the existing analog signal.

Several of the major industries have adopted HART as their sensor communication standard since it offers the benefits of digital communications together with compatibility with existing analog systems. In the longer term, however, if the FIELDBUS standards are universally adopted, HART will become redundant.

7.1.6 Trends

The reduction in size and cost of the transistors that make up our electronic subsystems has continued at the rate quoted in the beginning of Sect. 7 for many years, and there is every reason to suppose that it will continue for many years more. Talk is beginning about possible size limitations of a quantum nature, but it has to be remembered that we have only exploited planar structures and the possibilities of three dimensional structures are as yet largely unexplored. We are entering an era in which the silicon is of negligible cost. In these circumstances, development costs become even more important. Aids such as those described in Sect. 7.1.4, which enable development to be carried out on the device itself via an on–line computer, will make a major contribution in this area. We are now used to computer design aids, and digital simulators are now so good that a device that works in simulation can almost be guaranteed to work in practice, but unfortunately the same cannot yet be said of analog simulation.

Design and development costs will be moderated by the availability of tried and tested subsystems, and so it is important that a systematic approach is adopted rather than a piecemeal case–by–case one. Eventually manufacturers should be able to offer the sort of service provided by slide manufacturers, where the design is carried out on one's computer, sent over the network, and the product is received through the post in a few days.

One of the most exciting of recent developments has been microengineering, which turns the photolithographic techniques of circuit production to the manufacture of mechanical systems of micron dimensions. Subsystems as complicated as working millimeter–sized electrostatic motors have been demonstrated, which leads to the possibility of micro–robots working in environments such as the human body. The combination of microengineering and microelectronics on a single structure conjures up all sorts of possibilities (such as self–flushing gas microsensors).

In a little over a decade and a half, intelligent sensors have progressed from being a ridiculed academic pipe–dream to an essential component of modern technology, and there is much more to come.

7.2 Fiber Optic Sensors
W. Habel

At the beginning, optical fibers (lightwave guides) were created to transmit optical pulses over long distances with high transmission rates. Simple test systems consisting of a light–emitting diode, a fiber and a photodetector have been investigated. Since those days, fiber optic sensing techniques have grown significantly in number and type, and fiber optic sensors (FOSs) increasingly have become significant as smart sensing technology. The reasons are:

- their capability of being very sensitive, small, lightweight, chemically inert, and with no perturbing structural properties when embedded;
- their capability of being highly distributed;
- they withstand a few hundred degrees C during the curing process of composites;
- they are electrically passive and not disturbed by electromagnetic fields or by parasitic currents;
- they are network–compatible and amenable to multiplexing;
- they have small interface requirements (the opto–electronic elements and demodulation electronics are confined in the reading unit);
- there is a low risk of sparking because of the very low radiant energy emerging from the fiber optic system; and
- they are almost exclusively driven by standard photonics components.

A complete fiber optic sensor system consists of two parts:

1. the *sensing unit* contains the fiber optic sensing element equipped with a protective coating and/or an additional protective element (such as a pipe or a small tube) together with attachment material/clinge components
2. the *opto–electronic unit*, which contains the radiation source and a photodetector. Depending on the sensor type and on the size compatibility with the fiber, a semi–conductor laser diode (LD) or a luminescence diode (LED) is used. As photodetectors, PIN diodes or avalanche photodiodes (APD) can be used.

The intrinsic advantages have to be weighed against quite significant expense for the electronic equipment. Some fiber sensor types need very complex demodulation techniques, and this could lead to bias against fiber optic sensors. Nevertheless, the electronic devices within FOSs are continually being improved with respect to cost and customer–friendly use, and this should provide some compensation.

Basically, from the developer's point of view, two fundamental classes of FOS can be distinguished — the *intrinsic* fiber optic sensors and the *extrinsic* fiber optic sensors. Fig. 7.10 schematically points out the difference. An intrinsic FOS takes advantage of measurable changes in the transmission characteristic of the optical fiber itself; that means the sensing element is, at one and the same time, the carrier of information from and to the reading unit. Sensor types of this class are predestined for use in smart components because they avoid additional elements. Some extrinsic sensor types (such as micro strain sensors, see Sect. 7.2.4), where the fiber is not used as a sensor element but merely as a light guide to and from the sensing area, can also easily be used in smart structures.

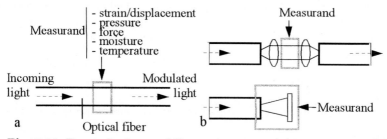

Fig. 7.10. Two main classes of fiber optic sensors: **(a)** intrinsic type; **(b)** extrinsic type. Both are shown in transmission mode (above) and reflection mode (below)

The quantity to be measured causes a variation of one or more physical parameters in the sensor. This variation must be detected, recorded, processed and should be re–transformed into a scaling unit of the measured quantity. The great challenge for the engineers is to separate the variations induced by the measured object from any variations induced by some other internal or external effects. Often faulty measurements are produced by an inappropriate application of the sensing element. Some aspects involved with these problems will be discussed.

Parameters to be varied by the measurand are the intensity, the wavelength or phase, and the polarisation state of light. Additionally used is, by means of optical time–domain reflectometry (OTDR) technique, the measurement of the travel–time of a light pulse launched at one fiber end and backreflected at markers. From a shortening or prolongation of the time of transit, the shortening or extension of the optical path length (contraction or extension of the interrogated fiber) can be assessed. It should be noted that a considerable number of fiber optic sensor types have been created in the past decade for measurement of almost all physical, and a lot of chemical, quantities. Of particular assistance was the development of components for communication links, such as connectors, splicers, couplers, and measurement devices.

In this section the examples are limited to FOS types for measurement of external disturbances such as strain, displacement, pressure, vibration, acoustics, and temperature, and for determining the location of damage along a fiber. It is important to point out that a large number of fiber optic sensors applied for temperature and vibration (acoustic) measurements are strain–based. However, their direct output signal fundamentally differs from conventional strain gauges, such as foil gauges, piezo films, piezoceramics or semiconductor strain gauges. The direct output of FOS depends on the principle chosen for interrogation. Ultimately, the effective output signal used by the engineers for detecting, evaluating and controlling is delivered from the data processing equipment. In the following subsections, as far as possible typical strain–related values, or values in terms of typically used parameters in mechanical engineering, are given.

7.2.1 Physical Principle of Fiber Optic Techniques

The basic element of a fiber optic sensor is a thin wire of glass or of plastic (polymeric) material. When light is ltransmitted into one end of the fiber surrounded by a fiber cladding with a lower refractive index than the core ($n_{\text{core}} > n_{\text{cladding}}$), it propagates through the fiber to the other end corresponding to the physical effect of total internal reflection [11]. Fig. 7.11 shows this effect. The acceptance angle Θ_A of the fiber (Θ_A is the maximum value of the range of the accepted angles Θ) defines the portion of light input at the fiber end. Only light that is input for $0 < \Theta < \Theta_A$ can be guided down the fiber. It is continuously reflected at the interface between the core and the cladding; the critical angle φ_c must not be exceeded. Depending on the diameter of the fiber core, 'modes' (the interference pattern within the core) are developed. Very small core diameters ($< 10\,\mu\text{m}$) allow only one mode to travel through the fiber (called single–mode fibers). The core–to–cladding index transition can change abruptly (step–index fiber type) or gradually with a parabolic profile (graded index fiber type). Thus, there are three main types of optical fibers used for different sensors; Table 7.1 explains the typical distinctions in their geometry. Because of the mass–production of single–mode fibers, they are cheaper than other types and therefore often preferred for sensor purposes. For special purposes, such as pressure, temperature, strain mesurements, so called high–birefringent (Hi–Bi) polarisation–maintaining (PM) fibers are used because the polarization state of the output signal is definitely effected by external perturbations.

Other specially designed fibers form the evanescent sensors. The core of such sensors can (locally) be coated with a cladding that modifies the refractive index in the core–cladding interface region. In the case of a variable environment (e.g. a change in the index of refraction between the uncured and the cured state of composites), the absorption coefficient of the fiber can alter. Such sensors are widely used for the detection of changes of chemical or biological environmental parameters.

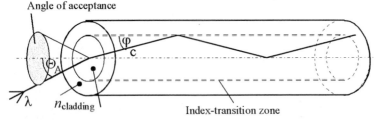

Fig. 7.11. Light beam propagation in a multi–mode fiber (λ = wavelength of the source)

7.2.2 Relevant Types of Fiber Sensors and Sensor Selection

The selection of sensors, including the demodulation technique used for solving a measuring task, is a very complex task. The primary aspects to be considered are:

- the number and type of the measurement points (i.e. sensor chain or sensor network, point or distributed sensors) and the parameters that need to be monitored;
- whether static or/and dynamic signals are to be recorded (i.e. short–term or long–term operation);
- any necessity to remove the reading unit from the sensors (i.e use of connectors);
- any temperature variations at the location of the sensor during the measuring time;
- the installation procedure (application during construction, during normal operation, during a maintenance break); and
- the need for reparability in the case of structure–integrated sensors (i.e. need of redundancy or acceptance of sensor failure after the occurrence of a detection event).

    Some aspects concerning these questions are described in [12]. From the user's point of view, an essential point in smart sensing is the length of the region to be evaluated. Long sensor fibers (area averaging sensors) or segmented sensor fibers (quasi–distributed sensors) allow the recording of deformations as well as cracks or other damages. They provide information about the structural health of extended structural components. The frequency range of such configurable sensors, based on commercially available measuring devices, is limited to a few Hz or less. However, vibration measurements, or even damage detection on the base of acoustic signals, are possible by means of interferometric short–gauge–length sensors. Distributed fiber sensors on the other hand, have the very desirable feature of being able to measure not only a physical quantity influencing the fiber, but also the position where the measurand is acting.

Table 7.1. Overview on the most common types of silica optical fibers used for sensors

fiber type	Multimode step–index fiber	Multimode graded index fiber	Single–mode step–index fiber
Light propagation (schematic)	Beam is reflected	Beam is refracted	Beam is guided
Geometry	Core (n_1), Cladding (n_2), Protective coating		
Typical dia– meters ⇒	core: 50 μm cladding: 125 μm coating: 140 μm to 250 μm	core: 50 μm cladding: 125 μm coating: 140 μm to 250 μm	core: 6 μm (870 nm) 9 μm (1300 nm) cladding: 125 μm coating: 140-250 μm
Refractive index profile	steplike from cladding to core	continuously from cladding to core	steplike from cladding to core

Fig. 7.12 shows a possible arrangement of different classes of fiber sensors for evaluation of the shape or stiffness parameters of a smart structure's component. The sensors can be embedded or attached to the surface.

7.2.3 Integrating and Quasi–Distributed Long–Gauge–Length Sensors

Sensors for Detection of Impacts and Fatigue Cracks. The crack sensor is the simplest form of a fiber sensor. It provides a bistable information: light is transmitted when the material is uncracked or undamaged; light transmission is disrupted at the position of a crack or damage. The use of fibers for crack detection and qualitative assessment of structures requires different treatment of the fiber depending on the kind of contact with the object to be detected (adhesive, or bonding to the host material). As optical fibers

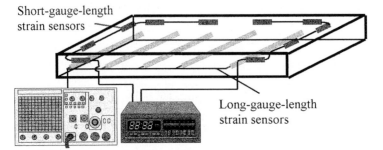

Short-gauge-length
strain sensors

Long-gauge-length
strain sensors

Fig. 7.12. Different fiber optic sensors (embedded or attached to the surface) as an integrated part of a smart structure

have a tensile strength of up to 12 GPa and assuming that no slippage exists between the sensing fiber, its coating and the measured object, a fiber break occurs for strains in the range of 0.85 mm/m to 8.5 mm/m because the coating prevents the formation of stress concentrations on the glass surface. Depending on the crack formation mechanism in the measured object (brittle or elastic material), the coating of a fiber crack sensor must be mechanically or chemically removed. A weakening of the coating (i.e sensitizing) should be preferred, because an uncoated fiber tends to break during application.

Experiences collected with fiber crack detectors applied on steel and concrete surfaces as well as embedded in composites has proved that simple crack detectors allow indication of the crack formation or provide qualitative information on the developing or stabilizing of crack activities. Fig. 7.13 shows results obtained on a steel and concrete structure's components by using surface–attached crack sensors, indicating the crack activity of a fiber loop used for detection of fatigue cracks during cyclic loading of a part of an old steel bridge [13]. The fiber has only been sensitized in the area where crack formation would be expected. The crack sensor starts to pick up detection readings in the formation phase of the crack, in other words in the plasticization of steel material round the crack tip. The location of the crack is possible by using an OTDR device. The first order of the attenuation signal of the fiber loop (centre curve in Fig. 7.13) could be used to trigger an emergency signal or an actuator. On steel structures, a detection threshold for cracks of about 10 μm to 30 μm can be achieved.

The same method can be used for crack detection of other materials. In concrete components, cracks opening in the range from 80 μm to 300 μm can be observed via sensors —for example during an intermittent loading of a large concrete beam, the start of crack damage can be detected. In the case of a sufficiently dynamic range in the sensor system, an estimation concerning stabilizing or opening of the cracks is possible. However, a correlation of the output signal to the crack width is hardly possible. More information, e.g. about the speed and direction of crack propagation, can be obtained by a suitable choice of arrangement in/on the structure (if the direction of

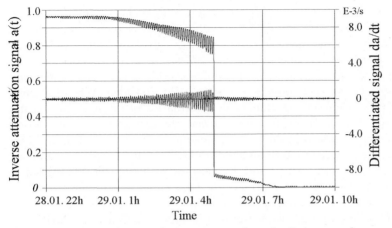

Fig. 7.13. Crack detection during propagation of a fatigue crack on an old steel bridge (centred curve is the differentiated attenuation signal of the fiber)

the propagation of cracks is approximately predictable). Using simple standard multimode fibers embedded in, and surface–attached to, high–density polyethylene pipes, dynamic measurements of the length of the fibers allows the evalution of the propagation speed of cracks in the range of 100–200 m/s [14]. The investigations in [14] confirmed the necessity to study further the dynamic behavior of the bonding region of embedded or attached fibers.

Recently, the use has been reported of embedded Hi–BI optical fiber sensors for the control of the health of multilayered structures. A spatial resolution of impact location of $\sim 30\,cm$ (accuracy $< 1\,cm$) could be achieved for an impact energy detection threshold of $5\,J$ [15].

Fiber Optic Sensors for Strain Measurements. Corresponding to the effect of decreasing the transmitted power due to local bends in optical fibers, the microbend sensor principle is the basis for a commercially available strain sensor, the 'optical string' [16]. This sensor consists of three stranded fibers. In the case of extension (or contraction) of the sensor, this arrangement produces microbends in the sensing fibers and those microbends lead to changes in intensity of the light transmitted. The optical string enables the measurement of the strain in the range of 0.5 % of the sensor length (max. resolution $4\,\mu m$, precision $\pm 0.1\,mm$ for long–term measurements).

The application of this sensor requires a free deformability of the fibers between their fixing points. Hence, hollow pipes are used for its embedding. This fact could be a drawback for integrating these sensors in smart structures. Intensiometric principles should be preferred for rather short–term measurements (construction–accompanying, proof loading monitoring, etc).

For long–term measurements, line–neutral methods are beneficial. Suitable techniques are based on low–coherence interferometry and backscattering. A long–length interferometric strain sensor present for long–term measurements on large structures [17] acts as a double Michelson interferometer: a sensing interferometer uses two fiber arms —a measurement fiber that is prestressed and in mechanical contact with the structure, and the sensor's reference, which acts as reference and compensates for the temperature dependance of the measuring fiber. The reference fiber must not be strained and needs to be installed loose in a pipe near the first fiber. When the measurement fiber is contracted/elongated, deformation of the structure results in a change of the length difference between the two fibers. By the second interferometer contained in the portable reading unit, the path length difference of the measurement interferometer can be evaluated. This procedure can be repeated at arbitrary times and, because the displacement information is encoded in the coherence properties of the light and does not affect its intensity, precision and repeatability of measurements are high without the need for continuous monitoring (as is required in other interferometric methods) (see Sect. 7.2.4). Typical parameters of this long–gauge–length sensor version are given in Table 7.2.

Table 7.2. Typical parameter of long–gauge–length sensors

Measuring length:	85 cm to 100 m
Measuring range:	up to 2% (for $< 160°$C)
Precision in measurement:	10 μm (error of measurement: $\Delta\epsilon = \pm 1.25 \cdot 10^{-5}$)
Proportionality factor between the measured delay and the applied deformation:	$(128 \pm 1)\,\mu$m/ps

Since strain measurement at different locations of very extended structures would require separate sensors for each section to be measured, an in–line multiplexing scheme is used [18]. This allows interrogation of long sensing fibers provided with reflectors that are positioned in pairs along the fiber. In this way, a long–term stable quasi–distributed measurement of up to ten strain areas along one fiber can be achieved. The precision of measurement reported is better than 1%; the optimization of the reflectors with regard to an adjustment of the power distribution is being improved.

The double Michelson interferometer principle also requires the strain–free installation of the reference fiber. This necessity does not normally lead to problems in large concrete structures (for which the method has been developed). However, smart structures are likely to be perturbed. Two–arm sensor principles basically have the drawback of requiring a hollow pipe for the loose installation of the reference fiber; one–arm interferometers should therefore be preferred. Hence, an alternative to two–arm long–gauge–length sensors is one long fiber containing fiber sections separated by reflectors. By measurement of the time of flight of a short pulse transmitted into the fiber and backscattered in these areas or on markers (splices, photoinduced reflectors, or squeezing points) at the end of these sections, the measurand can be determined at definite locations along the fiber. Fig. 7.14 shows a possible arrangement. An elongation (compression or contraction) of a measuring section, determined by two reflector sites on the fiber, changes the travel time of the pulse: $\Delta\epsilon \sim \Delta t_\mathrm{p}(c/2L_\mathrm{o} \cdot n)$, where c is the speed of light, n the index of refraction. Based on this relationship, the changes in the average strain of a chain of marked sections along the fiber can be interrogated by an OTDR device.

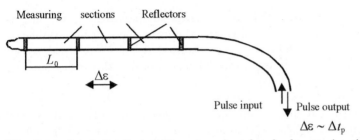

Fig. 7.14. Quasi–distributed fiber sensor based on backscattering signal evaluation

This method allows the evaluation of strain profiles in large components. The strain resolution achievable is determined by the OTDR device used. A modern, high–resolution picosecond OTDR device enables the resolution of elongation to 0.1 mm, assuming a minimum distance of the reflectors of 100 mm in the measuring section [19]; the maximum strain is about 2.5%. OTDR devices are very expensive and, because of the necessary integration time of the device, one scanning run takes between one and some ten seconds, depending on the expected precision. Moreover, the sections have to be interrogated one after another. In order to be able to fully exploit the performance of the backscattering technique, a new method for interrogating the optical path length in multiple fiber sections has been published [20]. A noise–limited spatial resolution of 20 μm over a section length of 5 m at a response time of 0.5 s can be achieved. In this way, continuous measurements over several sections can be expected.

The length of the fibers evaluated by conventional OTDR technique is limited by the permissible power loss along the fiber (in other words by the quality of the markers), and can reach up to several hundred meters, or tens of sensing sections.

Fiber Optic Sensors for Force/Pressure Measurements. The optical fiber types normally used show a small pressure sensitivity owing to variations in the guided waves. A correlation between pressure and inducing events is hardly possible, even if appropriate coatings that enhance the disturbing effect are used. In contrast to this, polarimetric fiber optic sensors, based on

high birefringent fibers, respond to pressure with a change in the polarisation state of their output light. Although the use of high birefringence for
distributed measurements in Hi–Bi fibers is accompanied by some difficulties
(high precision–alignment requirements when splices have to be made, and
the high cost of polarisation–preserving fibers), recent studies has been reported on sensing arrangements that allow the detection of the position of a
force and an estimation of its intensity [21]. The length of the sensing fiber
should, if possible, be in excess of 100 m. The margin of resolution of the
relevant force and its position has not been reported.

**Fiber Optic Sensors for Moisture, Temperature and Chemical Pa-
rameters.** Distributed sensors for chemical sensing have an important role
in the fabrication phase and to monitor corrosion. A conceivable fiber sensor
type is based on evanescent wave–type fibers coupled with the surrounding
material (see Sect. 7.2.1). However, the sensitivity of such a sensor type may
be a limitation for some chemical detection requirement. It should therefore
rather be expected that any sensors that act as chemically induced microbend
types, are promising for use. One sensor arrangement presented for physi-
cal/chemical measurements in conjunction with smart structure applications
is a microbend arrangement in which microbends are applied to the (usual)
optical fiber by expansion of a chemically reacting and water–swellable polymeric layer (called a hydrogel). For *distributed moisture detection*, a special
hydrogel layer that expands by up to 400% on the presence of moisture has
been used [22]. In combination with the OTDR technique, the sensor is capable of distinguishing water ingress points as small as 50 cm over a length
of more than 100 m. A subsequent absence of water (by a drying process)
can also be observed. The gel system used was primarily a water detector;
however, by varying its composition the sensor could be tuned to respond to
chemical parameter changes as well.

For measurement of *temperature distribution*, a backscattering–based sensor is used. When light propagates through an optical fiber it is scattered
by spontaneous fluctuations in the dielectric constant of the glass material:
molecular vibrations (called Raman scattering), propagating density fluctuations (Brillouin scattering), nonpropagating entropy fluctuations (Rayleigh
scattering). Brillouin scattering depends on both temperature and strain,
whereas Raman scattering only depends on temperature. By determining
the intensity and spectral composition of small parts of a backscattered light
pulse, external pertubations can be evaluated. In order to measure only temperature influences along a fiber, the Raman backscattering components are
determined. The temperature is determined as an integral value for a short
section of the optical fiber (e.g. 1 m) and, analogous to the known OTDR
method, the space coordinate is determined from the travel time of the propagating light pulse. In this way, the temperature can be measured simultaneously along the total length of the fiber. The measuring time for a fiber length
of up to 8 km is about 1 min with a spatial resolution of 1 m; a resolution of

0.5 m or 0.25 m requires an integration time of 2 min, or 4 min respectively. The measuring range of this distributed temperature sensor is $-50°C$ up to $300°C$, with a temperature resolution of less than 0.05 K. Because the Raman scattering depends only on physical constants and on parameters of the light transmitted into the fiber, the absolute temperature can be obtained directly from the backscattered signal. Owing to different material qualities of the fiber used, a calibration function must be determined before measurement starts. Application of this technique is commercially focused on geotechnical and environmental fields [23].

Due to the very low Raman scattering coefficient (measuring the dependence of the backscattered signal on temperature) these systems are fairly expensive. Additionally, at high temperatures, the Raman ratio approaches units, which means that the sensitivity decreases. That is why Rayleigh/Brillouin temperature–dependent scattering is used for distributed temperature measurements, although the very strong Rayleigh backscatter component is unfortunately dependent on temperature and strain. From investigations made by Wanser and co–workers [24], it can be concluded that fused silica fibers show an essentially reversible temperature dependence up to $650°C$. The spatial resolution of this method is as high as the spatial resolution of the OTDR, e.g. for a 0.01 dB OTDR a temperature–sensing resolution of 15.4 K can be achieved. Temperature differences between sections of less than a meter in length are resolvable. Another application of distributed temperature sensing that also uses the Rayleigh backscattering signal has been reported for diagnosing overheating conditions within electrical equipment. In a temperature range of 0–$150°C$, temperatures along a 20 m to 100 m fiber could be measured with a resolution of ± 5 K and a length resolution of less than 10 cm [25]. The strain dependence of the Rayleigh backscattered signal can advantageously be used for high–temperature strain measurements because of an increasing strain dependence at high temperatures. This temperature–dependent sensitivity effect has to be taken into account in any measurements made.

7.2.4 Short–Gauge–Length Sensors

Numerous types of short–gauge–length optical fiber sensors for strain measurements in materials research and structure evaluation have been proposed, but very few sensor techniques are commercially available. In contrast to fiber sensors with long measuring bases, short–gauge–length fiber optic sensors, based on interferometric and spectrometric principles allow the measurement of local deformations with a very high resolution. The most well–known micro strain sensor configurations are the Mach–Zehnder, the Michelson, the Fabry–Pérot and the Bragg sensors. In this section, the most widely used sensor types from that list are described: Fabry–Pérot interferometer sensors and Bragg grating sensors for the measurement of strain, temperature and pressure. Because of the general interest and their increasing importance, a short discussion is also included on their properties and some sensor–typical as well as application–technological problems.

Fiber Fabry–Pérot Interferometer sensors. The basic transduction mechanism is as follows: A measurand induces a phase change in the optical signal, and is detected as an intensity change in the output interference signal. Such a phase change is caused by force–induced or temperature–induced strain. Interferometric sensors allow local measurements of strain with a resolution of $< 0.1\,\mu\mathrm{m/m}$, and they can also be used for detection of mechanical vibrations and acoustic waves. With regard to manufacturing and applicability, fiber Fabry–Pérot interferometer (FPI) sensors are the most often applied interferometric sensor types for structural assessments. They do not need a reference arm or sophisticated stabilization techniques as the Mach–Zehnder or Michelson types do.

There are two arrangements of FPI sensors. A simple in–fiber FPI (called IFPI) consists of two mirrors at locations in the length of the fiber. The first mirror is formed by fusing two ends of an optical fiber together; as the second mirror the end of the fiber can be used (Fig. 7.15a). The maximum distance of the mirrors (i.e. the cavity length) can reach some mm. In this sensor type, the cavity length corresponds to the gauge length.

In another design, the cavity is produced by positioning a fiber end–face opposite to another, with a small gap of usually some microns. Such an extrinsic FPI (EFPI) sensor is shown in Fig. 7.15b. The most widely used design is to fix into position the two fiber ends in a hollow tube. In this arrangement, the gap between the fiber and the inside of the tube usually contains air. The input/output fiber is fixed at one end of the tube. Another fiber end–face, which acts as a mirror, is inserted into the other end of the tube. With the help of both ends a very small gap (about $10–100\,\mu\mathrm{m}$) can be aligned. The very small gap between the mirrors causes, in contrast to an IFPI sensor, a very low transverse influence.

The functional principle of an EFPI sensor is as follows. The incoming light reflects twice: at the interface glass/air at the front of the air gap (the reference [Fresnel] reflection) and at the air/glass interface at the far end of the air gap (sensing reflection). Both reflections interfere in the input/output fiber. The measuring effect is induced when the hollow tube is axially deformed and the two fiber ends move longitudinally inside the tube (because they are only fixed at the ends of the tube), which results in changes on the air–gap length (i.e. the gap width). There is a phase difference between the reference reflection and the sensing reflection, which changes due to gap variations. This change modulates the intensity of the light monitored at the end of the output fiber.

For this two–wave interferometer configuration, the observed output intensity I_{out} is a sinusoidal function of the gap width s. Small values of strain variations (less than $300\,\mathrm{nm}$ end–face displacement related to the measuring base of $\approx 10\,\mathrm{mm}$) can be measured directly, because the output signal

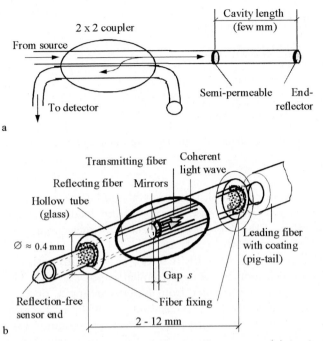

Fig. 7.15. Types of Fabry–Pérot interferometers: **(a)** intrinsic type; **(b)** extrinsic type

can be defined as linear between the peaks and troughs of the sinusoidal function. The measurement of larger end–face displacements, when a lot of periods are cycled, requires the counting of the interference fringes because the sensitivity decreases near, or becomes zero at, the maxima and minima values of the sinusoidal output signal. In this case, the resolution achievable with simple commercially available devices is unfortunately worse than that of conventional strain gauges. Because of the continuous oscillation due to strain/compression, the sensor signal provides only a relative measurement of strain induced in the sensor head.

There are two slightly more expensive ways of achieving a continuously high sensitivity, again at the maxima and minima of the output signal:

– using a two–gap sensing element; or
– illuminating the sensor with light of two wavelengths.

The two–gap sensor (Fig. 7.16) contains two input fibers positioned side by side in the hollow tube with a different end–face separation relating to the reflecting fiber.

This double–sensor configuration (called quadrature phase–shifted EFPI) ensures that at least one of the sensing units is sensitive and the direction of displacement change can be recognized. Another effect can be exploited.

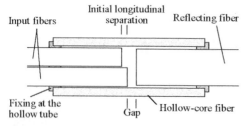

Fig. 7.16. EFPI design for unambigious measurement of axial and lateral strain [26]

Using such a single alignment tube with two input fibers, lateral as well as axial strain can be measured. In the case of axial strain load, the signal can be recorded from either of the two sensor elements. For measurement of two–dimensional strain changes, both signals are recorded.

By illuminating the simple one–fiber EFPI sensor (Fig. 7.15b) with two wavelengths, an absolute measurement of displacement is possible [27]. The changes of the gap width s can be computed from either scanning λ_1 or λ_2 or from measuring the phase difference for two constant wavelengths. Because the scanning device needs a limited time of a few seconds for scannning, these absolutely measuring sensors (AEFPI) are only capable of measuring static strain. It is worth mentioning here that the absolute strain measurement by using AEFPI sensors is a self–calibrating method with a large dynamic range. Using a spectrum analyser with a wavelength measurement resolution of $0.1\,\text{nm}$, an absolute gap–width measurement with a resolution of $5\,\text{nm}$ is possible; this corresponds to a strain resolution of $5\,\mu\epsilon$ even if the sensor is disconnected from the optoelectronic unit for a long period of time. The effective strain resolution can be matched by the choice of the measuring basis (see the following section).

A number of applications of EFPI sensors have been published. One of the first interesting applications has been the mounting of quadrature phase–shifted Fabry–Pérot sensors to the underside of the wing (near the root) of an F–15 aircraft for the purpose of fatigue testing its construction [28]. Another example of performance of EFPI sensors is shown in Fig. 7.17. Strain development in a laminate–coated concrete beam was observed in order to recognize delamination defects when the beam was loaded. By adhering EFPI sensors to the concrete surface and by embedding similar sensors in the coating laminate, static strain behavior at definite points in the different materials could be observed simultaneously when load was applied. More examples of application of EFPI sensors for the combined recording of static and dynamic signals are given in Sect. 7.2.4.

There are some interesting alternative Fabry–Pérot interferometer sensor configurations currently under investigation. One is the in–line fiber etalon (ILFE). ILFE sensors have a better mechanical strength and geometrical

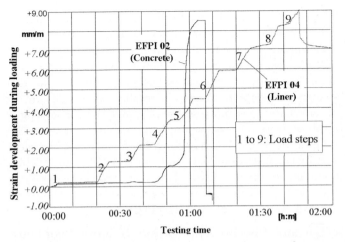

Fig. 7.17. Strain development in a composite–coated concrete beam under load. The step–like curve shows the strain in the fiber's reinforced–resin coating; the other curve shows the strain on the concrete surface exactly below the other EFPI sensor [29]

stability of the gauge length than other fiber sensors. Another alternative is the use of white–light–based sensor arrangements. Its important advantages are that the phase of the light propagated in the cavity is independent of environmental drifts from temperature or any transverse strain, and that the sensor operates in a completely output–independent manner. However, the reading equipment is quite expensive, and so this type should only be preferred if extreme demands on stability exist. In other cases, simpler and more inexpensive alternatives, such as intensity–based micro–displacement sensors can be used.

Badcock and Fernando [30] have investigated the static and dynamic behavior of a sensor configuration comparable to the sensor configuration of an EFPI type. Here, the strain ϵ can be calculated unambiguously from the measured output light intensity. This type of strain sensor was used for tension–tension and compression–tension loading. They embedded sensors in composite panels fabricated from Ciba–Geigy type pre–pregs. The attenuation of the transmitted light was recorded for evalution of the strain changes. Because they used a coherent light source, the effect of Fabry–Pérot interference could be observed. Both signals (interference fringes and intensity modulation) helped to avoid a loss of the fringe reference position.

There are a few devices for interrogating short–gauge–length fiber strain sensors offered by F & S Technologies Inc. (USA) [31] and Canadian Marconi Company, Canada [32]. Simple one– or two–channel devices enable the connection and interrogation of EFPI sensors and other short–gauge–length sensors for measuring strain, temperature and pressure. The two–channel system enables the recording of vibrations up to 20 kHz. However, when static strain is measured (max. resolution of $1\,\mu\mathrm{m/m}$) the measuring information is lost when the sensors are disconnected or the power supply is interrupted. For disconnectable measurements the Absolute Fiber Optic Support System (AFSS) based on AEFPI sensors (see above) is offered. This more expensive one–channel system enables the measurement of absolute (static) strain values even if the system is switched off, the strain resolution being $25\,\mu\mathrm{m/m}$ (measuring basis of the sensor: $\sim 4\,\mathrm{mm}$). Measuring systems for getting a strain resolution (in absolute terms) of 2–$5\,\mu\mathrm{m/m}$ are being developed. The four–channel Fiber–Optic Strain Sensor System (plus four analog electrical data channels) offered by the Canadian Marconi Company (Canada) enables absolute strain measurements up to $100\,\mathrm{Hz}$. The measurement accuracy is better than 1% of full scale. The sensors have a gauge length of 5–$15\,\mathrm{mm}$ and allow the measurement of strain as high as $5\,000\,\mu\mathrm{m/m}$.

Fiber Sensor Design and Special Application–Related Problems. Basically, fiber optic sensors are applied in two ways: (a) surface–mounted on metallic and non–metallic surfaces; or (b) embedded in polymeric, mineralic or low–melting materials. A number of questions referring to the host material and to the sensor design have still to be solved before the fiber optic measuring technique becomes an established one. These usually relate to the physical parameters and the external design of the units. These and other aspects require intensive cooperation between user, sensor designer and supplier.

The first step in planning a fiber optic sensor application is the selection of the sensor type (IFPI, EFPI, AEFPI or ILFE) depending on different requirements:

– *Sensitivity, Measuring Range and Resolution of the Measurand.* Micro strain sensors have a different gauge sensitivity. Fabry–Pérot interferometer sensors show a high sensitivity (in terms of phase changes); other interferometric sensor principles have sensitivities that are about two orders of magnitude less. It is important to note that, for in–fiber Fabry–Pérot sensors, the strain sensitivity is not a function of the gauge length. All types of in–fiber sensors have a limited ability of deformability. Photo–induced IFPI sensors can withstand strain of more than $20\,000\,\mu\mathrm{m/m}$; EFPI sensors normally survive strain values of more than $10\,000\,\mu\mathrm{m/m}$. The strength values decrease when dynamic loading with high amplitudes appears. Extrinsic sensors enable more flexibility: first, the measuring basis is defined by the length of the hollow tube, not by the cavity length; second, for high strain requirements, one fixing point can be placed outside the tube (by adhering the fiber to the material to be measured) in order to allow free movement of the fiber inside the tube, and in this way a sufficiently large displacement is possible. In the same way, the resolution of the sensing arangement can be matched.

– *Direction of Loading, or Strain Field.* In the case of measurement of a strain field that is not parallel to the sensor axis, at least three measurements are required. Analoguous to conventional resistive foil strain rosettes, an optical measurement (Fig. 7.18) can be used.

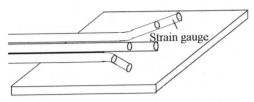

Fig. 7.18. Fiber optic strain rosette [33]

EFPI sensors have, in contrast to IFPI sensors, a distinct advantage because of their sensitivity to axial strain only. This is also important when fiber sensors are to be embedded, because in this case the measuring element is always affected by the surrounding material. The measurement of axial changes by using an IFPI sensor is disturbed owing to their sensitivity to radial strain components, and the strength is normally quite small.

– *Dimensions of the sensing element.* Especially for the embedding of sensors in composite materials, the problem concerning dimensions that can be accepted without running the risk of delaminating has to be considered. In–fiber sensors are just as smooth in the sensing region as the leading fiber itself; EFPI sensors, however, have a slightly larger diameter in the sensing region (usually 250–300 μm), which can strongly affect the features in fiber–reinforced composites. On the other hand, they could adhere better to homogeneous materials.

– *State of the Host Material Concerning Curing/Hardening Phase, Stiffness.* Since all embedded fiber–based sensors have a significant stiffness, described by Young's modulus for the glass material and the geometry, they are able to measure only deformations of such materials that develop sufficient bonding forces. When the curing process of, for example, polymeric panels or the hardening process of mineralic materials are to be observed, stiff fiber sensors are not suitable for measuring deformation at very early ages (such as at the interphase between rheological and solid states) because no, or only a small bonding force has time to develop between the sensor surface and host material. In only a few cases, a free movable EFPI sensor configuration (as also used for a large measurement range) could be helpful, because soft particles of the host material could burst into the tube. In order to prevent damage of EFPI sensors thereby, Habel and co–workers [34] designed an embeddable stress–free strain sensor configuration.

– *Expected Temperature Loading.* Temperature variations around the sensor lead to a change of the gap size due to thermal expansion of the glass material. Assuming that the sensor operates on/in the measuring object without slippage, the deformation is transferred linearly to the end faces, meaning that the effective temperature–induced gap variation is very small.

This is only valid when the sensor is attached to a glass surface or is freely movable; the false strain drift of $-20\,\mathrm{pm/K}$ can be neglected. In contrast to this, for an EFPI sensor with a measuring basis of $10\,\mathrm{mm}$ and a gap of $40\,\mu m$ attached to a steel material, a slow temperature change of $30\,\mathrm{K}$ leads to a mismeasurement in strain of $-0.16\,\mu m/m$ $(= -5,4\,\mathrm{nm/K}$ for application). This value cannot be neglected and has to be taken into account.

– *Surface Application or Embedding of Fiber Sensors.* The decision as to whether the sensors are to be applied to a surface or to be embedded in a material determines strongly the treatment of the fiber sensor and the choice of coating. The most important question concerns the load transfer provided by a durable coating. The attachment method on surfaces determines, on the other hand, the quality of load transfer; it should be carried out by experts.

– *Fiber Optic Strain Gauges for Quick Installation.* A relatively vulnerable bare glass tube that contains the fiber sensing element can be difficult to handle on site, especially when the application procedure must be carried out very quickly or under adverse environmental conditions. For such cases, optical strain gauges similar to resistive foil strain gauges are available [34]. The glass tube containing the EFPI sensor is embedded in a specially modified resin. These fiber optic strain gauges (Fig. 7.19) were, among other applications, fully adhered inside a box girder of a steel bridge in Rouen (France).

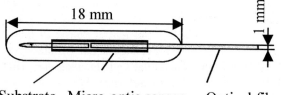

Substrate   Micro-optic sensor   Optical fiber

Fig. 7.19. Fiber optic strain gauge tailored for surface adherence

Bragg Grating Sensors. The discovery of the photosensitivity in germanium–doped fibers by Hill and co–workers in 1978 was the basis for fabrication of in–fiber reflective Bragg grating filters with a narrow–band, backreflected spectrum. When ultraviolet (UV) light is incident upon such a fiber, the refractive index n of the fiber increases. Meltz et al. [35] demonstrated that gratings can also be formed in the core of an optical fiber by illuminating it from the side by overlapping a pair of coherent UV beams. In the meantime, improved fabrication techniques allowed the manufacture of a grating in the core of a single–mode fiber (an in–fiber grating) at special wavelengths, such as $830\,\mathrm{nm}$ or $1300\,\mathrm{nm}$, with a given periodically changing refractive index (grating period or pitch Λ) and length L as the measuring basis.

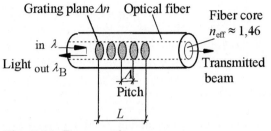

Fig. 7.20. Bragg grating sensor

A grating sensor can be manufactured as an integrated component of the fiber. The distance between the grating planes can vary. Bragg gratings for sensor purposes are primarily referred to as uniform gratings: the grating along the fiber has a constant pitch and the planes are positioned normally to the fiber axis, as shown in Fig. 7.20. In other types of Bragg gratings, the grating planes are tilted at an angle to the fiber axis (blazed gratings) or grating planes have a monotonically varying period (chirped gratings). The latter gratings are primarily used in long–haul transmission lines for telecommunication purposes.

The principle of the Bragg grating function is as follows. When a broadband light signal travels down the optical fiber that contains a Bragg grating, it will be reflected in accordance with the periodic variations of the refractive index of the grating. However, the reflected light will be out of phase and will tend to cancel out the incoming light signal. Only one wavelength of the light spectrum which satisfies the Bragg condition

$$\lambda_B = 2n_{\text{eff}}\Lambda \tag{7.1}$$

is reflected back (n_{eff} is the effective refractive index of the fiber core). The value of the Bragg resonance wavelength λ_B is determined by the grating pitch Λ manufactured and corresponds to twice the period of the refractive index modulation of the grating. The grating periodicity is relatively small, typically less than $1\,\mu m$.

From (7.1) it follows that the Bragg resonance wavelength λ_B will change when n_{eff} changes (for example by temperature variation) or Λ changes (through pitch changes by fiber–grating deformation). That means changes in strain or temperature (or both) will shift the reflected centre wavelength. In general, λ_B increases when the fiber is strained ($\Delta\epsilon > 0$) and decreases when the fiber is compressed ($\Delta\epsilon < 0$). The shift of the resonance wavelength can be monitored by using a spectrum analyser; in this way, one can determine strain variations (for constant temperature) or temperature variations (without any deformation of the grating).

When a Bragg grating sensor is to be used as a strain sensor and when the temperature varies under normal operation, a single measurement of λ_B makes it impossible to differentiate between strain or temperature changes.

This undesirable temperature–sensitivity of fiber grating sensors requires taking special measures in order to achieve a separation of the strain and temperature results. Assuming uniform axial strain changes in the grating area, the strain seen by a grating can be computed by a simple linear equation:

$$\epsilon = K \frac{\Delta \lambda_B(\epsilon_z)}{\lambda_B} + \xi \Delta T. \tag{7.2}$$

K can be estimated by a calibration procedure and is about 1.3 for single–mode fibers illuminated with 1300 nm. The 3M Company produces Bragg gratings with a temperature–sensitivity of $\xi = 6.3 \cdot 10^{-6}$ /K (for a source wavelength of 1550 nm) [36], and ξ decreases for lower wavelengths.

Table 7.3. Typical parameters of commercially available fiber Bragg gratings

Parameters	3M [36]	Northern Photonics [37]
Bragg wavelength band	800 nm, 980 nm, 1300 nm, 1550 nm	1550–1560 nm (in 1 nm increments) 980–1600 nm possible
Grating length	0.5–20 mm	10 mm (5–100 mm possible)
Bandwidth	0.1 nm to > 1.0 nm	0.25 nm
Temperature coefficient (referred to wavelength shift)	10 pm/K (1550 nm)	13 pm/K

The detectable strain resolution depends on the Bragg grating bandwidth and on the resolution with which the wavelength is measured. With standard gratings (typical parameters are shown in Table 7.3) and not too expensive spectrum analysers, a wavelength shift to about 0.01 nm, which then corresponds to a strain resolution of 9 $\mu\epsilon$, can be detected. By using more expensive spectrum analysers, especially in laboratories, strain sensitivities < 1 $\mu\epsilon$ can be achieved. There are, however, limits in the choice of the grating length.

Bragg grating sensors possess a number of advantages that make them attractive compared with other micro sensor arrangements:

– *Linear Response.* The Bragg wavelength shift is a simple linear response to the sensor deformation ($\epsilon = K \Delta \lambda_B / \lambda_B$). In contrast to interferometric micro sensors, there is no ambiguity for strain–compression subjected to the fiber.

- *Absolute Measurement.* The strain or temperature information obtained from a measurement system is inherently encoded in the wavelength (strain and/or temperature, index changes due to cladding affection); an interruption in the power supply does not lead to a loss of measurement information.
- *In–fiber Manufacturing.* In–fiber manufacturing enables low–cost fabrication of a large number of gratings.
- *Line Neutrality.* The measured data can be isolated from noisy sources, e.g. bending loss in the leading fiber or intensity fluctuations of the light source.
- *Potential for Quasi–Distributed Measurement with Multiplexed Sensing Elements.* Because a number of gratings can be written along a fiber and be multiplexed, a quasi–distributed sensing of strain and temperature is possible by serial interrogation of a limited number of gratings.

A few disadvantages should be considered. They can be overcome by using special sensor arrangements and special demodulation techniques:

- *Relatively Short Gauge Length.* At present, Bragg sensors normally have gauge lengths less than 15 mm. For sensing applications longer gratings are preferred.
- *Small Measurand–Induced Optical Signal Changes.* Because the strain–induced shift of the Bragg wavelength λ_B is very small, high resolvable spectrometers or monochromators are necessary. Such devices are large, heavy, expensive and can lose their function on–site. For this reasons, more or less complex demodulation systems are being developed. Long–period grating sensors with higher sensitivity in strain could become alternatives.
- *Temperature–Sensitivity in Case of Field Applications.* Strain measurements on–site are perturbed by temperature variations. In order to compensate for this, in general a second sensor is used that is decoupled from the strain that measures the temperature changes. This is not a good solution, especially when a large number of fiber gratings at different locations are to be interrogated, and there are several ways for overcoming the cross–sensitivity between strain and temperature. One way is the use of two superimposed grating elements with spatial periods Λ_1 and Λ_2 [38]. The static strain sensitivity of this method is reported to be $0.8 \, \mu\epsilon/\sqrt{\text{Hz}}$.
- *Weakening of the Sensor Area by Manufacturing.* Because the fiber coating must be removed at the location where a grating is to be created, and due to irradiation with UV–beams and the following annealing of the grating, the properties of the glass material that determine strength and fatigue can be expected to change. In comparison with the strength of untreated fiber zones, a lower strength for Bragg gratings should be assumed.
- *Vulnerability of the Sensors During Application.* For applying the gratings, the sensing zone, recoated after completion of the grating's creation, must be decoated again. Even if the removal procedure of the coating is carried out carefully, damage on the fiber surface should be expected.

– *Stiffness.* The stiffness of the fiber —and the corresponding grating area— makes it impossible to measure curing processes at very early ages. For such purposes, stress–free extrinsic Fabry–Pérot sensors prevail against Bragg gratings [39].

A promising approach worthy of discussion is the multiplexing of Bragg grating sensors. According to the parameters of sensor and device, a limited number of gratings (normally eight to ten) can be addressed by a single output connection. By using a wavelength–tuned multiplexing technique, a selective interrogation of Bragg grating sensors (at different centre wavelengths) is possible. Using this technique, a strain mapping system with nine sensors partitioned in two Bragg grating arrays (with serial architecture) can be configured [40]. The output signal obtained from the scanning–electron device enables the resolution of strain at the μm/m level. This strain mapping arrangement was used in [40] to monitor strain changes in a 0.15 cm–thick aluminum plate caused by deflecting forces applied to the plate. The gratings responded well to localized tensile and compressive strains.

The similar WDM technique, which uses additional lengths of doped fibers between the gratings, is the basis for a four–channel Bragg grating sensor demodulation system (with parallel architecture) [41]. The system has been tested for multipoint strain measurements on the Beddington Trail bridge in Calgary, Canada. Gratings with the same λ_B can be used; the number of driven gratings only depends on the available pump power of the laser diode. The demand for disconnecting the demodulation unit from the sensors is met by using special connectors with polished angled end–faces. This multiplexing system was designed to allow a strain dynamic range of $\sim 5\,000\,\mu$m/m ($10\,000\,\mu$m/m can be achieved, as reported). Using a 12–bit ADC, a strain resolution of $\sim 1\,\mu$m/m can be achieved. The electronics procedure developed enables a dynamic response; however, its use for long–term strain measurements is to be preferred, especially as the long–term stability for measurements on–site is reported to be $3\,\mu$m/m.

In order to detect delamination damage in the composite hull of a catamaran it was monitored by approximately 100 Bragg grating sensors embedded in the skin of the hull [42]. Strains could be detected and then that information correlated with the hull damage.

Short–Gauge–Length Sensors for Vibration Measurement and Acoustic Emission (AE) Detection. Distributed dynamic measurements, which deliver input signals for active vibration suppression, is one of the important areas of interest in smart structure engineering. Another no less important effort is the detection of local or partial loss of integrity. It has been shown to be difficult to sense vibrations or acoustic emission signals by using long–gauge–length distributed fiber sensors. A promising technique is that of interferometry based on short–gauge–length sensors. Interferometric sensors have been shown to possess the sensitivity required to respond to extremely small displacements of a material that are caused by dynamic events or acoustic

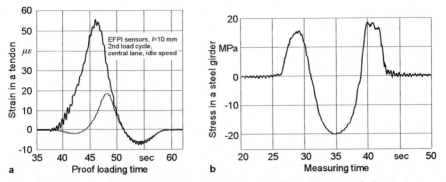

Fig. 7.21. Strain signal of proof–loaded bridges with superimposed dynamic content [43]: **(a)** pre–stressed concrete bridge in Berlin, Germany, proof–loaded by a motor–crane. From the superimposed oscillations follow the bridge's natural frequency and the amplitude spectrum; **(b)** steel bridge in Rouen, France, proof–loaded by a tram–car. The oscillations on the second positive peak clearly reflect the bridge response to crossing a rail joint.

waves. Apart from several sensor arrangements for noncontact interrogation of vibrating surfaces, which are based on reflective types of fiber sensors by means of a focused laser beam (utilized for surface velocity and length measurements), two discrete types of structure–integrated fiber sensors have been used for acoustic emission detection: two–arm Michelson interferometers and highly sensitive EPFI sensors. With regard to use for smart structures, EFPI sensors provide the better sensing method for vibration measurements and acoustic emission detection (see Fig. 7.15b).

In order to measure acoustic wave propagation, an interferometer sensor can be embedded or surface–attached, and then interrogated by using an AE detection system [31]. By using an active alignment process, the phase modulation sensitivity is very high during the interaction between the elastic stress wave fields in the material to be evaluated and strain induced in the sensor. Such a sensing technique not only allows the detection of strain waves up to 2 MHz but also the measurement of low–frequency strain. Low–frequency strain from the natural frequency of (say) an in–service load superimposed on a static strain can be detected. Figures 7.21a and 7.21b show examples of the noise–free detection of the harmonic content in a strain signal obtained by proof–loading certain structures. From this signal, typical dynamic parameters can be derived and the structural health can be assessed because the natural frequency is known to be related to it. Fig. 7.22 shows the strain response detected by a steel surface–attached sensor due to an impact pulse from a pencil lead broken on the surface [44].

By using embeddable sensors for continuous or periodic AE detection, fatigue cracks or overloading–induced cracks can be detected early and reliably so that a repair can be made at minimum cost, and routine inspections can be reduced. Increasingly, the future health monitoring of structures by

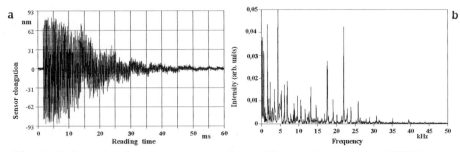

Fig. 7.22. Impact–induced strain waves detected by a surface–attached EFPI sensor (stress–free type): **(a)** sensor elongation (gap variations); **(b)** fast Fourier transform of the response shown in **(a)**

smart systems will use acousto–ultrasonic techniques. By introducing ultrasonic stress waves into the structure and detecting stress waves at definite points of the structures, changes in material damping characteristics due to damage can be recognized by using structure–integrated fiber sensors.

A similar excitation method can be exploited by using movable fiber optic microphones to detect structural inhomogenities or to produce proof of structural integrity. Another important potential application is the measurement of the velocity of acoustic waves transmitted through curing materials by a set of sensors: Depending on the state of curing, embedded sensors are able to measure different wave spectra, and after completion of the curing process the same sensors can be used for determining the in–service strain and vibration state of the structure.

Fiber Optic Temperature Sensors. Temperature sensors form the largest class of commercially available fiber optic sensors. Refraining from the advantage of electromagnetic immunity —the main reason for its use— some of these sensors can be embedded in or attached to tiny samples without perturbing or heat sinking them. Because of their small mass, they are able to respond rapidly to temperature variations. On the other hand, glass temperature sensors can be designed with a very low heat conductivity. This condition can be important for such applications, where the leads must not conduct heat. There needs to be drawn a distinction between thermometers based on the principle of thermal expansion, and fiber optic pyrometers.

One type of the latter category consists of a probe linked to a fiber optic cable. The tip of the probe contains a layer of transparent material sandwiched between two layers of reflective material. The transparent material, illuminated by a spectrum–stabilized LED source, varies its refractive index significantly with temperature. The corresponding variation in the reflection conditions modulates the colour of the light signal reflected back along the same fiber to the measuring instrument, which in turn translates the light spectrum data into temperature values. Sensor probes are commercially available for a temperature range from −40°C to 300 °C; up to four independent fiber optic sensors can simultaneously be interrogated for temperature (and also pressure) [45]. The resolution is 0.1 K (response time > 0.2 s). There are two types of temperature probes, a metallic type for harsh environments or where resistance to organic solvents is necessary, and a non–conductive type that has a probe diameter of 0.8 mm (length = 10 mm) and is specially designed for embedding into composite materials monitoring a curing process (maximum pressure load: 0.7 MPa).

Another group of commercially available temperature sensors are based upon the measurement of the temperature–dependent fluorescent decay time of photoluminescent sensor materials. The photoluminescent sensor is attached to the end of an optical fiber connected to an instrument. There are exchangeable probes for contact or noncontact sensing of temperatures in a range from −50°C to 200°C [46]. The probe tips can be replaced by means of a slip connector. One type of remote–sensing probe allows measurement in the temperature range −195°C to 300°C. There are also replaceable probe tips for measurements up to 450°C (intermittent) and for installation in an oil–polluted environment. The response times vary from 0.4 s to a few seconds; the leading fiber can be as long as 300 m. The measurement can be recorded with an error of ± 2 K RMS (uncalibrated), or ± 0.1 K RMS (temperature measurement at the point of calibration). Four channels can be interrogated simultaneously.

In order to measure temperatures higher than 350°C, sensor probes with the usual polyimide–based coatings cannot be used over a long period of time. The most suitable materials for sensors to be able to measure continuously from sub–room temperature to temperatures of about 500°C seem to be crystals (ruby or sapphire) or Cr:YAG.

The recording of very fast temperature variations, such as heat pulses for heat transfer studies, is a difficult measurement task. Normal fiber optic configurations suffer from a limited heat conductivity caused by the cladding of the sensing fiber. By using an in–fiber Fabry–Pérot interferometer sensor, and by heating up the core of the fiber directly at the cut end–face, a time response of ∼ 2.5 μs can be achieved [47]. Such a fiber optic thermometer responds to heat pulses with frequencies around 300 kHz. In comparison with the fastest conventional thermocouple (Pt–Au thin–film gauge, with a typical frequency response of ∼ 80 kHz), this technique offers a superior bandwidth in conjunction with the miniature format.

Contact sensing of temperatures above 1000°C is not possible by means of silicon–based fiber sensors and pyrometers. As an alternative, sapphire fiber sensors, with a melting point of more than 2000°C and appropriate optical properties, can be used for high–temperature fiber–based sensors. Such sensors are preferably designed as extrinsic sensors; the sensing mechanism is based on a Fabry–Pérot cavity elongation induced by thermal expansion/contraction of the materials employed. By using a sapphire fiber sensor, analoguous to a normal fiber EFPI sensor, a temperature resolution of 3.5 K can be obtained in a temperature range up to 1500°C [48].

7.2.5 Sensor Networks

When a large number of sensors are to be spatially distributed in an area and interrogated by a central processing unit, multiplexing techniques (by time division, wavelength division or frequency division) in conjunction with serial or parallel (or combined) network architectures are beneficial. In that case, the system costs and overall reliability are optimized owing to reduced expenditure of input/output interfacing. Depending on the measuring task, simple serial or parallel architectures of long–gauge–length sensors, or a combination of short–gauge–length and long–gauge–length sensors simultaneously driven and interrogated (e.g. by the OTDR–technique), or highly sophisticated interferometric sensor networks can be designed [49].

Fiber optic sensors are also a very attractive sensor base for neural network processors because of their capability of high processing speed and almost instantaneous response. Two– or three–dimensional sensor systems for real–time monitoring or determination of damage location in smart structures can be combined with an optical neural network–based processor. A pre–trained neural network is able to process the outputs of a number of sensors and then (in the case of a multiple–damage situation) indicate the damage locations.

7.2.6 Research Problems and Future Prospects

Although a remarkable number of very different fiber sensor applications have been carried out, a lot of field–application–related questions cannot yet be answered definitely. In general, the application of cylindrical sensitive elements completely differs from adhering or embedding flat foil gauges.

Fiber sensors should be built–in in newly–created structures or composites in the manner of structure–integrated neurosystems. Introduction of the fiber sensors into the manufacturing cycle of advanced materials must also be sought. For the purpose of creating really adaptive structures, a close interaction between control and actuation units have to be projected. However, when fiber sensors are embedded in a laminate material, such as glass fiber reinforced plastic (GFRP) or carbon fiber reinforced plastic (CFRP), they could reduce the tensile (or compressive) and fatigue strength of the composites. Normally, the fiber diameter is considerably larger (by up to ten times) than that of the reinforcement material. Several investigations have revealed that optical fibers that are not exactly positioned at the ply direction produce resin cavities that perturb the microarchitecture of the composite. In order to minimize this possible reduction of strength in the laminate to be provided with optical fibers, care has to be taken with the orientation and the adhesion characteristics determined by the coating material.

One more problem, already addressed above, concerns the perturbation of the material zone to be evaluated. All stiff sensors influence the measuring object especially in the plastic material stage or in low–modulus materials and this effect makes the material stiffer than it would be without sensor. A promising way round is the use of stress–free sensor designs (as described in Sect. 7.2.4).

In adaptronic structures, reliable data must be delivered from sensors over a long period of time. The user has to pay attention to three main points:

- a durable coating or covering material has to be chosen;
- a reliable load transfer from measurement object into the sensing element has to be arranged, that is free of perturbing effects (e.g. temperature, transverse pressure); and
- the installation method must not obstruct the construction process and the long–term functionality of the object being interrogated.

It has been shown that the coating materials usually used can fail under raw environmental conditions. The load transfer characteristics can be perturbed or a long–term bonding to the measuring object cannot be achieved. Alternative paths must be trodden.

There are worldwide activities to throw light on these problems. Extensive investigations are carried out in interdisciplinary research programs supported by national governments or the European Community (e.g. the Optical Sensing Techniques in Composites (OSTIC) (BRITE sponsored) program under the auspices of the EC [50], and the LINK Structural Composite Programme in the UK [51]). Intense research work addressing these questions has also been carried out for years in North America at the University of Toronto [33], the University of Maryland [52], and the Fiber & Electro–Optics Research Center, Virginia Polytechnic Institute and State University, Blacksburg, VA, USA (FEORC) in collaboration with the Center for Intelligent Material and Structures of the Virginia Tech [27].

The practical use of embedded and surface–mounted fiber sensors will increase strongly over the next few years; more and more fiber sensor techniques will be made fit for long–term stable operation under field conditions. This objective requires more intensive interdisciplinary cooperation with closer collaboration between potential user, sensor developer and electronic engineers. The main problems to be solved are dictated by the structural interface and the interconnection of the sensors to the host materials (with the aim of creating a standard interface for fiber sensors). Greater cooperation will then lead to results being more quickly available for practical applications, and finally to a better price–to–perfomance ratio in comparison with established sensors: low fabrication costs are essential because a large number of sensors are required.

References

1. Brignell, J.E.: Quo vadis smart sensors?. Sens. Actuators, A37–38, pp. 6–8 (1993)
2. Brignell, J.E. and White, N.M.: *Intelligent sensor systems.* (Institute of Physics Publishing, Bristol, 1994)
3. Brignell, J.E.: Digital compensation of sensors. J. Phys. E: Sci. Instrum., **10**, pp. 1097–1102 (1987)
4. Brignell, J.E.: Sensors in distributed instrumentation systems. Sens. Actuators, **10**, pp. 249–261 (1986)
5. Brignell, J.E.; White, N.M. and Cranny, A.W.J.: Sensor applications of thick–film technology. IEE Proc. I, 135, **4**, pp. 77–84 (1988)
6. White, N.M. and Brignell, J.E.: A planar, thick–film load cell. Sens. Actuators, **26**, (1/3), pp. 313–319 (1991)
7. White, N.M. and Brignell, J.E.: Excitation of thick–film resonant structures. IEE Proc.– Sci. Meas. Technol., 142, **3** pp. 244–248 (1995)
8. Taner, A.H. and Brignell, J.E.: Aspects of intelligent sensor reconfiguration. Sens. Actuators, A46–47, pp. 525–529 (1995)
9. Taner, A.H. and Brignell: The role of the graphical user interface in the development of intelligent sensor systems. Man–machine interfaces for instrumentation, IEE Digest No: 1995/175, 3/1–3/6
10. Taner, A.H. and Brignell, J.E.: A graphical user interface for intelligent sensor ASIC reconfiguration. Proceedings of Sensors & their Applications VII, Dublin, 10–13, pp. 365–369 (Sept. 1995)
11. Sensors - A Comprehensive Survey. (edited by W. Göpel, et al.). Chapter Optical-Fiber Sensors VCH 1992
12. Inaudi, D.: Fiber Optic Smart Sensing. (pre–print) *Handbook of the Methods in Optical Metrology* (to be edited by P.K.Rastogi, Wiley).
13. Habel, W.; Hofmann, D.: Application of fibre–optic sensors for measurements on for damage detection on structures. Symposium on Non–Destructive Testing in Civil Engineering. Berlin '95, Proc. **2**, pp. 1025–1032.
14. Brönnimann, R.; et al.: Measurement of Crack Propagation in Polymer Pipes with Embedded Optical Fibers. Smart Structures: Optical Instrumentation and Sensing Systems Conf. 1995, SPIE **250**, pp. 12–19.
15. Noharet, B.; et al.: Impact detection on airborne multilayered structures. Smart Structures and Materials 1995, SPIE **2444**, pp. 460–468.
16. Globale Bauwerksüberwachung mit Optischer Saite. Fa. DEHA–COM GmbH Köln.
17. Vulliet, L.; et al.: Development and laboratory tests of deformation fiber optic sensors for civil engineering applications. Laser, Optics and Vision for Productivity in Manufacturing I. Besancon June 1996. (pre–print)
18. Inaudi, D.; et al.: In–line coherence multiplexing of displacement sensors: a fiber optic extensometer. Smart Structures and Materials San Diego 1996, SPIE **2718**, pp. 251–257.
19. OTDR–Prospekt Opto–Electronics Inc. Canada (1995) and Information by L.O.T. Oriel GmbH Darmstadt, 02/96.
20. Geiger, H.; et al.: Multiplexed Measurement of Strain Using Short and Long Gauge Length Sensors. Pre–print by the author.
21. Campbell, M.; et al.: Optimisation of Hi–birefringence Fibre Based Distributed Force Sensors. Smart Structures: Optical Instrumentation and Sensing Systems Conf. 1995, SPIE **2509**, pp. 57–63.

22. Michie, W.C.; et al.: Distributed pH and Water Detection using Fiber–Optic Sensors and Hydrogels. J. of Ligthwave Technology 13(1995)7, pp. 1415–1420.
23. Hurtig, E.; et al.: Fibre optic temperature sensing: application for subsurface and ground temperature measurements. Tectonophysics 257(1996) pp. 101–109.
24. Wanser, K.H.; et al.: High Temperature Distributed Strain and Temperature Sensing using OTDR. in: Applications of Fiber Optic Sensors in Engineering Mechanics (Ed. by F. Ansari). ASCE 1993. New York.
25. Boiarski, A.A.: Distributed Fiber Optic Temperature Sensing. in: Applications of Fiber Optic Sensors in Engineering Mechanics (Ed. by F. Ansari). ASCE 1993. New York.
26. Bhatia, V.; et al.: Optical Fiber Extrinsic Fabry–Pérot Interferometric Strain Sensor for Multiple Strain State Measurements. Smart Structures and Materials, SPIE **72444**, pp. 115–126.
27. deVries, M.; et al.: Applications of Absolute EFPI Fiber Optic Sensing System for Measurement of Strain in Pre–Tensioned Tendons for Prestrained Concrete. Smart Structures and Materials, SPIE **2446**, pp. 9–15.
28. Murphy, K.;A; et al.: Fabry–Pérot fiber optic sensors in full–scale fatigue testing on an F–15 aircraft. Fiber Optics Smart Structures and Skins IV Conf. 1991, SPIE **1588**, pp. 134–142.
29. Unpublished Research Report. No.: 4 – 06/96 IEMB e. V. Berlin/Germany, 1996.
30. Badcock, R.A.; Fernando, G.F.: Fatigue damage detection in carbon fibre reinforced composites using an intensity based optical fibre sensor. Smart Structures and Materials, SPIE **2444**, pp. 422–431.
31. F & S Technologies Inc. (USA), Product Data and Technical Specifications, 1996.
32. Canadian Marconi Company (Canada). CMA–2026 Fiber–Optic Strain Sensor System Specifications. 1994.
33. Measures, R.M.: Fiber Optic Sensing for Composite Smart Structures. in: AGARD Conf. Proceed. 531(1992), pp. 11–1 to 11–43.
34. Habel, W.R.; et al.: Fiber sensors for damage detection on large structures and for assessment of deformation behavior of cementitious materials. Proc 11th Engineering Mechanics Conf. Fort Lauderdale 1996, Conf. **1**, pp. 355–358.
35. Meltz, G.; et al.: Formation of Bragg gratings in optical fibers by a transverse holographic method. Optics Letters 14(1989)15, pp. 823–825.
36. 3M Specialty Single–Mode Fiber Products. (1996), via AMS–OptoTech. GmbH, Martinsried/München.
37. Technical Specifications: Bragg–Photonics Inc. and QPS Technology Inc. (1996), via Döhrer–Elektrooptik GmbH, Karlsbad/Germany.
38. Farahi, F.: Simultaneous Measurement of Strain and Temperature Using Fiber Grating Sensors. Proc 11th Engineering Mechanics Conf. Fort Lauderdale 1996, Conf. **1**, pp. 351–354.
39. Habel, W.R.; et al.: Deformation measurements of mortars at early ages and of large concrete components on site by means of embedded fiber optic microstrain sensors. Cement & Concrete Composites (Special issue on Fiber Optic Sensors) 19(1997)1, pp. 81–102.
40. Davis, M.A.; et al.: Structural Strain Mapping using a Wavelength/Time Division Addressed Fiber Bragg Grating Array. 2nd. European Conf. on Smart Structures and Materials. Glasgow 1994, SPIE **2361**, pp. 342–345.
41. Technical Specification: Fiber Optic Strain Measurement System. ElectroPhotonics Corp. 1995, and private communication (1996).
42. Troy, Ch., T.: Fiber Optic Smart Structures. Photonics Spectra, May 1997, pp. 112–128.

43. Habel, W.R.; Hillemeier, B.: Results in monitoring and assessment of damages in large steel and concrete structures by means of fiber optic sensors. Smart Structures and Materials, SPIE **2446**, pp. 25–36.

44. Hofmann, D.: Unpublished research report. Fibre Sensor Laboratory of the IEMB e.V. Berlin/G, Nov. 1996.

45. Technical Specifications: Fiber Optic Probe Guides und METRICOR 2000–System. (1996). via Photonetics GmbH Stuttgart/Germany.

46. Technical Specifications: Fiberoptic Probes and Accessories for Model 790 Systems. Luxtron 03/93. via Polytec GmbH, Waldbronn/Germany.

47. Farahi, F.: Fiber Optic Sensors for Heat Transfer Studies. SPIE-proc. **1584**, pp. 53–61.

48. Wang, A.; u.a.: Sapphire optical fiber–based interferometer for high temperature environmental applications. Smart Materials & Structures 4(1995), pp. 147–151.

49. Kersey, A.,D.: Multiplexed Fiber Optic Sensors. Distributed and Multiplexed Fiber Optic Sensors II, Boston 1992, SPIE **1797** (1992) pp. 161–185.

50. Michie, W.C.; et al: Optical Fiber Techniques for Structural Monitoring in Composites. in: AGARD Conf. Proceed. 531(1992), pp. 8–1 to 8–9.

51. Eaton, N.C.; et al: Factors Affecting the Embedding of Optical Fibre Sensors in Advanced Composite Structures. in: AGARD Conf. Proceed. 531 (1992), pp. 20–1 to 20–14.

52. Sirkis, J.S.: Electro-opto-mechanical Design of Fiber Optic Smart Structures. 2nd. European Conf. on Smart Structures and Materials. Glasgow 1994, SPIE **2361**, pp. 330–337.

53. Gardner, J.W. and Bartlett, P.N. (eds): *Sensors and sensory systems for an electronic nose.*(Kluwer Academic Publishers, Dordrecht, 1992)

8. Adaptronic Systems in Engineering

8.1 Adaptronics in Space Missions —
An Overview of Benefits
B.K. Wada

Often the primary objective of space exploration and utilization includes in–situ measurements and observations, and then the transmission of signals back to Earth. Precision mechanical systems are necessary to help meet this objective. In the 1980s, one of the technical challenges was the control of large (up to 100 m in dimensions) and precise (sub–micron level) space structures. These challenges motivated the initiation of research in adaptronics to provide alternatives to the approach of adding hundreds, if not thousands, of sensors/actuators and their controllers in order to control the vibrations of 'linear' structures. Among the difficulties with the proposed Multiple Input Multiple Output (MIMO) approach were reliability of the control system, processing demands, and the necessary accuracy of knowledge of the structural system characteristics in an operating environment in space.

Successful experimental results by [1] and others at several universities in adding active–add damping to small cantilevered structures, using piezoelectric actuators at its base near the area of maximum strain energy, provided the initial motivation. One early definition to identify its applicability to a broad range of space challenges is 'a structural system whose geometric and inherent structural characteristics are beneficially changed to meet mission requirements either through remote commands or automatically in response to adverse external stimulation' [2]. This definition expands the benefit of just adding damping to the structure to include solutions to other sustaining challenges including ground validation tests, deployment, system identification, linearization, static adjustment, reliability, and cost reduction. The significant point is that the definition focuses on the potential benefits and not just the technologies.

The simple approach to implementing the ideas of adaptronics into successful laboratory and space experiments stimulated interest. From small beginnings with the first technical session [3] and the first annual conference [4], a large number of meetings, two journals and several technical committees currently exist.

8.1.1 Basic Principles

Prior to adaptronics, the approach to the dynamics of space structures is analytically and experimentally to determine the eigensolution, to determine the *inertial* displacements and accelerations due to the force functions, and then to evaluate the design forces or displacements. The control of large structures moved in parallel with the approach of dynamics by placing sensors and actuators at locations of maximum motion and attempting to properly phase a subset of actuators in order to cancel the structural motion. Cancellation necessitates an accurate prediction of pending structural motion, and one difficulty in the approach includes the inability to predict reliably the motion of a structure, and to the necessary accuracy.

The application and integration of adaptronics represent a return to the fundamentals of structures, namely to sense and actuate the basic parameters of structures, which include internal forces, relative deformation, and internal strain energy. A structure itself provides communication between the sensors and actuators by its ability to transmit force magnitude and distribution information.

The challenges using the basic parameter of inertial acceleration (an approach prior to adaptronics) for control of space structures are as follows.

1. Proof–mass actuators must provide the external inertial forces because ground does not exist to react against the forces.
2. Difficulties exist in the design of proof–mass actuators with high internal resonance to avoid its coupling with the structure.
3. The sum of the external forces will result in a net excitation force in all six rigid–body degrees–of–freedom due to errors in the location and direction of the external forces relative to the center of gravity.
4. Accelerometer output includes both the rigid and elastic structure motions. The elastic motion to be controlled is determined by subtracting the rigid body motion from the total motion, potentially resulting in large errors for very small elastic motions.
5. An extremely accurate real–time knowledge of the structural characteristics is necessary to control the structure. The controlling forces are potentially destabilizing.
6. High–speed processors are necessary to simultaneously control all actuators and detect erroneous signals or output from faulty sensors and actuators.
7. The inertial actuators are not suitable for quasi–static adjustment changes.
8. The system is not robust.

The advantages in using the basic parameters of internal forces and strain energy (adaptronics) are thus.

1. Actuators and sensors are generally at locations with high strain energy. Each actuator–and–sensor collocated pair is capable of independently

withdrawing strain energy from the system. Since strain energy is an algebraic quantity, the energy dissipation is a direct function of the number of actuator–and–sensor pairs. Redundant actuator/sensor pairs are a part of the system to add redundancy.

2. An actuator integrated into a structure results in very high stiffness design with high internal resonance.
3. The control algorithms are local, simple, and require simple processors.
4. The control algorithms are very insensitive to structural uncertainties, including nonlinearity.
5. The actuators as excitors for on–orbit system identification only excite the elastic deformations because they subject the structure to equal and opposite forces.
6. The actuators can provide for structural softening or stiffening so as to adapt the structure to make in more controllable.

8.1.2 Application of Adaptronics for Space

Although the motivation of adaptronics for space is to actively damp structures, many other benefits exist that provide both benefits and future challenges. The challenges and potential for their solution using adaptronics are as follows.

Robust Design. A significant benefit of adaptronics is the ability of designers to develop robust structural designs [5]. A robust design does not depend upon stringent control of the individual design and process parameters to assure its operational success. The conventional approach to design, fabrication and testing of large precision structures requires an overwhelming number of precisely controlled parameters and processes; thus the design is not robust; furthermore, the requirements for the controlled parameters and processes require extensive analysis and laboratory experiments. Adaptronics, in contrast, provides for a robust design by adjusting the structure during its operation to compensate for the allowable errors in the parameters for the various designs and processes, and thus provides for a lower–cost design applicable for large and small spacecraft.

Ground Testing. [6] identify the difficulty and importance of ground testing to validate large precision structures. The difficulty in ground testing is the earth's gravitational force that induces upwards of three orders of magnitude more preloads and internal strain energy into the spacecraft structure than it experiences in its operational state. Thus very often significant strain energy sources, small gaps in structural joints, and the structural precision information important to the success of the structure are not detectable on the ground. Using adaptronics, the role of ground testing is to bound the uncertainties in the significant characteristics of the structure and to establish that the range of the adaptability encompasses the uncertainties and includes the spacecraft's operational design conditions.

Deployment and Space Assembly. Reliability of deployable structures continues to be a problem. The external forces necessary to deploy the structure do not have the external work needed to overcome any unexpected internal strain energy within the structure, leading to incomplete deployment. Adaptronics provides approaches to remote control of the level of strain energy buildup [7] in the deploying structure by including active members in appropriate locations that extend or contract at user–controlled levels of tensile or compressive loads respectively. For structures assembled in space, problems related to the installation of members between joints are similarly resolvable by controlling the magnitude of internal strain energy.

Linearization of the Structure. Gaps in joints are difficult to detect on the ground because of gravitational loads. Their potential presence in space is troublesome because it creates a non–deterministic structure. Experiments indicate such structures respond chaotically when subjected to a harmonic input. To attain, retain and control these structures at the displacement levels equivalent to the magnitude of the gaps becomes an insurmountable problem. One approach to attaining controllable preloads in carefully designed indeterminate structures by preloading selected members using active members is suggested by [8]. Preloaded members preload the joints, leading to a linear structure.

System Identification. Adaptation of the structure in space requires knowledge of the structural characteristics prior to its modification. The active adaptronics members in the structure become excellent exciters and sensors to perform modal surveys of the system [9, 10]. The direction and magnitude of force application are compatible with the motions of response. The equal and opposite force application of each active member excites only the elastic modes, not rigid body motions.

Controllable Structures. Linearization of the structure belongs in the category of controllable structures [11]. Other approaches include the capability to change the dynamics of the structure in order to avoid multiple eigenvectors, eliminate modal localization and change the eigenvectors to enhance the observability of the sensors on the spacecraft.

Static Adjustment. Adjustment of the quasi–static geometry of the structure is a major challenge in precision structures that is often more significant and difficult than controlling the dynamics. [8] and [12] establish approaches that allow the optimal correction of the structural surface errors with a finite number of active members. 'Inchworm' actuators exist that are capable of large displacement and high resolution, and they accurately maintain the position without power. One example shows that the use of active members in 10% of the truss elements of a structure is capable of reducing the errors by about 80%.

Vibration Attenuation. Adaptronics continues to add reliable, active damping to structures — the original objective that initiated research in adaptronics for space. A local feedback controller [13] using a wide variety of active members provides damping to the structure and damps selected modes to allow high–performance MIMO controllers [14]. A vibration attenuation system [15] in conjunction with vibration isolation and active optical path-length control has successfully reduced errors by over a factor of 5000. The pathlength was controlled to about 10 nm. Flight experiments demonstrating this technology include damping of a 12–meter–long truss beam [16] during 10–20 s in a zero–gravity environment on a KC–135 aircraft in 1990, damp-ing of a 1.8–meter–long truss on the MIR space station [17] in 1996, and an ACTEX–I experiment [18] on a spacecraft flown in 1996 to demonstrate damping using embedded piezoelectric actuators.

Vibration Isolation, Suppression, Steering, and Pointing (VISSP). Adaptronics provides the capability to develop a subsystem between a pre-cision instrument and a spacecraft that provides isolation from spacecraft vibrations, cancels motion from forces applied directly to the instrument, and precisely steers and then points the instrument. The capability of six active members to provide vibration isolation and suppression has been suc-cessfully demonstrated in [19]. In 1994, during the STRV–!b space mission, [20] validated the technology to cancel static and dynamic motions at the tip of a cryo–cooler finger for the five harmonics, starting at 60 Hz. In [21] the feasibility of VISSP was established. An experiment to space–validate VISS by the USAF Phillips Laboratory and the Jet Propulsion Laboratory is in development for a planned launch in 1998.

8.1.3 Summary

The objective of this first section of Chap. 8 is to describe the benefits and the current status of adaptronics to meet the challenges of current and future space missions. Many of the components to integrate adaptronics into a space system exist. At the time of writing the challenges are to identify the benefits to the appropriate designers, and to utilize and demonstrate the value of adaptronics.

8.2 Adaptronic Systems in Aeronautics and Space Travel
C. Boller

8.2.1 Implications and Initiatives

Implementation of sensors, actuators and possibly even control into a struc-ture or material is already a major issue in aeronautics and space travel. Air-craft nowadays all contain a large number of sensors, such as for temperature, speed, acceleration, brightness, humidity, volume flow or electromagnetic signals, and actuators allowing the movement of components, the exertion of forces, the injection of a fluid or a gas, or the reduction in light intensity. All actuation is performed by either human or electronic control, the latter requiring the type of sensors mentioned before. Within this chapter, however, these systems will not specifically be described. The intention here is much more to describe systems that go beyond this stage, being still generally related to research and development. The focal areas for adaptronics have been determined to be [22, 23](cf. Sects. 8.2.1 and 8.2.2):

- aircraft health and usage monitoring, which is a sensory system mainly related to the identification and possibly validation of damage in structures or of situation awareness (e.g. a foreign object, or friend–and–foe recognition);
- active and adaptive structures, which is related to shape control, vibration damping and light permeability of structural components, using materials with sensing and actuation capabilities such as piezoelectrics, SMAs, ER fluids or electro–chromatic glasses;
- adaptronic skins, being load–carrying structural elements with integrated avionics (antennae), which can be either a sensory, active, adaptive or even intelligent structure depending upon its term of use.

So far fiber optics, piezoelectric materials, shape memory alloys (SMA), electrorheological (ER) fluids or electrochromatic glasses are the sensing and actuation materials and elements used today (cf. Sects. 8.2.3 and 8.2.4). These elements are added to a structural system made out of passive materials and are linked together with a controller, thus resulting in what can be called today a realized adaptronic system. This is the initial conceptual idea for enhancing the performance of actual and next–generation aerospace structures and has resulted in a variety of major activities such as the space activities at NASA–JPL (e.g. Controls and Structures Interaction (CSI) program, Precision Segmented Reflector (PSR) program, etc.), the US Air Force smart structures program, SDI or research and development programs going on in the area of adaptive wings (NASA/Lockheed, Northrop–Grumman, DASA/DLR/ Daimler–Benz, BAe/Dowty) or in adaptive rotorblades (e.g. Eurocopter/DLR/Daimler–Benz, MURI–Programme between Univ. of Maryland, Cornell Univ., Penn. State Univ. and Univ. of the District of Columbia in the USA). In damage monitoring, the USAF Wright Laboratory Smart Metallic Structures Program and its relation to the USAF Aircraft Structural Integrity Program ASIP has been a major initiative [24]. Another ongoing EC–funded project entitled 'Monitoring On–line Integrated Technologies for Operational Reliability' (MONITOR) has been started. The use of fiber-optic sensors for monitoring load sequences and determination of suitable techniques for on–board damage monitoring in aircraft are the topics to be covered in that project. Techniques to be developed will be evaluated in simulations as well as on real components and partially also in a flight test.

Within most of these programs, a three–step approach is considered, which can be summarized for an adaptive–wing concept as follows:

1. Take the existing structure (e.g. a wing) with its conventional sensors and actuators and improve control and data processing, which allows extension of the use of existing flaps.
2. Implement sensors and actuators based on appropriate materials with known characteristics (e.g. piezoelectrics, SMAs, ER–fluids) and start to adapt the structural design according to these materials.
3. Use an adaptronic composite system for design of the wing.

So far the status of these programs is such that work is mainly performed within the first step, adding conventional sensors, actuators and control to a structure that is conventionally designed on the basis of conventional passive materials. Progress in data processing and handling is specifically taken advantage of. The latter two steps stated above, however, strongly require more knowledge in what can be called sensory, active, adaptive, intelligent or generally speaking adaptronic materials.

Since activities in adaptronics have mostly originated from aerospace applications, a relatively large amount of publications has come from this sector. A first workshop on 'Smart Materials, Structures, and Mathematical Issues' was presented at the U.S. Army Research Office in 1988 [25]. In 1992 AGARD's Structures and Materials Panel organized a meeting on 'Smart Structures for Aircraft and Spacecraft' [26], which has possibly been one of the first conferences on smart structures being fully related to aerospace. In 1994 the American Institute for Aeronautics and Astronautics (AIAA) and the American Society of Mechanical Engineers then arranged the 'Adaptive Structures Forum' [27], which has been mainly related to aerospace. A second forum of that kind took place in 1996.

Another conference being very much related to aerospace is the International (originally U.S./Japan) Conference on Adaptive Structures [28], which was established in 1991 and has been held since every year. Other conferences where adaptronics applications for aerospace are partially presented include the annual 'Fiber Optic Smart Structures and Skins' [29] organized by SPIE since 1988, the annual 'Adaptive Structures and Materials Conference' and the biannual 'European Conference on Smart Structures and Materials', both being held since 1992.

A variety of papers being related to aerospace can also be found in the two international journals on adaptronics, namely is the journal *Smart Materials and Structures* (now published bimonthly by the Institute of Physics Publ. since 1992 and edited by R.O. Claus, G. J. Knowles and V.K. Varadan) and the *International Journal on Intelligent Material Systems and Structures* (published quarterly by Technomic Publ. Co. since 1990 and edited by C.A. Rogers and recently by J.S. Sirkis). Another important source is the *AIAA–Journal* .

A lot of motivation has been spread through the ideas generated and successfully realized for space applications (e.g. [30]–[32]). In the meantime, applications are also considered with respect to military and civil aircraft, which will be described throughout the following section.

8.2.2 Condition Monitoring

Condition Monitoring is related to sensing aerospace structures. This includes sensing flight and/or position parameters, recognition of foreign objects or even monitoring the health of structural and propulsion systems. The former can be considered to be established nowadays and will therefore not be discussed. The latter is, however, related to sensing systems that also take over load–carrying functions, such as conformal antennas.

For space structures, monitoring can help to know the status of the mission and to which degree these structures can still be used after environmental degradation, thus providing information on the space structure's effectiveness. Monitoring of discriminating hostile threats play an additional role in defence.

Another important aspect are adaptronic skins (mainly denoted as smart skins). They are designed for control of aerospace structures such as acoustic noise and vibration, drag and skin friction using advanced polymeric smart materials, MEMS (Microelectromechanical Systems) and built–in antennas [33]. The objective is to develop something being denoted as 'smart wall papers'. Applications include smart helicopter rotorblades with microstrip patch antennas and detection and discrimination of hostile threats resulting from laser, radio–frequency and X–rays, such as might occur in the satellite attack warning and assessment flight experiment (SAWAFE) for actively filtered transparencies and conformal antennae [34]. Wireless remote and continuous telemetry for application to rotorcraft and smart skin aerospace structures are further areas discussed [35].

Damage Monitoring. Aerospace structures have been traditionally designed for safe life. Improved knowledge (i.e. in fracture mechanics), better design tools (i.e. fatigue life evaluation concepts) and better technology (i.e. in nondestructive testing) has, however, allowed these structures to be designed fail–safe, thus leading to lighter structural weight and/or enhanced performance.

To allow for fail–safe conditions, monitoring procedures and systems have been introduced. Many of them — and especially those being used during flight — are not able to monitor damage itself but only a damage–related parameter such as loading. This therefore requires the additional inclusion of a theoretically based procedure, which converts these loads into a likelihood of damage occurring or progressing. Knowing more precisely the incident of damage thus allows transfer of costly manpower–related off–service inspections into in–service automated inspection without compromising operational reliability. Benefits are therefore seen as a result of automation and increased availability.

Condition monitoring has become of real value with composite materials, where major concern exists with respect to barely visible impact damage (BVID) that occurs inside the structural material and can often not be seen from the outside. To minimize the effort required for inspecting the material or component with conventional Non Destructive Testing (NDT) technology, the NDT–technology has to become an integral part of the material itself [36]. This can be done by integrating or adapting a sensor network into the structural material. The types of sensors mainly considered are fiber optics [37, 38] and piezoelectrics [36]. These sensor networks allow either the monitoring of the damage–inducing load (i.e. the impact load) or a defined signal, which has been sent through the material. The latter case ends up as a material or structure with integrated NDT–capabilities, being directly able to detect damage. Other uses with such a material include monitoring any condition which is required as an input for an adaptive or intelligent structure (e.g. monitoring the pressure profile for an adaptive wing).

Built–In Non–Destructive Monitoring. A summary of NDT–techniques being mainly used today is listed in Table 8.1; these techniques have been validated with respect to their potential of being integrated into a material or structure. Apart from visual inspection, use of ultrasonics and eddy currents are the procedures generally applied. Procedures such as computer tomography, holography, shearography, thermography, Barkhausen noise or magneto optical eddy current have recently gained more attraction because of improved availability of these techniques. Major applicability and experience gained with these techniques has been with metals. A limited number of them is also applicable to composite materials.

Future aircraft health and usage monitoring systems to be designed as built–in systems must be able to monitor damage on–line. Major damage to be detected includes fatigue and corrosion in metallic structures, and delaminations, especially BVID, in composite structures, resulting either from material degradation (aging) or mishandling (including misrepair). Damage has to be detected, localized and monitored in accordance with damage tolerance design principles.

Under these conditions strain, chemical sensing, modal analysis, Lamb waves, acoustic emission, and acousto–ultrasonics are the monitoring technologies worth considering. An overview of the different techniques is also given in [36].

Modal Analysis, which is based on the fact that vibration modes change due to a change in structural integrity, is a further NDT procedure which is widely applied for monitoring space structures. Hickman et al. [39] have shown on a demonstrator aircraft that modal analysis could well be used for monitoring damage such as lost rivets or bolts. Analytical and experimental results published in various papers [40] show that the delamination's size

294   8. Adaptronic Systems

Table 8.1. Technology selection for health monitoring systems

Monitoring Technology	In-Flight Applicability	Degree of Development	Loads Monitoring	Damage Identification	Damage Propagation	Fatigue Crack	Impact (BVID)	Delamination	Corrosion
Visual/Borescope	No	high				Yes			Yes
Strain	Yes	high	Yes			(Yes)			
Flight Parameters	Yes	high	Yes			(Yes)			
Magnetic Particle	No	high		Yes	Yes	Yes			
Eddy Current	No	high		Yes	Yes	Yes	(Yes)	(Yes)	
Penetrant	No	high		Yes	Yes	Yes		(Yes)	
Paintings	No	low		Yes			Yes		
Chem. Sensing	Yes	low		Yes					Yes
Radiography	No	high		Yes	Yes	Yes	Yes	Yes	Yes
Modal Analysis	Yes	high	Yes	Yes	Yes	Yes	Yes	Yes	?
Acoustic Emission	Yes	high	Yes	Yes	Yes	Yes	Yes	Yes	(Yes)
Holography	?	low		?	?	?	?	?	?
Ultrasonic	Yes	high		Yes	Yes	Yes	Yes	Yes	?
Acousto-Ultrasonics	Yes	high		Yes	Yes	Yes	Yes	Yes	?
Lamb Waves	Yes	high		Yes	Yes	Yes	Yes	Yes	?
Shearography	?	low		?	?	?	?	?	?
Thermography	No	low		Yes	Yes	Yes	Yes	Yes	
Barkhausen Noise	?	low	?	?		?			?
Magneto. Opt. Eddy Curr.	No	low		?	?	?			?
Comp. Aid. Tomography	No	low		Yes	Yes	Yes	?	?	?

must be at least 10% of the surface monitored by a single sensor to be reliably detected by that technique.

Acoustic Emission (AE) has been successfully used for monitoring discontinuities, fatigue failures, materials behavior, welds (including welding processes) or stress corrosion cracking in pressure vessels, aerospace vehicles and engineering structures. Acoustic emission is the elastic energy being suddenly released when materials undergo deformation. The F–111 fighter/bomber aircraft has been tested in a chamber where the aircraft was periodically chilled to $-40°C$ and stressed between $+7.3g$ and $-3.0g$ and an AE system was used to locate sources of structural failure [41]. Techniques were developed to eliminate loading noise, and a strategy was established to identify locations where sensors should be placed to obtain optimized signals. These developments have become feasible since handling, processing and interpretation of data has been improved through better computer technology and freshtrials [42]. Acoustic emission can also be well applied for determining damage in polymer–based composites. However the high ratio of acoustic signal damping in these materials must be kept in mind, which can lead a sensor to be required every 10 cm, depending on the frequency to be monitored.

Acousto–ultrasonics is a technique which has been proved to be even more sensitive than acoustic emission [43]. Fig. 8.1 illustrates the principle of this technique. It requires two probes, one of which is used to introduce ultrasonic stress waves into the structure (actuator) and the other to pick up these stress waves at another position (sensor), where sensor and actuator can both be piezoelectric elements. As soon as the damaged area lies between the two probes, the shape of the received acoustic signal changes because of change in material–damping characteristics due to the damage (crack, delamination) that has occurred. Initial tests performed by Keilers and Chang

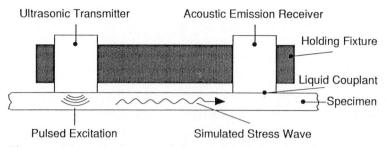

Fig. 8.1. Schematic diagram of the acousto–ultrasonic technique

[44] on composite plates and joints have shown that the method works and is worthy to be pursued.

Lamb waves is another method, being also related to the actuator/sensor principle which is based on Lamb wave theory (e.g. [45, 46]). A longitudinal and transverse plate wave is emitted into the structure. The ability of these waves to propagate over long distances is highly advantageous. However, much care is required to find the right angle for inducing the Lamb waves, especially when the structure to be monitored is of a geometric shape significantly different to that of a plate.

Sensors and Data Processing. As has been mentioned before, strain gauges or piezoelectric sensors are possible options for being used as sensors. Another type of sensors being often mentioned is fiber optics. Measures [37] performed a variety of laboratory tests in integrating optical fibers into the leading edge of a composite aircraft wing for damage detection, and an overview of the scheme is shown in Fig. 8.2.

Even though this leading edge seems not to have been flight–tested, fiber–optic sensors have been flight–tested under other circumstances, as reported in [47, 48], and showed good performance. The equipment required for signal data processing is actually still comparatively large and costly, but progress in data processing and electronic units leads to the expectation that this can be significantly reduced over the next few years.

Neural networks. For handling the data generated by a large number of sensors, neural networks have become a viable technique. So far, a lot of numerical work has been done for damage detection (e.g. [49]), which shows promise with regard to future experimental work.

Situation Monitoring. Besides monitoring damage, a lot of other things can be considered as well, such as the position of flaps or landing gear, the recognition of foreign obstacles, meteorological conditions, or the pressure distribution along an aerodynamic profile. This monitoring is done using smart load–bearing antennae (smart skins), where adaptronics plays a major role.

Antennae used today in aerospace systems are mainly applied for communication between the aerospace vehicle and the ground, between different

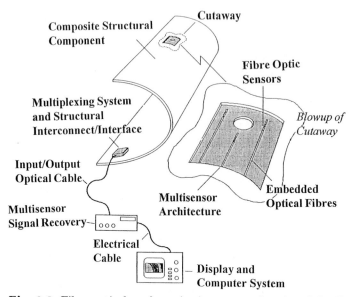

Fig. 8.2. Fiber optic–based monitoring system for aircraft leading edge [37]

vehicles, or for determining any other environmental conditions. Whenever considering the integration of various sensors into an aerospace vehicle, antennae could very much help to collect the information generated by these sensors and thus avoid a possibly large amount of wiring. Varadan and Varadan [50] have made suggestions on how to remotely transmit the sensor signal information via electronically steerable antennae (Fig. 8.3).

However, care is required with regard to the situation because other electromagnetic effects may affect the sensor signals being transmitted. Recently such a system has been proposed for three–dimensional monitoring of flaws in materials [51].

Within the Satellite Attack Warning and Assessment Flight Experiment (SAWAFE), different types of sensors have been tested for viable on–orbit health and status monitoring [34]. The technology goals were to define the nature of an attack, provide awareness of tampering with the space platform or its primary sensors, and provide collateral information for failure analysis. Pyroelectric polymer films such as PVDF were considered for infrared laser detection. Si/Au thermocouples have been selected for continuous wave response indication. Multiple–layer coatings to be deposited on the polymer for providing selective absorption at specific wavelengths have been considered. Electronics for preamplification is provided by surface–mounted packages on flexible printed circuit board substrates to maintain conformability. For X–ray sensing, charged couple devices (CCDs) and special types of fiber optics were developed where the latter was glass–darkening fibers and scintillation fibers respectively. Both sensor types had to be integrated onto an RF an-

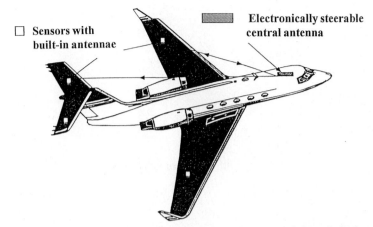

Fig. 8.3. Wireless telemetry for health monitoring of aircraft [50]

tenna structure. An overview of the design showing laser sensors being placed between the arms of a spiral antenna and wound layers of scintillating and darkening fiber sensors is shown in Fig. 8.4. The mechanical design of the avionics system has been based on a 'rack' architecture and has been positioned inside the spacecraft.

Ongoing Activities. For over 20 years the Aircraft Structural Integrity Program (ASIP) [52], as well as others, has been ongoing within the United States Air Force (USAF). ASIP itself requires that an airframe be capable of withstanding the growth of an assumed initial flaw under normal operational usage over a prescribed time interval. This can be well done by use of a monitoring system. To this extent a Structural Health Monitoring System (SHMS) has been designed for the USAF on a modular basis [53] (Fig. 8.5). The system includes sensors, local preprocessors, a central processor, and software capable of making aircraft maintenance and logistics decisions. Individual sensors track strain, acceleration, temperature, corrosive environment, and structural damage. Due to the modular architecture of the SHMS with physically distributed units being logically centralized, excellent flexibility is achieved, allowing for system growth and ease of replacement without impacting the baseline design.

In Europe, the MONITOR–programme's objective is to determine potential reductions in operating cost of aerospace structures by implementing either a structure–integrated loads or damage monitoring system. MONITOR aims to (1) understand transport operators requirements and translate these into a health and usage monitoring system specification, (2) develop a monitoring system that will allow structural usage to be effectively monitored and assessed, (3) develop a damage detection system that will allow structural health and integrity to be effectively monitored and assessed, (4) confirm the performance of prototype health and usage monitoring systems

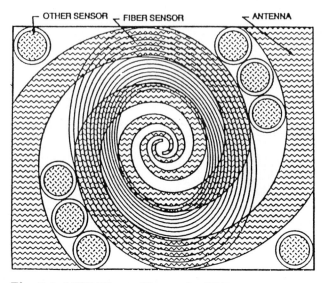

Fig. 8.4. SAWAFE panel layout for STEP–3 flight experiment [34]

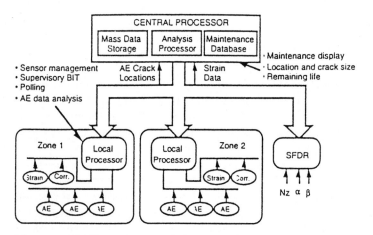

Fig. 8.5. SHMS architecture [53]

in ground and in–flight evaluation, and (5) provide guidelines to best practice in the design, manufacture and qualification of structurally integrated sensing systems.

8.2.3 Shape Control

Fixed–Wing Aircraft: The 'Adaptive Wing'.
Lift and drag as well as the velocity of the airflow and the ambient gas temperature/pressure are very sensitive functions of an aerofoil such as an aircraft wing. Much care is therefore spent on the design and especially the camber of these aerofoils, where drag should always be kept at a minimum. As long as the camber is kept constant, this minimum of drag is only possible for a specific amount of lift. Whenever more lift is required (e.g. for a maneuver), drag may increase significantly. Fig. 8.6 shows a selection of lift–drag relations for different cambers.

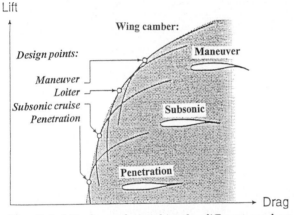

Fig. 8.6. Lift–drag relationships for different cambers [53]

A minimum in any lift–drag relationship of an aerofoil can be achieved when the camber can be varied according to different operating conditions. One solution to this problem is a system of rotary actuators and linkages to hinge a flexible wing (Fig. 8.7) [54]. This enables the camber to be adaptively controlled and the aerofoil to follow the minimum lift–drag relationship, thus leading to what is called the 'mission–adaptable wing'.

The solutions having been considered initially were mainly based on standard mechanical systems, which resulted in a complexity and weight penalty of the designs and actuation mechanisms. Much emphasis is nowadays placed to overcome these limitations by determining the use of adaptronics in this field. In Northern America, as well as in Europe, a lot of work is performed to replace the heavy slow mechanical system shown in Fig. 8.7 with adaptive panels using multifunctional materials and elements that can be rapidly flexed over a wide range of angles [87]. A weight reduction is expected allowing 30% higher payloads, 50% greater range, and 30% increased maneuvrability. Improvement in stealth performance can also be expected due to the ability of continuously varying curvature [55].

A key project called the 'Adaptive Wing' has been recently established between DASA Airbus, the German Aerospace Establishment DLR and Daimler–Benz Research and Technology, where pressure distributions around the aerodynamic profile will be monitored in–service using an integrated sensor network that will then allow to control 'intelligent' flaps being integrated into the wing.

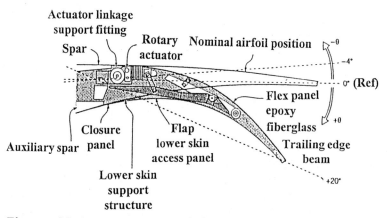

Fig. 8.7. Mission–adaptable wing [53]

MEMS (see Sect. 8.2.5) is an area which significantly gains importance with regard to locally influencing the shape on aerodynamical profiles, where more reference will be made in one of the following sections.

Rotorcraft, Spacecraft and Missiles. Most of the adaptronics work related to rotorcraft considers vibration damping. Some of the principles that will be described in Sect. 8.3.2 can also be used to statically bend or twist a rotorblade.

Most of the adaptronics work started with space applications, where various overviews have been given for different countries and organizations (i.e. [56, 31]). A variety of analytical studies have been performed [57], where shape control and vibration damping of antenna structures and solar arrays has been a major issue. Proof of concept has been made experimentally by bonding numerous piezoceramic patches onto a parabolic antenna for space applications, thus allowing the adjustment of the antenna's shape acccording to varying thermal conditions [58].

Missiles are a good testbed for applying adaptronics. Their aerodynamic profiles are usually much smaller than real aircraft, they only undergo one mission, and technology is less cost–sensitive. Misra et al. [59] fabricated a winglet where preconditioned SMAs were incorporated in the facesheet. Three electrical circuits were created that permitted control of four SMA wire lengths independently. This allowed an 'antagonistic' control approach, generating a 'balanced' or symmetric force response for an externally applied load, which was important with regard to a submarine missile changing from air into water or vice versa.

Another example is a missile model being downscaled by a factor of three, where the wing has been powered with a piezoelectric bimorph bender [60]. Good correlation has been shown between theory and experiment where deflections in excess of $\pm 14°$ have been achieved. This allowed the missile load factors to be increased by up to $12.6g$ and the turn radius to be reduced significantly. Improvement in development has been shown on a larger 4–inch span winglet where deflection amplitudes have been increased by up to $\pm 16°$ when using a shell offset of up to 20% [61].

8.2.4 Vibration Damping

Fixed–Wing Aircraft.

Mass Vibrations. A first example of a typical high–performance aircraft wing was built out of graphite epoxy laminate with piezoelectric actuators distributed over 71% of its surfaces [62]. Actuators were arranged in three banks, as shown in Fig. 8.8 and were wired so as to induce bending in the laminate. Three tip displacements were used for displacement feedback with the control objective being gust disturbance rejection and flutter suppression. From the tests performed with the model an increase in approximately a factor of three in stiffening the structure through the application of a closed–loop control was achieved. The overall RMS response in bandwidth up to 100 Hz was reduced by about 15.4 dB, with the first mode being completely eliminated. *Acoustic Vibrations.* Noise reduction in aircraft fuselages is a major issue with turboprop aircraft, where proof–of–concept has been shown [63]. Microphones have been used as sensors and loudspeakers as actuators. Linking the microphones and the loudspeakers with a controller finally succeeded in achieving the goal. A next step is to bond piezoelectric patches as sensors and actuators onto the inner panels of the fuselage at analytically well–determined locations [64].

Rotorcraft.

Mass Vibrations. With regard to expanding the flight envelope of helicopters and increasing speed, active control of rotorblades is a major issue actually in rotorcraft research and development, which mainly includes blade impedance control or individual blade control. In a feasibility study related to adaptronics performed by Strehlow and Rapp [65], impedance control using blade

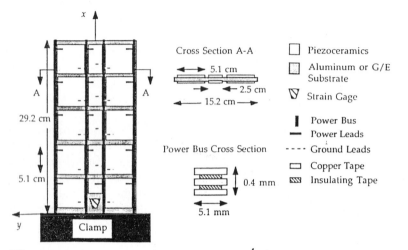

Fig. 8.8. Active aeroservoelastic wing model [62]

neck actuators and individual blade control using blade translation actuators or active flaps, both for active blade twist, have been considered. From the various materials available with actuation capabilities, piezoelectric ceramics were determined to be the most useful for rotorblade applications at this stage. It has, however, been stressed that active strain of these materials has to be improved by a factor of ten without reducing their other properties. Assuming an allowable strain of 0.6% and a maximum electrical field of 1000 V/mm, neither the active blade neck bending for lead–lag damping augmentation,

nor the active blade neck twisting for individual blade control, nor a trailing–edge flap being coated with a piezoelectric fulfilled the operational requirement, which were about one order of magnitude larger than the capabilities of these actuation concepts. Only the hinged flap concept proposed in [66] was able to allow adequate twisting of the rotorblade.

The hinged flap shown in Fig. 8.9 has been recently improved by using a piezoelectric bender and a new flexure mechanism to deflect the trailing edge flap [67]. This has helped to eliminate hinge friction and backlash. Flap deflections of 11.5° or more were demonstrated at operating frequencies up to 100 Hz on a test bench. Deflections of the flap greater than 5° while operating at 90% span location are expected when scaling the test bench up to a real–sized rotor blade.

Similar conclusions regarding improved sensing and actuation properties of adaptronic elements have been drawn by Straub [68] when analysing the McDonnell Douglas helicopter AH–64 rotorblade. A servoflap control with discrete actuators driving hinged control surfaces (Fig. 8.10) has been proposed as the actual optimum solution. In a follow–on study [69], hinged control surfaces have been analyzed and optimized in more detail. It turned out

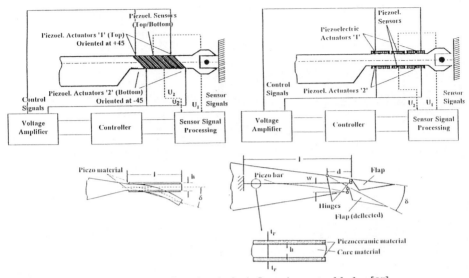

Fig. 8.9. Different concepts for adaptively influencing rotorblades [65]

that the servoflap rotor root torsional stiffness and the precollective angle are the most powerful parameters for optimizing actuation work and rotor power. An adaptive blade–root torsional spring has been proposed to further reduce required actuator work by 55% and rotor power by up to 10%. Optimization of the overhanging nose balance and the radial extent finally led to reduction in required actuation work by 86%.

Ben–Zeev and Chopra [70] developed a dynamically scaled helicopter rotor model featuring a trailing–edge flap driven by piezoceramic bimorph actuators. Flap deflections decreased significantly with increasing rotor speed, which was mainly caused by frictional forces created at the junction where the flap is supported during rotation of the blades. The use of a thrust bearing was found to alleviate this problem.

Another actuation type based on magnetostrictive actuation technology has been proposed in [71]. A blade–mounted Terfenol–D actuator was developed for the high–weight–penalty army UH–60A helicopter, where vibration reductions greater than 90% have been predicted. Experience gathered by using a piezo stack actuator for actuation of a flaperon, which consisted of a small movable surface to trip the boundary layer, located on the top surface of a wing model, and an electrostrictive stack actuator for moving a leading–edge droop flap hinged at 25% chord of a wing model, has been reported in [72]. Adaptive rotor blades using SMA–based active camber control have been proposed in [73]. The SMA actuator pulls a lever arm forward, which introduces torsion on a rod allowing to control a flap.

In Germany a key project entitled 'Adaptive Rotor' has also been launched, which includes Eurocopter Deutschland, DLR and Daimler–Benz Research

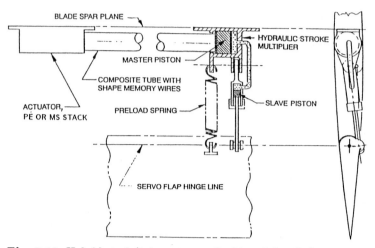

Fig. 8.10. Hybrid stack/tube actuator for hinged flap [68]

and Technology. The following issues are considered: (1) individual blade control with an aeroelastic servoflap; (2) active variation of the rotorblade's shape and thus adaptation of the blade to various flight conditions; and (3) active gear–support for isolating the vibrations being transfered to the rotor.

Acoustic Vibrations. The activities being performed for noise reduction of turboprop fixed–wing aircraft can be well transferred to helicopter applications. A 'smart wallpaper' to be used for monitoring the noise in cabins and thus being used for control has been proposed in [75].

Space Vehicles. An overview of USAF activities and perspective has been given by Das et al. [74]. Early work in the 1980s has been mainly related to active control algorithms that resulted in vibration suppression techniques. In parallel, NASA funded some activities in Large Space Systems Technology (LSST) being mainly related to large antennas and space platforms. After that, NASA funded follow–on efforts which were focused on hardware demonstrations.

Vibration damping (or even suppression) systems combined with condition monitoring are required to allow space structures to be permanently retuned according to changes resulting from aging, damage, dynamic behavior, etc. [34].

The USAF Phillips Laboratory, in cooperation with the Strategic Defense Initiative Organization (SDIO), initiated the Advanced Composites with Embedded Sensors and Actuators (ACESA) Program, being related to design, fabrication, and demonstration of composite components containing embedded sensors and actuators. It turned out that piezoelectric ceramics were the most suitable to be used for sensors and actuators and being integrated or adapted to graphite–epoxy composites. Three 17–ft long, 5–in diameter composite tubes comprising the secondary mirror support metering truss for the

ASTREX facility were manufactured as active control elements for a system–level demonstration. These tubes were equipped along their length with sections of piezoelectric actuators, colocated and nearly colocated sensors on each of their four flattened sides (Fig. 8.11).

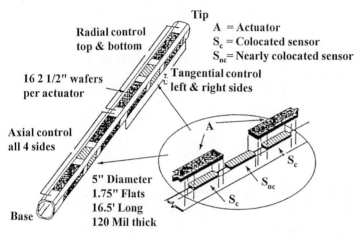

Fig. 8.11. Embedded piezoceramic sensors and actuators used on ASTREX [74]

Size and power of the system still needs to be significantly reduced if the system should be beneficially applied in a spacecraft environment. Furthermore, incompatibility of the sensor/actuator and host material under specific environmental conditions can lead to problems still to be solved.

A test structure (Fig. 8.12) consisting of a cantilevered graphite–epoxy tripod, with embedded piezoceramics, approximately $60 \times 30 \times 25 \, \text{cm}^3$ in size, has been used within the Advanced Control Technology Experiment (ACTEX) for demonstrating active vibration suppression. The system was designed to be fully exposed to the space environment and was well instrumented with accelerometers and thermistors, allowing determination and influence of the dynamics of the structure and evaluation of the performance of the control system.

Work related to avoiding sensor jitter was performed in [76] using a truss experiment which contained two active piezoelectric struts. Each strut had a collocated displacement and force feedback. Machinery noise was imposed to the structure and could be successfully damped.

In real space structures, sources of that jitter are typically cryocoolers, which have to be used to keep infrared space surveillance sensors at their operating temperature. Fig. 8.13 shows a schematic view of such a system and the two alternate load paths allowing transfer of the vibrations. These include the cryocooler 'cold finger' and the common structure. Piezoceramic actua-

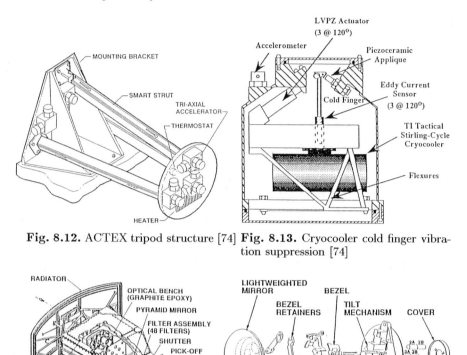

Fig. 8.12. ACTEX tripod structure [74]

Fig. 8.13. Cryocooler cold finger vibration suppression [74]

Fig. 8.14. Correcting the spherical aberration of the Hubble Space Telescope [32]

tors were therefore placed in both of these load paths in order to significantly reduce the vibrations originating from the cryocooler.

A major source of jitter in spacecraft is also solar array systems, where vibrations can be caused by either stepper motors or satellite housekeeping activities such as attitude thruster firings. Structurally integral active and passive vibration damping techniques have therefore been applied. Active damping has been provided via embedded piezo ceramic sensor/actuator technology. Viscoelastic passive damping has been used in the joint between the solar array drive assembly and the spacecraft bus.

Finally, a structure operating successfully is the Articulating Fold Mirror (AFM) which forms part of the optical scheme for correcting the spherical aberration of the Hubble Space Telescope [32]. Each AFM utilizes six electrostrictive multilayer ceramic actuators which also contain the required sensor unit while control is performed for the six actuators in a central unit (Fig. 8.14).

8.2.5 Smart Skins and MEMS

Smart skins in aerospace mainly means the integration of antennae into the structure or making the aircraft as much as possible electronically invisible ('stealthy') by using materials that are especially electronically steerable. Structural integration of antennae means that the antenna may also take over load–carrying functions, which may result in weight savings. This may require vibration–isolating features, allowing up to 50% of the aircraft's surface to be used for integrating antennae. Development in microelectronics and microsystems has allowed the compilation of more functions into one antenna. Today 66 antenna apertures are located at 37 sites on an F–18 fighter aircraft, covering a frequency band from 200 MHz to 18 GHz [77], and these are intended to be reduced to only nine apertures in nine sights. This will allow a reduction in the weight by a factor of two and in the cost by 30% [78].

Another initiative with the USAF Wright Patterson Laboratory is to design, develop and test a conformal structural load–bearing communication navigation and identification (CNI) antenna in the 0.15–2 GHz range [79]. Other attempts and ideas are thermoadaptive and electrochromic adaptive antennas developed by McDonnell Douglas [80, 81] or a conformal spiral antenna developed at Penn State University [82]. Phased array antennae are a further result of miniaturisation efforts, leading to weight savings and structural conformality.

MEMS is a collection of microsensors and microactuators being linked with a microcircuit control. It is thus an adaptronic structure on a microscopic level. Fabrication is based on using silicon micro–machining for manufacturing the electromechanical devices, sensors, and actuators and also integrating antennas by conventional microelectronics packaging.

One of these systems is a remote Surface Acoustic Wave (SAW) sensor to be used to study the deflection and strain of 'flex–beam'–type structures of helicopters [50]. It consists of remotely–readable passive SAW sensors and a microwave reading system. It has been implemented in rotating blades. The strain and deflection were measured remotely by an antenna communicating with the sensors (via a built–in antenna), first while the blades were stationary and then when the blades were rotating at nearly actual speeds.

Another widely discussed application of MEMS is measurement of both pressure and shear of the fluid flow on aerospace structures which is ideal for controlling vibrations or drag [83]–[85]. This may allow detection of the point of transition from laminar to turbulent flow and the active transmission of acoustical energy into the boundary layer thus allowing the low energy fluid particles to accelerate in the transverse direction and mix with the high–energy flow outside the boundary layer. A schematic arrangement of such a device is shown in Fig. 8.15.

The sensing and actuation part is mainly countered through polymeric piezoelectric materials. Actuation is induced locally on the basis of microri-

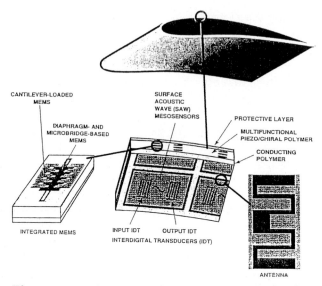

CANTILEVER-LOADED
MEMS

DIAPHRAGM- AND
MICROBRIDGE-BASED
MEMS

SURFACE
ACOUSTIC
WAVE (SAW)
MESOSENSORS

PROTECTIVE LAYER

MULTIFUNCTIONAL
PIEZO/CHIRAL POLYMER

CONDUCTING
POLYMER

INTEGRATED MEMS

INPUT IDT   OUTPUT IDT
INTERDIGITAL TRANSDUCERS (IDT)

ANTENNA

Fig. 8.15. Aerofoil section with MEMS and actuators [Varadan and Varadan 1996]

blets. Since the sensing and actuation elements are highly thin and small, they can be flush–mounted on an aircraft's surface. The waves being generated on the structure may even be monitored at a remote location, which avoids wiring. Application of a SAW MEMS sensor has been experimentally proven in a test performed in the 16–inch low–speed unsteady–boundary–layer wind tunnel at Penn State University/USA. Three–dimensional MEMS has been proposed to be an ideal solution to produce active microriblets for the control of drag [86].

MEMS technology can also be considered for active structural noise control. A MEMS sensor of comparatively small size has been recently tested and compared to a conventional piezoelectric (PZT) sensor [50] where the MEMS sensor showed a comparable and sometimes even better performance. Finally, it has to be mentioned that work is also going on in combining nondestructive evaluation with MEMS for health monitoring of aircraft structural components [50].

8.2.6 Control

Progress in computer science, and especially digital control, has given this area recently significant importance. Although it is still mainly considered for military or high–performance applications, a transfer to civil applications will occur as soon as maturity is proven. One of the initial algorithms to be developed is flight control of a battle–damaged aircraft, allowing the aircraft to return safely to base. Within the aforementioned adaptive wing concepts, more extensive use of existing controls is one of the initial steps tackled.

8.3 Adaptronic Systems in Automobiles
H. Janocha

Adaptronic systems in automobiles contain multifunctional elements possessing capabilities of sensing, actuation and signal processing and incorporate features of self–monitoring and self–repairing, for example by shifting important functions from defective to functional system components. In the case of an accident, this self–repairing feature could be the deciding factor for reaching a workshop and preventing the automobile from becoming stranded.

There are currently no complete adaptronic systems in automobiles. However, possible applications are foreseeable particularly for driver assistance, safety mechanisms (e.g. seat belts, airbags) and lighting systems. Some new information systems (telematics) also provide possible application of complete adaptronic systems. A few examples of adaptronic systems already implemented for driver assistance will be described from the viewpoint of actuation. Other systems will be touched upon shortly in the outlook (Sect. 8.3.7).

8.3.1 Preamble

A normally equipped automobile typically comes with electronic engine control, antilock braking (ABS) and an airbag. Table 8.2 lists the most important actuators involved in these systems. The electronics is typically not integrated into the actuators, but, rather, located in the control unit. All actuators, with exception of the airbag igniter, are electromagnetic in nature. Piezo-electric and magnetostrictive operating principles (solid–state actuators) are not present.

Actuators are subject to vastly different environmental conditions, depending on where they are mounted in the automobile. The actuators involved in engine control are typically mounted in the engine compartment and subject to a broad range of temperatures ($-40°$C to $+150°$C), vibration (up to about $30\,g$) and a wide variety of chemicals (salt spray, fuel, oil, etc.) [88].

Various requirements with respect to dependability are placed on the actuators: failure of an airbag or an ABS actuator can be fatal. In contrast, the failure of a central locking actuator device may be aggravating but does not lead to functional defects relevant to safety.

Most actuators must be capable of operating in real–time, i.e. they must exhibit short reaction times relative to typical system time intervals (as measured, for example, by the period of crankshaft rotation). An injection valve must inject a precisely defined amount of fuel into the combustion chamber at a fixed point in time. Solid–state actuators could be implemented to fulfill this function in the future.

Table 8.2. Actuators in automobiles [88]

Actuator application	Operating principle	Electrical load	Real–time capability
Engine control			
injection valve	solenoid	inductive	yes
ignition	spark gap with high–voltage transformer	inductive	yes
idle control	solenoid	inductive	yes
exhaust return valve	solenoid	inductive	yes
fuel tank relief valve	solenoid	inductive	no
throttle	electric motor	inductive	yes
ABS			
valves	solenoid	inductive	yes
pumps (accumulator)	electric motors (relays)	inductive	no
Airbags			
priming cap / gas generator	igniter	resistive	yes
Locking system			
power locks	electric motors (relays)	inductive	no
Transmission control			
valves	solenoid / regulator	inductive	yes

The automotive industry is constantly under pressure to reduce costs and simultaneously improve performance. There are various approaches to solving this problem with respect to actuators: control devices can be networked, actuators can be implemented to fulfill multiple functions, and control units can be integrated into actuators. Legislation is also increasingly demanding diagnosis capability. The state of development will be discussed in more detail next.

8.3.2 Networking of Control Devices

The basis for exploiting the multifunctionality of actuators is the flow of information between the control devices involved. Various options are available, as set out hereafter.

- A point–to–point connection of control devices offers a viable solution when little information is to be exchanged.
- In contrast, networking of control devices within a standardized automobile network is a considerably more flexible approach that becomes more economically attractive with increasing demand for information exchange. The

current division of the market with respect to data bus standards (CAN, ABUS, VAN, J1850 and more) presents limitations to implementing this approach.

The reasons for introducing a data bus are fourfold:

- All control devices can access the same information simultaneously.
- A piece of information or a sensor signal can be made available to more than one control device at the same time.
- The signals from the most different control devices could be combined to create new functions.
- The wiring costs would reduce dramatically despite an increase in system networking.

The architecture of networked systems is based on powerful control devices with functional groups for individual or even multiple units (system control devices). For the time being, one device generally controls the engine torque and another the use of the automatic transmission. Units containing both functions are already being implemented in individual cases (see Sect. 8.3.4).

Actuators and sensors are normally included in the system control devices but are also available for other systems. It can also be meaningful to combine components to form subsystems. The integration of certain functions forms the basis for assemblies with high dependability and the ability to be diagnosed. This is an important aspect concerning the already–existing and the yet–to–be–realized requirements for on–board diagnostics (see Sect. 8.3.5).

As an example, Fig. 8.16 illustrates the structure of a system for managing the drive train of a vehicle. The engine control unit includes all engine–related functions, such as idle air control and those necessary under load conditions. An E–gas throttle body with integrated electronics is implemented to control air–flow. This and the integrated pedal sensor form the E–gas sensor/actuator subsystem. Its functions include determining the driver's intention, controlling the air–flow (or vacuum pressure), self–monitoring and diagnosis of components and fulfilling certain safety tasks for the drive train.

8.3.3 Multiple Usage of Actuators

The driver's control signals of 'braking', 'accelerating' and 'steering' can cause problems in vehicle dynamics in unusual situations occurring, for example, during winter driving or panic reactions. Feedback control systems have been developed — of which ABS is the most well known — to control vehicle dynamics and the drive system during such situations. Forthcoming solutions involve networking systems that have been already established, such as ABS and engine control systems, so as to offer greater functionality and to reduce the number of actuators (such as the extra throttle valve positioner, used till now to regulate drive slippage).

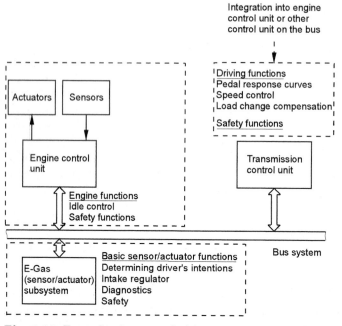

Fig. 8.16. Example of a networked management system for the drive train [89]

The multiple usage of actuators resulting from networking is demonstrated in Table 8.3. For example, the engine torque can be reduced to zero nearly continuously by 'shutting off' the cylinders one–by–one by preventing injection of fuel to them and eliminating the corresponding ignition. Anti slip control systems (ASR) and electronic transmission controls (ETC) make use of this possibility.

Vehicle Dynamics Control System. Electronic systems for driving safety, such as ABS and ABS/ASR, are intended to maintain the usual driving behavior of the automobile even in critical dynamic situations involving forward motion during full braking or acceleration. This is achieved by limiting slip of the tires so that the automobile remains manoeuvrable. As soon as the driver gets into situations involving lateral movement of the automobile — such as during a drastic steering manoeuvre, when changing lanes or avoiding an obstacle — support systems supplementing ABS and ABS/ASR become necessary in order to maintain control. Exceptional driving skill is not even necessary. Creating such a system was the goal of Bosch in the development of the vehicle dynamics control system (VDC). The concept relies upon proven ABS and ABS/ASR components.

Fig. 8.17 shows the complete concept schematically, with the automobile as the controlled member, sensors for determining the input parameters to the closed loop, actuators for influencing the braking force and drive power, and a hierarchically structured feedback controller consisting of an overriding ve-

Table 8.3. Examples of actuators with multiple uses in automobiles [88]

Actuator	System	Extended system functions
selective cylinder injection	engine control	ASR: reduction of engine torque in case of traction loss ETC: reduction of engine torque during shifting
retardation of ignition timing	engine control	ASR, ETC: see above
brake torque control	ABS	ASR: electronic differential blockage, drive stabilization: active control of yawing moment
automatic transmission	ASR	ATC: driver assessment, road–type identification, load recognition

hicle dynamics feedback controller and an underlying slip feedback controller. The overriding feedback controller provides reference values of the slip to the underlying feedback controller. The controlled state variable, the floating angle, is determined by an observer unit. The reference behavior is obtained by evaluating the signals describing the driver's intentions from the steering angle sensor (desired turning angle), the inlet pressure sensor (desired delay) and the engine management system (desired driving torque). The coefficient of friction and the vehicle speed, which are approximated from the wheel revolution counter, lateral acceleration sensor, roll–speed sensor and pressure sensor, are included in the calculation of the reference behavior.

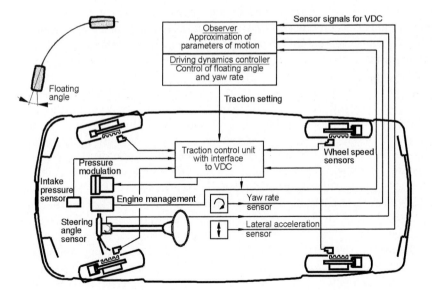

Fig. 8.17. Concept of the vehicle dynamics control system (VDC) [90]

The actual behavior of the automobile is determined primarily from the yaw–rate signal and the floating angle approximated by the observer. The vehicle dynamics controller controls both the state variables of yaw rate and floating angle, and it calculates the transmission torque necessary to compensate for the difference between the actual state variables and the reference state variables. To generate the reference transmission torque, the necessary reference changes in slip are determined for the appropriate wheels by the vehicle dynamics controller. The reference changes are set via the underlying braking and anti–slip controllers and the hydraulic braking and engine management actuators. Should the rear axle of the vehicle press outward too strongly, for example during fast cornering, the system first reduces the driving torque leading to an increase in the lateral traction of the rear wheels. If this intervention in engine operation does not suffice, the VDC system then also brakes the outer front wheel until any skidding motion lets up. A simultaneous reduction in speed offers additional safety.

Electronic Transmission Control . Electronic transmission control (ETC) has been implemented successfully in automobiles with automatic transmissions for more than ten years. The interface to peripheral control units or systems within a vehicle is a focal point in the development of ETC. In this scheme, the ETC control unit communicates with — in addition to the automatic transmission — other control units or computers belonging to the various systems, such as engine, ABS, instrumentation and VDC.

A common strategy has been established among most producers since the introduction of electronic control of hydraulic transmissions: a control switch is used to select from among various gear–shifting programs. Successive development encompasses adaptive or self–adapting transmission control (ATC) with which driveability is optimized. Gear–shifting programs and special gear–selection strategies are chosen from a memory bank depending upon predefined driving conditions.

Implementation of ABS functions does not place any new fundamental demands on the control electronics. System expansion is accomplished by combining information that is already available within the transmission control system, engine electronics, the control unit for the anti–lock braking system, or the anti–slip control system. In particular, no additional sensors need be implemented. Fig. 8.18 illustrates the functional structure of ATC in BMW automobiles with 5–speed automatic transmissions. The 'intelligent' control system constantly monitors the driver's behavior and selects the corresponding shifting characteristic that is optimally adapted to the current driving behavior — from economical to extremely sporty.

Recognition of both the driver type and any special driving conditions leads to the selection of an appropriate gear–shifting program that provides suitable shifting points. A program is also available for recognizing and selecting the most suitable gear for operating conditions with high driving resistance due, for example, to a mountain ascent or trailer load. A lower gear

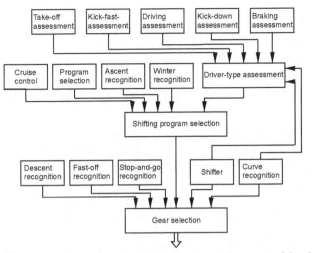

Fig. 8.18. Structure of adaptive transmission control by function [91]

is normally selected in comparison to level driving with low load. The winter program is activated when slippage of the drive wheels is recognized. The often surprising encounter with slippery roads is almost always counteracted by the winter program through a shift to one higher gear, providing for an increase in driving safety, particularly in automobiles without VDC [91].

All powerful, 'intelligent' transmission controllers must attempt to perform gear selection as closely as possible to that of a driver shifting manually. The development goals of ATC are

– a reduction in average shifting frequency;
– a reduction in average fuel consumption;
– an increase in driving safety;
– a decrease in driver intervention; and
– more enjoyment of driving automatic vehicles.

8.3.4 Integration of Control Units

Similar to sensors with integrated microelectronics, which are hardly established in automotive applications, actuators with integrated power (or even signal–processing electronics) are still little–used in automobiles. Nowadays, electronic transmission control units are generally independent units based on printed circuit boards. The units are mounted, for example, in the passenger compartment electronic box or behind the splashboard. A multiwire plug–and–socket connection forms the interface to switches, other control devices, sensors and actuators in the transmission.

Two basic solutions exist for integrating control units:

- Integration of several independent electronic control devices to form a single control unit (e.g. integration of driving–speed control into the engine–control electronics).
- Integration of electronics into or onto the controlled mechanical automobile subsystem ('intelligent' actuators, mechatronic systems). A current example comprises the integration of ABS electronics into the hydraulic ABS control block.

Both variants are also conceivable for ETC, for example.

Integration of ETC in the Engine Control System. The engine– and transmission–control functions can be summarized in the form of centralized drive–management electronics. The main advantage of this approach is the potential for cost savings through the reduction of housings and electronic components, as compared with the solution with two separate devices. An additional advantage of the combined unit is the certain degree of transparency with which the close communication between engine– and transmission–control devices takes place: the use of a single microcontroller within the combined unit provides immediate access of shared information by the corresponding functions.

In addition to the problems associated with the realization of such a complex unit, the integration of the two control systems exhibits a substantial disadvantage: suppliers of automatic transmissions are bound in business terms to the suppliers of integrated electronics. The electronics supplier is determined by the engine producer and therefore most often by the automobile producer. The responsibility for the transmission supplier's 'automatic transmission system' is thereby dissolved.

This discussion makes clear that the integration of engine– and transmission–control electronics would be particularly popular in markets in which automatic transmissions are well established and the engine or automobile and transmission producers are one and the same. This description primarily fits the market in North America.

Integration of ETC into the Automatic Transmission. New approaches to rational automobile production are carried forward in the concept of production modules. Only one system supplier is responsible for the development of such production modules, their production and maybe even their implementation on the assembly line. The complete creation of an automobile front–end, including radiator and lights, or a complete brake module (brake pedal, power brake servo–unit, main braking cylinder, and the ABS hydraulics with its electronics) are actual examples of production modules.

From a similar point of view, the integration of electronic control directly into or onto the automatic transmission proves attractive. The automatic transmission is thereby transformed into a complete production module. Siemens refers to the ETC unit involved in this integration as a 'prodmod'. The feasibility of incorporating a prodmod into a transmission is highly dependent upon the miniaturization of the actual transmission electronics and upon meeting the demands of the harsh environmental conditions. One solution to the problem — based on 'low temperature cofired ceramics' (LTCC) — is beginning to establish itself. This technology involves a multilayered ceramic substrate to which the active electronic components, such as the microcontroller or the final power stage are affixed. A large portion of the passive electronic components, such as the electrical resistances, are attached directly to the ceramic substrate using silk–screen printing methods.

New concepts involving the division of functions also focus on integration with the transmission. The gear–shifting strategy can be integrated into the engine–control system (or other module of the automobile) as a component function, while the prodmod mainly fulfills functions specific to the transmission (such as the control of the shifting process itself). The automatic transmission becomes an 'intelligent' actuator that receives commands to shift from a centralized drive–management system (engine–control system plus gear–shifting strategy). This approach corresponds to the division of tasks already assumed nowadays by automobile producers (shifting strategies) and transmission suppliers (shifting control). Prodmod offers the following advantages.

- simplified wire harnesses (particularly when implementing an automobile network);
- fewer variants (one hardware/software variant for each transmission component);
- reduced logistical complexity and simpler application;
- delivery of the transmission and its control unit to the automobile assembly line as a single unit;
- integration of sensors and actuators with electronics; and
- the chassis/automobile need not provide for room or connector to accommodate the transmission control unit.

Effects on cost, dependability and other aspects will also be presented. A comparison is made based on one transmission type (5–speed automatic) with a constant transmission–to–electronics interface while varying the connection to the wiring harness of the automobile. Table 8.4 describes the variants compared, together with their features. On the whole, the normal 'stand–alone ETC' solution is compared with two variants of the prodmod.

The comparison of costs for electrical and electronic components includes the costs for control electronics, the wiring harness (that portion of the wiring harness for the transmission and wiring internal to the transmission), revolutions counter and temperature sensor in the transmission (monitoring tumbler switch not included). Particularly variant III/Prodmod 'Full CAN', in which the electrical interface to the transmission consists of only four wires (supply line and CAN bus), offers substantial potential for cost savings. This potential comes about as a result of a reduction in the wiring harness, which

Table 8.4. Comparison of costs for ETC electronic/electrical components for various degrees of integration [92]

Variant	I	II	III
Name	Stand–alone ETC	Prodmod	Prodmod/ 'Full CAN'
Interface with transmission	12 leads to sensors/valves	12 internal connections to sensors/valves (within transmission)	12 internal connections to sensors/valves (within transmission)
Interface to automobile	17 leads (some conventional, remaining CAN)	17 leads (some conventional, remaining CAN)	4 leads (V_{batt} + /−, CAN +/-)
ETC mounting location	E–box	Inside transmission	Inside transmission
ETC design / version	Standard (PCB)	Prodmod	Prodmod–CAN
Notes	Standard design as 'housed circuit board'	ETC production module; sensors and valve terminals are integrated	ETC production module; sensors and valve terminals are integrated; transmission plug with 4 pins (only CAN interface)
Relative costs	1.0	0.9	0.8

more than compensates for the increased cost due to integration of the control unit [92].

8.3.5 Self–Diagnostics

Two factors support the use of self–diagnostics. On the one hand, US legislation demands system diagnosis, and on the other hand, system diagnosis is absolutely necessary in order to achieve the correct monitoring of the functioning of systems with respect to safety. The diagnosis of actuators depends upon the analysis of control signals as well as the analysis of sensor signals within the system. The general problem will be presented on the basis of two examples: the monitoring of exhaust behavior in the engine control system, and the monitoring of the airbag system function.

The US's California legislation demands continuous control of all components influencing exhaust emissions. Defects must be recorded in memory chips and displayed to the driver. In the future, similar requirements will also be fixed by legislation in Europe. Relevant actuators will therefore not only require self–monitoring but also monitoring within the system. Recognition of misfiring serves as an example. Correct functioning of ignition is first measured by the crankshaft revolution counter based on the smoothness of engine running. The ignition current is also monitored so as to exclude any defects in the ignition coil or wiring harness. When misfiring is detected, the cylinder needs to be 'turned off'. This is accomplished by cutting off the supply of fuel to the cylinder. Monitoring of the ignition current is interesting from other aspects: the flow of current over time can provide additional information about the fuel ignition process in the cylinder. The actuator can be implemented as a sensor.

Diagnosis is essential for safety–relevant systems. The unnoticed defect of an airbag system can lead to unpredictable damage claims. The entire system is therefore not only diagnosed during initialisation when the power supply is switched on, but also the system is monitored constantly in normal operation. The resistance of the airbag igniter is measured repeatedly and compared with reference values. Any detected defects are also recorded and reported to the driver via control–panel indicators.

The VDC discussed in Sect. 8.3.3 serves as an example to show how a comprehensive safety concept can be implemented through consistent use of available sensors together with computerized intelligence. The actuator is tested actively by sending short pressure–modulation cycles and analysing the response measured by the pressure sensor. Defects in pumps, compression devices and solenoid valves — even ones external to the closed loop — can be recognised in this fashion. The sensors are monitored for broken wires and implausible signal behavior during the entire driving operation.

The most important sensors are tested actively in a second stage. The pressure sensor is tested for correct functioning during the active test of the actuator and during all active braking manoeuvres. The yaw rate sensor is tested by actively detuning the sensor element and evaluating the signal response. The steering angle sensor, with local intelligence has self–monitoring functions and sends any error messages directly to the vehicle's control unit. The transmission of digital signals to the control unit are also monitored constantly.

In the third stage, the sensors are monitored during the entire period of stationary operation through analytical redundancy. A model–based calculation tests whether or not the relationships between the sensor signals, as determined by automobile motion, are being violated. In the case of a defect, the system will be either partially or fully shut down, depending upon the type of defect [91].

8.3.6 Implementation of Functions

Within the first phase of application, a function was often guaranteed to operate correctly by a mixture of analogue and digital technologies (see Fig. 8.19).

The state of development is marked by a strong increase in the scope of operating system software in control computers. The memory limit of 64 kB

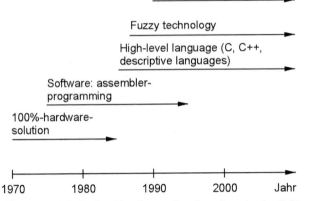

Fig. 8.19. Methods of implementing functions in the field of automobile electronics [93]

presented by 8–bit technology represents an increase by more than a factor of 8 over the 7.5 kB memory needed by the first serially produced controllers of automatic transmissions in 1983. This development will continue particularly with the implementation of real–time operating systems and powerful 16–bit or 32–bit microcontrollers. An increase to about 250 kB is expected for the new generation of transmissions and controllers by the turn of the millennium. There is a significant relative increase in the parts for gear selection and diagnosis software relative to the shifting functions [91].

The method of defining functions in high–level computer languages is still far from the formulation used by transmission experts (such as designers and application engineers). This barrier can be overcome through the use of descriptive and specification software (CASE tools) that is capable of generating C code directly with a precompiler. Presentation on a descriptive level also offers possibilities to implement graphical tools for increasing the transparency in describing the functions. Fuzzy logic is another approach to implementing functions. The transfer of expert knowledge into 'if..., then...' formulations is thereby simplified.

Knowledge–based systems such as fuzzy logic will be technologically advanced through 'mechanised learning'. Such systems do not require that the expert fully formulates his or her knowledge; it will be learned in a 'mechanised' fashion during a training phase. The knowledge–based system will need to be supplied with a stimulus (input) in the form of 'good examples' during the learning process. A road recognition system serves as an example. The system is to derive the type of road being traveled (e.g. motorway, highway, city street, etc.) from input quantities such as the average driving speed, average absolute lateral acceleration, average gas pedal position and other signals. The recognition system also offers information about the course of the road (level or hilly) and can therefore offer support in optimising driving functions.

The particularity of mechanised learning in this example lies in the generation of context in the form of fuzzy logic rules in an off–line training phase by means of typical measurement results for input data and the corresponding association with the type and course of the road. A phase involving the analysis of the measurement results and the definition of the rules by a human expert are no longer needed. The fuzzy rules are then implemented in the control unit and evaluated on–line in the automobile [93].

8.3.7 Outlook

Additional adaptronic systems for automobiles will be available in the near future. A few of these systems will now be described briefly (see also [94]).

Further Driver Assistance Systems. A system for automatic distance control (ADC) provides for a sufficiently safe distance to other automobiles using feedback control of speed and distance that themselves depend on the automobile's speed and road conditions. The distance is constantly monitored. A reduction in speed is automatically achieved through braking and adjustment of the throttle when a minimum distance is reached. The automobile must have an automatic transmission and ABS.

In addition to the gain in comfort compared to current cruise control systems, the ADC system eliminates the risk of unrestrained collision. In a traffic jam, the automobile with ADC even stops fully automatically and only continues forward when the driver takes over control. Accidents involving rear–end collisions are prevented and traffic is harmonized by the use of ADC. The system reduces fuel consumption and prevents stressful situations that commonly occur in dense traffic.

An ADC system consists of a sensory system (e.g. an infrared laser or radar sensor) for measuring the distance to the automobile in front, a measurement device for automatically determining the friction coefficient of the street, and an automobile computer. Determination of the coefficient of friction is derived from the difference in rotational speed between wheels being driven and those not being driven. The rotational speed of the wheels is acquired from the revolution counters located at the wheels of automobiles equipped with ABS.

Retaining Systems. Retaining systems include front airbags for the driver and front passenger, side airbags from the seats and door padding, safety belts with force blocking and tighteners for front and rear passengers, seats with head rests adapted even for extra–large passengers, and optimal cargo protection. One focus in current development is on the increase of effectiveness of retaining systems in real accidents for various passenger and seating positions using complex sensors. Components of future retaining systems include:

– seating and child–seat recognition for deciding whether or not to activate
  an airbag;
– recognition of any passenger's position, weight, and use of safety belt, as
  well as the gravity of the accident, for automatically achieving the optimal
  settings of airbag and safety belt systems;
– automatic adjustment of head rests and height of the safety belt;
– a combination of head and thorax airbags; and
– sensors for controlling, for example, knee and seat cushion airbags for re-
  lieving the extended leg performing braking and the pivoting of pedals to
  clear the foot area.

Lighting Systems. Gas–charged lamps deliver three times the effective
light as halogen lamps with about 30% less electrical power consumption.
However, reeling and rolling motion of the automobile can be blinding to
other road users in traffic. Automatic headlight range control works against
this blinding effect by maintaining a constant lighting distance independent of
the automobile load. This system consists of a sensor on each of the front and
rear axles, positioning motors for regulating the headlamps, a speed sensor
and a control unit. The headlamp range is readjusted within a few hundred
milliseconds.

Information Systems. Driver information (e.g. concerning the weather,
traffic jams, and accidents) and directing traffic with telematic systems to
prevent unnecessary delays and to provide for stress–free driving are important contributions toward preventing dangerous situations. One fundamental
component for intelligent information systems is an auto–radio for receiving traffic information, which is transmitted with RDS/TMC (Radio Data
System/Traffic Message Channel) via an infrastructure already existing. A
mobile telephone (Global System for Mobile Communication, GSM) is also
important as well as GPS (Global Positioning System) satellite positioning
for determining the location of an automobile on the ground.

Dynamic guidance is accomplished by a navigation system that calculates
the best route based on a digital road map and the current traffic reports
received via RDS/TMC. The driver is then guided to his goal by acoustic
signals. Dynamic guidance systems reduce insecurity resulting from problems
in orientation, thereby reducing accidents. The search for a parking place is
also reduced, along with the corresponding consumption of fuel. The route
computer monitors incoming TMC messages with respect to the individual
routes. In the case of traffic congestion, the router calculates a new route
leading around the traffic problem.

In the case of an accident, a crash sensor automatically triggers an emergency signal that is directed without delay to police rescue and traffic–
headquarters via mobile communications. The location of the accident is
determined by satellite using GPS and sent with the emergency signal. Other
drivers can be warned immediately, thereby preventing further accidents or

delays. A life–saving advantage can result from establishing contact with the rescue service when travelling on isolated roads.

8.4 Adaptronic Systems in Machine and Plant Construction
H. Janocha

8.4.1 Positioning Systems

Many tasks in machine and plant construction require positioning systems with which displacements with micrometer or nanometer precision can be achieved at high speeds. Solid–state transducers (see Sects. 6.2 and 6.3) are very well suited for this task as they are also capable of driving heavy loads.

Because of hysteresis, the absolute value of the displacement achieved with a solid–state transducer is not known with great accuracy. This property does not reduce positioning accuracy as long as relative displacements are to be carried out or the change of position is measured externally. Otherwise, absolute positioning is necessary. This can be realised by driving the transducer within a closed control loop, which requires measurement of the controlled output and a controller for setting the transducer voltage based on the difference between the control setting and the controlled output (see Fig. 8.20).

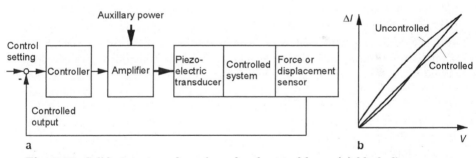

Fig. 8.20. Solid–state transducer in a closed control loop: **(a)** block diagram; **(b)** actuator response

The controlled output is derived from the measurement of force, displacement or strain. The sensor should be placed as closely as possible to the point at which the controlled output should be achieved. Predominantly strain gauges are implemented to achieve displacement resolutions in the range of several tens of nanometers. Inductive and capacitive displacement sensors are used when finer resolutions are necessary.

The advantages of closed–loop control are

- positioning free of hysteresis
- high absolute positioning accuracy
- compensation of drift
- stable positioning despite changes in load
- extremely high stiffness.

The state of the art with respect to precision and resolution is embodied in a positioning table with a displacement range of $100\,\mu$m along the x–axis. The displacement is achieved using a piezoelectric actuator and monitored by a capacitive sensor. Additional actuators and sensors control undesired deviations along the y and z axes. This 'concept of adaptive mechanics' (as described by the author) keeps to an ideal path within $0.3\,$nm [95].

Based on a precision table of this nature and the use of high–speed solid–state actuators, tool positioning units on turning lathes can be extended to achieve surfaces that are not rotationally symmetric ('noncircular turning'). The latest developments aim at achieving the same results in open control. These attempts make use of the smart actuator concept described in Sect. 6.1, the real system having first been identified and modeled. This approach enables the system to be modeled for example in a digital signal processor that is connected in series with the real transducer and used to compensate hysteresis [96]. The first products making use of such adaptronic systems will soon become commercially available.

Another example involving the use of solid–state actuators for positioning tasks is a system for correcting thermally induced tilt in machining equipment. Thermal deformation of machine tools is a significant factor determining the achievable manufacturing precision. The effects of thermal deformation are in the range of translational influence. An apparatus which can be attached to the machine table was developed to investigate tilt correction on a mid–sized drilling and milling machine. Position–controlled piezoelectric actuators were implemented to achieve the correcting displacements. These solid–state actuators are capable of achieving the necessary positioning accuracy, axial stiffness and positioning speed. A functional prototype of the tilt–compensated table was designed for a nominal load of $10\,$kN, tilt settings of up to $100\,\mu$m and a total stiffness of greater than $500\,$N$/\mu$m [97].

The following application, involving grinding, was already patented in the 1980s. The use of cubic crystalline boron nitride (CBN) permits grinding speeds much higher than those achievable with conventional grinding materials such as corundum. However, tighter precision requirements are placed on dressing the grinding wheel. Piezoelectric actuators are suitable for the dressing process. Their smart properties enable them to be used both as sensors and actuators. Fig. 8.21 illustrates one configuration for active dressing using piezoelectric actuators. In the first step of dressing, the dressing stone is positioned close to the grinding wheel along the NC axis of the grinding machine. The slide plate stops moving as soon as contact with the grinding wheel has been sensed by the piezoelectric transducer. In the second step,

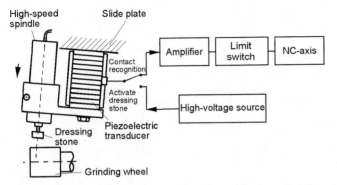

Fig. 8.21. Principle of adaptive dressing with piezoelectric actuators (Source: University of Hanover)

the wheel is dressed with the help of the piezoelectric actuator, making use of its high loading capability.

8.4.2 Damper Systems

A common challenge in the construction of machines and plants is the reduction or elimination of undesirable vibrations in mechanical structures. This is achieved using passive and active dampers.

Passive Damping Systems. Passive dampers absorb vibrational energy from a structure and convert it into a different form of energy or effect an opposing exchange of energy (dynamic absorber). Passive dampers would be attached to the structure at critical points, i.e. at points exhibiting undesirably high amplitudes of vibration. The dampers achieve their maximum effectiveness at a specific frequency. Examples can be found in milling and turning spindles, or in the supports of flat surface grinding machines.

The most predominant form of passive damper is that of the dynamic absorber, whose behavior can be modeled by a mass, spring and damping element. The design of such a system involves the selection of values for mass, stiffness and damping that lead to the greatest reduction of vibration.

Fig. 8.22 illustrates a dynamic absorber containing an electrorheological (ER) fluid. In each case, the auxiliary mass is spring–mounted in a housing filled with an electrorheological fluid. The damper housing is attached to the structure to be damped. Electrical control of the flow behavior of the ER fluid influences the coupling between the housing and the auxiliary mass. In this way the dynamic absorber can be adapted optimally to match the vibrational behavior of the structure to be damped.

In dynamic absorbers based on the shear mode, the housing filled with ER fluid forms one electrode and the auxiliary mass vibrating relative to the housing forms the other electrode. The vibrating motion results in shear loading of the ER fluid located between the electrodes. In the flow mode, the

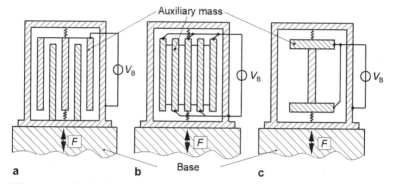

Fig. 8.22. ER fluid dynamic absorber: **(a)** shear mode; **(b)** flow mode; **(c)** squeeze mode

ER fluid flows between the electrodes on the auxiliary mass and the damper housing as a result of their relative motion. In dynamic absorbers based on the squeeze mode, the vibrating motion of the auxiliary mass causes a change in displacement between the electrodes. The result is an electrically controllable squeeze flow [98].

Fig. 8.23 shows an example of a multi–purpose vibration damper with electrorheological fluid [99]. The valve used to alter the resistance of ER fluid flow, thereby controlling the absorber force as a function of the control voltage, is integrated into the bypass of the double–cylinder damper. The outer cylinder, which is electrically insulated from the remaining housing, forms the outer electrode; the inner cylinder is the grounded electrode. The diameter of the piston is 22 mm, that of the piston rod 8 mm. The fluid displaced by the piston flows through the bypass, in which the electrode voltage produces a radial electrical field that influences the flow behavior of the ER fluid. A damping force of 320 N was achieved for a piston speed of 10 cm/s at an electrical field strength of 4 kV/mm. In the absence of an electric field, the damping force is 30 N. The dependence of the conductance of the ER fluid on the flow speed can be used as a sensor. This dependence results in a decrease in electrical current with increasing piston speed. The ER fluid is a multipurpose element: as a hydraulic medium, as a special material for which a valve can be created with no moving parts (ER effect); and as a sensor with which changes in the piston motion can be detected.

The squeeze mode is particularly suited for realizing electrically controllable dampers for large absorbing forces with low amplitudes of vibration. Squeeze–mode dampers can be implemented to reduce the radial vibration of rotating shafts. A rolling bearing encircling the rotating shaft and containing an ER fluid damper based on the squeeze mode has become widely known [100]. The radial resonant vibrations occurring for various rotational speeds of the shaft were reduced substantially with the ER fluid damper. The area of machine tools offers additional possible application of ER/MR

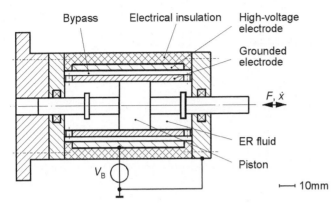

Fig. 8.23. Example of an ER fluid damper.

fluid squeeze–mode dampers in which high–force resonant vibrations can be reduced.

Active Damping Systems. Active dampers make use of actuators to produce a force opposing the disturbing vibration at the location of occurrence. This type of application takes place in closed–loop control since the disturbance must first be measured at the location of interest. Special process parameters can be identified with the help of a general process model. A control algorithm can be formed that supplies the appropriate control signal to the actuator. Active dampers are more complex than passive ones as they can adapt to changing process behavior and are therefore effective within a much wider band of frequencies.

Based on their ability to generate both high–frequency displacements and large forces, solid–state transducers are well suited as actuators for active damping of undesired vibrations in heavy mechanical structures. Fig. 8.24 illustrates this feature in the case of external plunge grinding, in which relative displacements between the tool and work piece resulting from dynamic instability (chatter) are compensated for or damped. The additional necessary mechanical energy is generated using a piezoelectric actuator integrated directly into the center points. A system applying this concept presented in the middle 1980s demonstrated broad–band improvement of dynamic behavior of the grinding machine [101].

A so–called 'ARMA' process model tracks the real process on–line, with a least–squares (LS) procedure for identifying the unknown parameters of the process — the undesired vibrations. Input and output process parameters are passed through form filters to the individual parts of the model (see Fig. 8.25). The difference e in the model outputs is a generalised error and serves as input to the LS procedure for adapting both model parts to the process. The estimation vector, which contains the identified process parameters, forms the basis of an adaptation algorithm used to adjust the controller according to

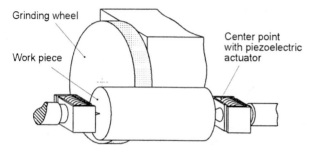

Fig. 8.24. Active vibration damping with piezoelectric actuators in external plunge grinding

appropriate strategies (minimal variance, dead–beat, PID, etc.) and follows the changing process behavior [101].

One variation, referred to by its creators as a 'semi–active damper' , is an interesting version [102]. Semi–active devices are those in which the 'passive' stiffness and/or damping properties are varied in real time based on a feedback signal. Semi–active elements have, like passive elements, the ability to dissipate system energy. Through implementation of an appropriate adaptive control law, semi–active elements are able to adapt to different vibration environments and/or system configurations. Another advantage over passive damping elements is their ability to utilize sensor information from other parts of the structure, as so–called noncollocated sensor/actuator architecture. Typically, semi–active elements have low power requirements and are less massive than their active counterparts.

One of the most significant sources of passive damping in built–up truss structures is the connection joint. Energy dissipation in joints occurs primarily as a result of impact and dry friction present at the sliding interface. With nonlinear, local control, the energy dissipated by the frictional joint can be maximized, using the normal force as control input. This concept of varying the normal force in a frictional joint to enhance the energy dissipation from a vibrating structure has been realized with a piezoelectric disc plate (see Fig. 8.26). Applied voltage at the stack disc tries to extend the piezoelectric material, which results in increasing the normal force. If the dynamical behavior of the piezoelectrical material is to be neglected, so the normal force is proportional to the input voltage. The goal of the controller is to prevent the semi–active friction damper from sticking.

8.4.3 Valves and Valve Systems

The development of pneumatic and hydraulic drives — in connection with the demand for smaller supply lines and actuators — requires entirely new concepts in the case of electrically actuated valves. Higher switching frequen-

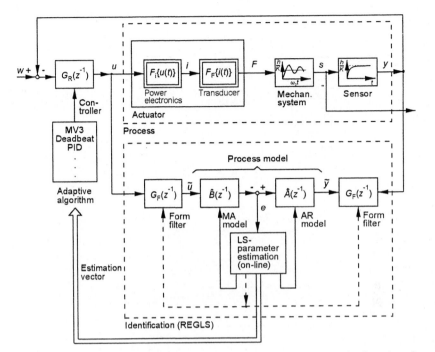

Fig. 8.25. General signal–flow diagram for real–time control of active damper systems

cies, analogue control of the flow throughput and miniaturization of the valve components are areas requiring more attention than ever.

Valves with Solid–State Transducers. One means of solving these problems involves replacing conventional electromagnetic actuators with solid–state transducers. Using piezoelectric and magnetostrictive components, blocking forces in the range of several thousand newtons and displacements of up to several hundred microns can be achieved. The high energy densities achievable using solid–state transducers enable the required force–displacement product to be reached with a comparably low transducer volume. At the same time, the achievable operating frequencies are much higher than those achievable with electromagnetic devices.

The actuator used in the laboratory model of Fig. 8.27 is composed of a piezoelectric stack transducer with a mechanical or hydraulic displacement–amplifying system. The maximum operating frequencies achievable with this valve arrangement lie between 100 Hz and 400 Hz. The size of the entire valve unit is considerably greater than that of the magnetically driven valve.

With the goal of approximating performance, a commercially available piezoelectrically operated valve with pneumatic pilot control was compared with a comparable magnetically operated valve with respect to static, dynamic and electrical properties [103]. The basic construction of the piezo-

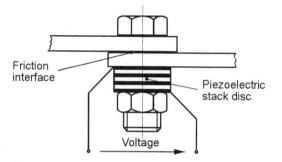

Fig. 8.26. Active connection joint in a built–up truss structure

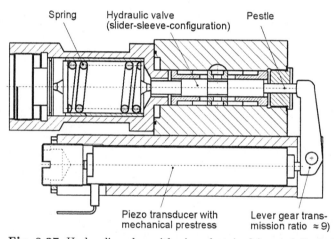

Fig. 8.27. Hydraulic valve with piezoelectric drive and displacement amplification

electrically driven valve is shown in Fig. 8.28. The valve has pneumatic pilot control that can be affected using a bimorph actuator. The on/off positions can be achieved by controlling the pressure on the pilot control membrane.

The piezoelectric valve demonstrates a remarkably greater dependence of the switching point from the system pressure but a much narrower hysteresis that the magnetic valve. Higher flow rates are achievable with the magnetically driven valve and the voltage necessary to switch and hold the valve are lower. The use of piezoelectric actuators in indirectly operated valves normally exhibits worse dynamic behavior than that of magnetic valves; this is because of the low actuating forces of piezoelectric bimorph elements and the resulting demand for small valve openings in the pre–stage. The piezoelectric valve studied, however, demonstrates advantages with respect to the electrical power demand: while the magnetic valve requires a power of 4 W to switch, the piezoelectric valve requires only about 5 mW; furthermore, the piezoelectric valve is much better suited for use in off–grid applications.

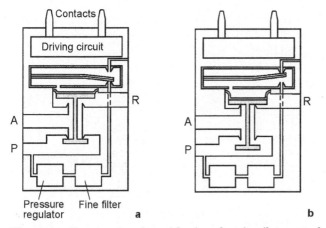

Fig. 8.28. Pneumatic valve with piezoelectric pilot control:
(a) closed position, (b) open position

An important result of the study states that direct valve designs are preferable in the development of highly dynamic valves. Based on this design aspect, switching frequencies of about 400 Hz currently appear realistic.

Valves with ER/MR Fluids. Servo valves without movable mechanical parts can be realized using an ER or MR fluid as a hydraulic medium in the flow mode. Valves for electrorheological fluids are usually constructed of coaxial cylindrical electrodes or as an arrangement of parallel plates (electrodes). MR fluid valves are realized in the form of a gap in a magnetic circuit. The flow resistance, and therefore the pressure drop across the valve, are controllable by the electric/magnetic control field. A sufficiently high control field amplitude can close the valve entirely, reducing the fluid flow to zero. Such an ER/MR fluid valve can be built compactly, driven with low power, and operated without wear. Conventional proportional valves consist of a large number of precisely machined parts; ER/MR fluid valves in comparison are less costly and less susceptible to wear. An additional advantage of servo valves based on ER/MR fluids lies in the achievable switching frequencies.

One possible application for an ER fluid valve is the positioning unit shown in Fig. 8.29 [104]. The design consists primarily of a hydraulic cylinder, four ER fluid valves arranged in the form of a full bridge, a four–channel high–voltage source, and a controller. The position x of the piston is measured by a displacement sensor and the force F working against the piston by a force sensor. A pump operating at a constant speed generates the pressure in the hydraulic circuit. Continuous control of the pressure drop across the ER fluid valves enables continuous control of the pressure on the piston and therefore its direction and speed of travel as well as its holding force.

An ER fluid servo drive based on this principle has recently become commercially available. According to the author, the unit can be used in automation, the automobile industry, testing, or even vibration isolation. A testing device equipped with this drive unit reached operating frequencies into the range of kiloHertz during preliminary tests.

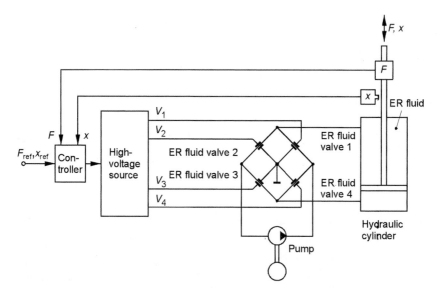

Fig. 8.29. Example of a positioning unit making use of ER fluid valves

8.4.4 Brakes and Clutches

Many solutions applied in the construction of machines and plants demand quick and compact brakes and clutches. LORD Corp. offers a commercial disc brake based on magnetorheological fluid. In comparison with eddy–current brakes, the MR fluid brake generates high torque even at low rotational speeds. The torque is adjustable over a wide range independently of the speed via the control current. The design and the braking torque M as a function of the control current I are shown in Fig. 8.30 [105]. Both gaps between the housing and the brake disc are filled with MR fluid. The magnetic flux generated by the coil is perpendicular to the direction of shear in the MR fluid. The brake has an outer diameter of 92 mm and requires a maximum driving power of about 10 W at a maximum current of 1 A. The brake is capable of converting a maximum of 700 W mechanical power into heat.

Compact clutches can be realized using ER and MR fluids for transmitting torque up to about 20 Nm with switching times on the order of several milliseconds. Small design changes can enable the brake described above to be modified to a disc clutch. One area of application of MR fluid clutches is for transmitting power to small machines such as generators, ventilators or

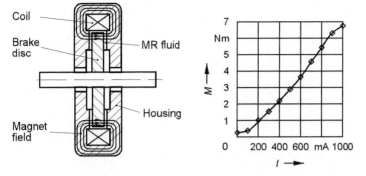

Fig. 8.30. MR fluid brake: **(a)** basic construction; **(b)** dependence of the braking torque M on the control current I within the range of speeds $200\,\mathrm{min}^{-1}\,to\,1000\,\mathrm{min}^{-1}$.

air–conditioning compressors. The application of electrorheological fluids in clutches is limited, since the ER fluid heats up while in operation, leading to a substantial increase in electrical control power.

8.4.5 Additional Applications

In remote controlled actuators for generating forces and/or displacements (manipulators, teleoperators), the desire could arise for a tactile response at the operating panel to the actual forces or displacements. This sense of touch can be achieved with the help of ER fluid actuators. This is achieved by arranging ER fluid actuators in an array. Small actuators enable compact arrangements, resulting in finer resolutions. Fig. 8.31 shows an example design conceived at Daimler–Benz AG, Munich [104].

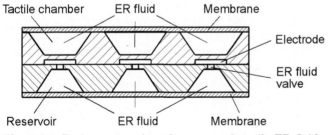

Fig. 8.31. Basic construction of an array of tactile ER fluid cells

Each actuator consists of an upper tactile chamber and a lower compensating chamber. Both chambers are filled with ER fluid and connected via an ER fluid valve. Each chamber is bounded by a membrane. Pressing on the membrane of the tactile chamber, e.g. with the finger, displaces ER fluid through the valve into the compensating chamber. The resulting resistive pressure can be altered by applying a voltage across the valve. The sense of touch can be simulated over a wide area using a matrix of several tactile actuators.

The squeeze mode offers a further possibility to realize a tactile array. The array is bounded by a membrane with a grounded metal foil on its underside. The base of the tactile field consists of several electrodes that are arranged in the form of an array and to which a high–voltage can be applied independently to each one. All high–voltage electrodes have a common, grounded counterpart, namely the metal foil on the underside of the membrane.

8.5 Adaptronics in Civil Engineering Structures
G. Hirsch

Civil engineering structures are exposed to dynamic loading from several sources, including high winds, earthquake ground motion, rotating and reciprocating machinery. While large deflections of tall buildings do not necessarily pose a threat to the safety of a structure, they can cause considerable discomfort and even illness to building occupants. The requirements for control of civil engineering structures concerning comfort are significantly less demanding than those for safety.

The concept of active control — a first step to adaptronics in civil structures — is an attempt to 'make structures behavior more like aircraft, machinery, or human beings' in the sense that they can be made adaptive or responsive to external loads.

Kobori and Minai [106] advocated the concept of 'dynamic intelligent buildings' capable of executing active response control when they are subjected to severe earthquakes.

In the U.S., Yao [107] marked the beginning of active control research when he proposed an error–activated structural system whose behavior varies automatically in accordance with unpredictable variations in the loading, as well as environmental conditions. A remarkable number of different systems, mechanisms and devices has been proposed by researchers in the past 20 years. Although each of them introduces a certain novelty, all the presented systems can be classified in three groups:

− active tendons;
− active mass dampers; and
− pulse control.

It should be noted that adaptronics in civil engineering structures is not making progress at present. Moreover 'structural control' is not the same as 'control theory', which has been developed in electrical engineering and applied mechanics, or the methods for control of space structures. The essence of structural control is the management of the performance of relative massive civil structures that require the application of large control forces but do not require a high degree of accuracy. Control of space structures have developed knowledge that, to some degree, provides information of value to structural control but does not solve the problems of civil structures control.

The control of earthquake response of structures is only one part of structural response control. The effects of wind, explosive shock, micro–tremors, etc., are also of concern. The control of the response of sensitive items, such as medical equipment, emergency power equipment, etc., must be considered.

The first international conferences on structural control were held at Waterloo University, Canada, in 1979 and 1985 [108, 109]. Although active control has been researched and utilized in many applications throughout the last decades, it has been only in the past several years that applications in civil engineering structures have been contemplated. Progress in development of structural control has been summarized within the scope of the International Workshop on Structural Control, Hawaii 1993 [111].The First World Conference on Structural Control, Pasadena 1994 was the first official act of the newly formed International Association for Structural Control. State–of–the–art reports, second generation of active control were topics of the conference.

Much progress has been made in research and development of active control technology in the U.S. and in Japan [112]. However, the true potential of active control has not been fully exploited. Concerning adaptronics in civil engineering structures, it should be noted that for the control of large motions, large uncertainties in the structural model exist since tests are not at amplitudes corresponding to building collapse, and time–varying non–linearities may exist in the foundations or soil that are significant to the lower vibration modes of interest [113].

This section is about the state of the art of active control in civil engineering, the second generation of active control, the results of experimental and full–scale tests in Japan and the U.S. and conclusions relating to the realization of adaptronics in civil engineering structures. It doesn't claim to be complete and is a selection of the author's experiences in passive and active control.

8.5.1 State of the Art for Active Control of Civil Engineering Structures

Concerning control of civil engineering structures, we have to distinguish between

– slender, tower–like structures;
– tall buildings; and
– long–span bridges (suspension and cable type).

The structural control depends from the dynamic properties (mass distribution, stiffness and damping) as well as the dynamic loading (wind, traffic, machinery, earthquake). From an engineering point of view, the different realities require adapted measures. Practical guidelines relating to vibration problems in structures are given by Bachmann et al. [114].

Tower–like structures. Tower–like structures are understood, in general, to be slender, tall structures (as television towers, lookout towers, chimneys, masts, and bridge pylons). Usually, gust–induced vibrations in the wind direction predominate, especially those at the fundamental bending frequency. The vibrations connected with vortex shedding that is transverse to the wind direction, however, can be more important. Particularly sensitive in this respect are steel chimneys (of welded construction, not insulated or lined with masonry, and with a fixed base). Vibrations of chimneys, masts and other low–damped tower–like structures lead to structural safety (fatigue) and serviceability problems. The occurrence of unacceptable vortex–induced cross–wind vibrations cannot be ruled out. As remedial measures in this case, both passive aerodynamic and mechanical aids should be mentioned. From an economic point of view, tuned mass dampers (TMD) are becoming more and more popular. The mass ratio (the mass of the TMD to the generalized mass of the structure) is often chosen to be 0.05.

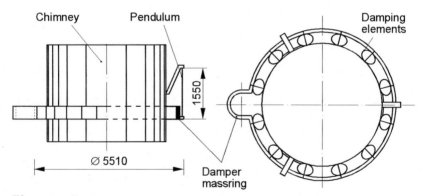

Fig. 8.32. Design of pendulum–suspended TMD for chimneys

An illustrative example (see Fig. 8.32) of the application of remedial measures to an existing group of steel chimneys when another chimney is added is given from Hirsch in [115]. However, in cases of transient loading by wind–gusts or earthquake, the effectiveness of TMDs will be reduced and therefore the active control will be particularly important in these cases of dynamic loading.

In practice, the optimum values of natural frequency and damping are usually not attained precisely. However, the sensitivity of such added systems with respect to deviations from the optimum is relatively small. Moreover, the TMD is not effective from the beginning of critical excitation of the main structure, as Fig. 8.33 shows.

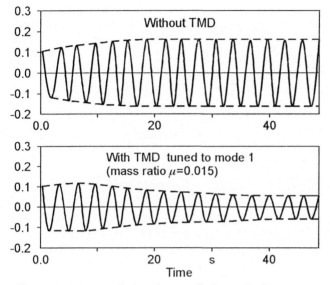

Fig. 8.33. Uncontrolled and controlled top displacement

The time delay between the beginning of a dynamic response and the counteraction of TMD is one of the most important topics in structural control (passive and active) from an engineering point of view. Therefore, Koshika et al. [116] have proposed an active–passive composite tuned mass damper (APTMD). The feasibility and the practicability of the theory was confirmed by demonstrating its control effect using an experimental model, as Fig. 8.34 shows.

The APTMD is one of the proposed active control device in civil engineering structures, especially of tall buildings against strong–wind effects and earthquake loading, described in the next subsection.

Tall Buildings. Tamura [117] shows several popular mechanisms of active vibration control systems for civil engineering structures. Mass damper systems, which use the inertia force of the auxiliary mass as the reaction force, are most commonly adopted. They need only a small space for installation and they can suppress the response of tall buildings very effectively during strong winds.

The mass damper systems are classified into four types from the energy supply's point of view: the passive type (TMD), semiactive type (SAMD), full active type (AMD) and hybrid (HMD). The TMD is effective if the natural period of the device is tuned well to that of the tall building (mostly

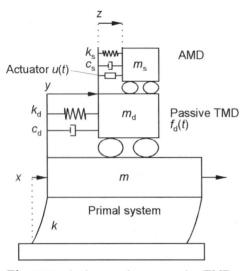

Fig. 8.34. Active–passive composite TMD

the fundamental mode). However, if the period of the device or the building gets changed, or the predominant period of the building response differs from the period of the device, the effectiveness of the TMD cannot be maintained. Therefore the other three types are developed to solve this problem. The SAMD is designed to ensure the optimum adjustment modifying the characteristics of the devices. Both of the AMD and HMD are the systems to use active control forces by supplying the energy from outside for suppressing the vibrations of the buildings. The AMD has no spring and no damping element; whereas the HMD is equipped with adequate values of spring and damping elements. In the case of a power blockout, the effectiveness of TMD will remain, and this is important in relation to reliability of control.

Active bracing or tendon systems can apply the control force directly to structures. This system is one of the most studied mechanisms. The reasons for favouring this mechanism is considered that the bracings and the tendons are existing structural members. This is important relating to realization of adaptronics in civil engineering structures. This system seems to be effective when the tall building structure is light. However, if the structure becomes larger and the external excitation level is higher, it will be difficult to apply the system because the required control force increases significantly. Moreover, the dynamic behavior of the bracings or tendons have to be considered.

Hybrid systems combined with base–isolated buildings have been proposed against earthquake loads. They have the possibility of decreasing the vibration of the structures drastically, if the adequate devices for supporting the buildings with low stiffness and the devices for generating the control force accurately are developed. Since the isolation devices have usually nonlinear properties, the control algorithm for such systems has been actively studied recently [106].

Bridges. The dynamic loading of bridges is of different kind:

- Traffic load (randomly);
- dynamic wind loads (gusts, buffeting, vortex shedding, galloping, flutter), occurring randomly, harmonically, or in a self–excited manner;
- earthquake (transient).

To avoid or suppress bridge–flutter, know–how transfer from other disciplines of engineering (ship and aircraft engineering) is possible and suitable in order to understand the control application. Domke (in [109]) reported on active deformation control of a 10 m test girder of Aachen University. As Fig. 8.35 shows, pneumatic cushions were supported on steel cables, which were connected to both ends of the girder. Pressures in the cushions were adjusted according to the deflection of the girder until the cable forces counteracted the sum of dead weight and live load.

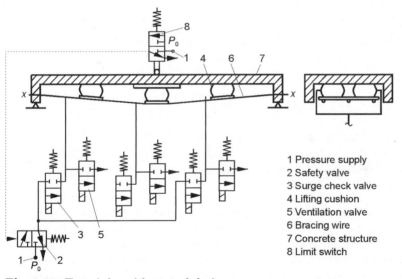

1 Pressure supply
2 Safety valve
3 Surge check valve
4 Lifting cushion
5 Ventilation valve
6 Bracing wire
7 Concrete structure
8 Limit switch

Fig. 8.35. Test girder with control device

Moving vehicles induce vibration of bridge girders by two mmechanisms: one is due to the moving force acting the girders in a finite time and the other due to the roughness of the road surface. In the case of cable–stayed bridges, this mechanism will be excited at the higher modes of the bridge deck and the supporting points of the cables. Parametric vibrations of the cables may occur (Bangkok Bridge for example), because vehicles are mass–spring–dashpot systems and the interaction between the girder and vehicles is not negligible in the girder vibration. The traffic–induced vibration causes the following problems: (1) runability of vehicles, especially of high–speed trains; (2) fatigue of girders and vibration–induced noise; and (3) excitation of buildings near bridges. The first problem occurred with the Honshu Shikoku Bridge and was reported by Y. Fujino in [111]. The third problem often occurs in elevated urban viaducts. Vibrations of the bridge piers induce waves propagating in the ground and this excited the buildings nearby. Because the occupants are sensitive to noise and vibrations, this may entail controversy.

Fujino and co–authors (in [112]) reported about active control of traffic–induced vibrations in highway bridges. Fig. 8.36 shows the vehicle, bridge girder with surface roughness, and an active TMD.

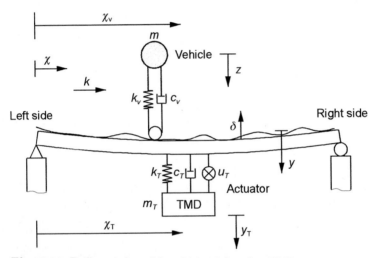

Fig. 8.36. Bridge girder with vehicle and active TMD

This simplified model should be self–explanatory. It is shown that both bridge response and support reaction forces can be significantly reduced. Linear quadratic (LQ) optimal control with full state feedback is used. The passive TMD (around 5% of the beam's modal mass) is tuned to the first mode of the beam. Fig. 8.37 shows some results. namely the dynamic reaction forces with and without control.

In order to effectively control wind–induced vibrations of long span bridges aerodynamic devices from the viewpoint of required energy are more attractive than mechanical control means. Fig. 8.38 shows the simplified model of bridge–section with aerodynamic devices (surfaces) to flutter–control.

Active aerodynamic measures are being used successfully in aeronautical engineering to suppress wing flutter. Using additional control surfaces, control forces that are anti–phase to the inducing forces are implied to prevent

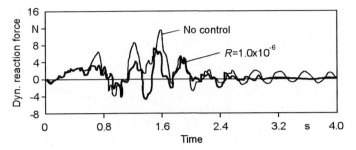

Fig. 8.37. Dynamic reaction forces with and without control

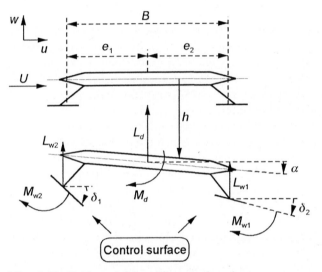

Fig. 8.38. Bridge section with control surfaces

the vibration of the structure in the critical flow. Sensburg et al. (in [108]) reported the results of flight tests with active flutter suppression. The suppression of bridge flutter by means of control surfaces will be based on the same idea. Fig. 8.39 shows some characteristic results.

8.5.2 The Second Generation of Active Control

Much of the theoretical basis in the development of active structural control over the last twenty years is rooted in modern control theory. For example, most of the control algorithms used in the current operating control systems for large civil structures are based on the principles of the linear quadratic regulator (LQR), as Housner et al. [112] showed. However, it needs to be recognized that control applications to civil engineering structures are unique in many ways and present a set of different challenges. For exam-

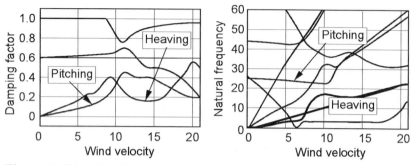

Fig. 8.39. Example of bridge flutter suppression

ple, in comparison with conventional control design as practiced during the first generation. Some distinguishing features of civil engineering structural control will be: structure (complex system with more than one eigenmodes, nonlinear behavior), sensing and actuation (few sensors and actuators, large control forces with higher speed), and control strategies (simple but robust and fault–tolerant control, suboptimal control, implementable control laws).

In incorporating active control into a structure, either as a new design or a retrofit, it is important to consider active control as a member of a family of innovative protection technologies which include, in earthquake loading for example, among others, base isolation and passive energy dissipation. For a specific application, technical merits and cost effectiveness of active control systems can then be evaluated more realistically in this context. Moreover, as stated in the recommendation of a working group on experimental methods [111], the testing of possible control devices that can deliver the required control force (for example) is necessary in order to assess the implementability of theoretical results: practical issues such as time delay and spillover effects can only be addressed after one learns of their magnitude and effects through experiments.

8.5.3 Application of Active Control from Practical Engineering Aspects

Although active structural control has been researched and utilized in many applications throughout the last several decades, it has been only in the past several years that applications in civil engineering structures have been contemplated. The acceptance of building applications has similar features in the U.S. and Japan, the leading states in active control of civil engineering structures. However, the acceptance procedures in Japan are expedited by direct involvement of each owner–construction company with the professional community. In the U.S., procedures were identified as to be geared by cost, insurance and performance criteria. While specific performance criteria should be established for individual devices to meet safety and operational requirements, the performance of buildings should meet current safety and service criteria. It was felt that the design engineer should be trusted with this engineering judgement and additional limiting standards should not be established. This will probably encourage and not limit necessary innovation.

The most recent implementations are combinations of add–on devices. While mass dampers provide the majority of implementation (i.e., tuned mass dampers, active mass dampers, hybrid mass dampers), active stiffening or bracing systems, energy dissipation/absorption dampers and hybrid isolations were also implemented in actual applications or full scale experiments. It was noted in [111] that most of the systems were considered as add–ons and the integration of the systems into structural design is not yet completely developed. That is important in relation to the application of active control from practical (engineering) aspects.

The following is a synopsis of the conclusions of several workshops on structural control concerning the development needs in the field of control engineering practice: (1) actuators, including modeling and identification; (2) performance robustness versus stability robustness, improving performance and improving stability for large civil engineering structures; and (3) the development of accurate, reliable and inexpensive displacement sensors would provide a substantial benefit to many approaches in structural control.

From an engineering practice point of view, the actuators for civil structures are most interesting. Simple and robust actuators are necessary. There are only few practical examples of actuators for civil structures. Fig. 8.40 shows an example of an actuator with variable damping showed by Kawashima and Unjoh in [111].

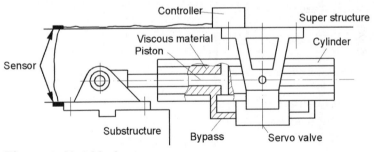

Fig. 8.40. Variable damper

8.5.4 Results of Experimental and Full–Scale Tests (in Japan and the U.S.)

Application of active control of civil structures from practical aspects has been pioneered in Japan. Multiple earthquake loadings and strong–wind effects on tall buildings enforced the realization of new techniques. Synopses of practical examples are given in the Japanese contributions in [111] and [112]. In this section, some practical solutions will be explained.

The Kyobashi Seiwa Building was built in 1989 and is the first building in the world that has an active control system. The active mass driver was installed to suppress dynamic responses caused by earthquakes or strong winds. It was reported from Yutaka Inoue et al. in [111] that the building had experienced several moderate earthquakes and strong winds during which ground acceleration, wind velocities and structural responses had been measured. The measured responses during the earthquakes are compared with the simulated responses by numerical analyses for an uncontrolled structure. Wind–response observations were performed every 30 min with and without control. From these comparisons, a remarkable decrease in amplitude due to the active mass driver system was confirmed.

An active tendon system has been examined by using a six–story steel structure [110]. Fig. 8.41 shows that a control force is transmitted to the structure through diagonal braces connected to the first floor by servo–controlled hydraulic actuators. The 600–ton symmetric building, as shown, has been erected in Tokyo, Japan. In fact, two control systems has been tested on the structure (a biaxial active tendon system and a biaxial active mass damper system).

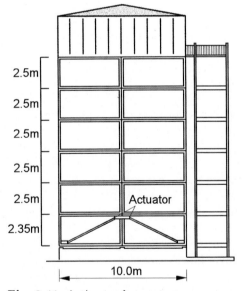

Fig. 8.41. Active tendon system

In relation to adaptronics in civil engineering structures, the active bracing system represents one of the best possibilities and therefore will be outlined in more detail. Reinhorn et al. [118] reported on the braces, the hydraulic actuators, the hydraulic power supply, the analog–digital converter and the sensors. The observed performance of the system under actual earthquakes and other artificial loadings will be presented.

The design of the braces was based on the maximum control force and the anticipated stiffness with the assurance that buckling will not occur under actuator actions. Circular steel tubes were used with 360 cm length, 165 mm diameter, 4.5 mm thickness and 564 kN strength. The measured stiffness of the braces is 98.4 kN/mm in the x–direction and 73.8 kN/m in the y–direction. Four units of Parker, heavy–duty hydraulic cylinder series 2H–style TC (NFPA style Mx2) were selected as actuators, with the following specifications: 735 mm length, 152.4 mm piston diameter, 63.5 mm rod diameter, ±50 mm stroke and 344 kN average capacity. Although the expected movement in the actuators was only ±12 mm, larger–sized actuators were chosen to enable length corrections during construction. In future applications, a much shorter actuator would be sufficient. The average capacity of the actuator was based on the working pressure (20.68 MPa) of the hydraulic oil and the average piston area. Two hydraulic actuators were coupled in series in each direction and are monitored by one servovalve and one servovalve controller of type MTS 458. The inner control loop for the hydraulic actuators is used for position feedback. The servovalve MTS 252.2x can supply up to 55 l/min at a pressure drop of 6.89 MPa.

The final design of the hydraulic system allows the active system to remain ready for full power controlled operation, while requiring the hydraulic pump to operate for only a few seconds each hour to keep the system full charged. An analog–digital converter was chosen based on the requirements that the analog controller must be compatible with the hydraulic service manifold and with the servovalves, and be capable of simultaneously controlling the two sets of servovalves.

The microcomputer executed the control algorithm, monitored the status of operation of various hydraulic components and monitored the status of the structural system. The system consisted of a PC computer with an INTEL 80386/25 MHz processor equipped with an INTEL 80387/25 MHz math coprocessor. Two analog–to–digital and digital to analog conversion boards provided interface for up to 16 channels of differential inputs from sensors and four channels of analog outputs to controllers. In addition, 16 digital logic channels were available on the computer boards.

The control system had four servovelocity seismometers of type Tokyo Sokushin VSE 11 for each principal direction of the structure, with an output range of ±100 cm/s. The velocity sensors were located on the ground, at the first, at the third and at the sixth floors of the building. The same sensors could provide acceleration information up to ±1000 cm/s². Additional transducers were mounted at each floor to monitor building behavior. Each actuator was equipped with a displacement transducer (LVDT) having a range of ±12 mm, which is used to adjust the length of the brace via the servovalve loop.

Some recorded samples show in Fig. 8.42 the structural response under 32% El Centro earthquake (uncontrolled and controlled structure). Also from the top–floor deflection (Fig. 8.43) one learns that the control measure will be effective with time delay.

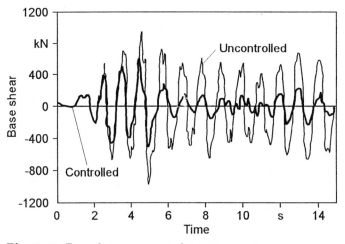

Fig. 8.42. Base shear response of structure

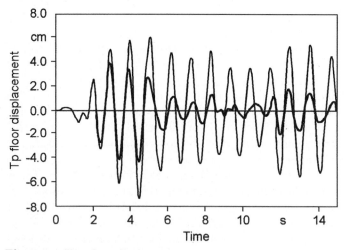

Fig. 8.43. Top–floor displacement

The results presented in the Reinhorn paper [118] demonstrate the following:

- The concept of an active tendon or bracing system, originated almost 20 years ago, has led to the successful development of the device for civil engineering structural control.
- The success of the full–scale active brace system performance is the culmination of numerous analytical studies and carefully planned laboratory experiments involving model structures.
- The control system can be implemented with existing technology under practical constraints such as power requirements and under stringent demand of reliability.
- The use of the control measure in existing structures can be a practical solution for retrofit, as demonstrated by this full–scale experiment. Note that the active braces were added only after the structure was completed.
- The experience gained through the development of this system can serve as an invaluable resource for the development of active structural control systems in the future.

A new type of tuned active damper (Mitsubishi) for high–rise buildings is shown in Fig. 8.44. Because the high–rise buildings are being put up everywhere and these days are put to many uses (high–class hotels, offices, apartments), whatever the purpose, one of the more unpleasant characteristics of such buildings is the degree of sway which occurs in high winds or earthquakes. The pendulum–type tuned active damper is a result of experiences with tuned mass dampers in towers, stacks and bridges.

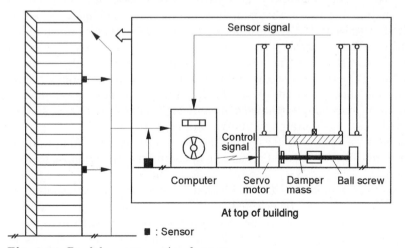

Fig. 8.44. Pendulum–type active damper

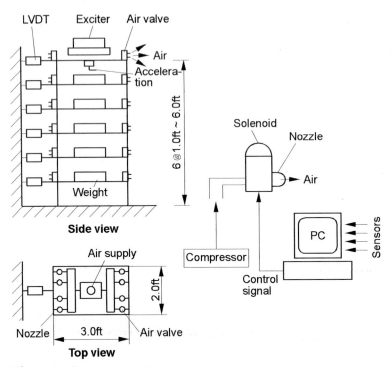

Fig. 8.45. Pneumatic active control device

The pendulum has the same natural period as the building. It has a multi stage suspended damper mass that can be housed in a single storey. When the building sways, the damper counteracts the building's movement and tends to absorb it. The sensors detect the movement of the building and the computer controls servo motors driving ball screws to position the damper so that it makes the optimum use of its natural tendency to absorb the motion. In case of winds, the damper immediately goes into operation. It is effective against vibrations of only 1 gal, but it works equally to counteract strong sway. Concerning earthquake, it cannot eliminate the effects of a major earthquake but it can bring residual vibrations under control.

Finally a pneumatic active control device may be an interesting example of active control. The schematic diagram of control configuration is shown in Fig. 8.45.

It has been demonstrated by [119] that a sequence of force impulses applied to a structure can be selected in such a way that the power spectral density of the displacement at a particular point within the structure matches the spectral density produced by earthquake ground motions as closely as desired. This result naturally suggests the possibility of using the force pulses to counteract or reduce displacements produced by earthquakes.

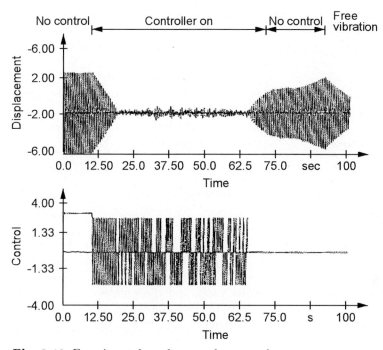

Fig. 8.46. Experimental results on pulse–control

Some experimental results on control pulse and top–floor relative displacement are shown in Fig. 8.46.

8.5.5 Conclusions

The placement of large masses at the top of a structure (a tall building, for example) appears to be an expensive solution, and the approach may be limited by the size and stroke of the active mass damper. In both the active and passive state, the active member is a functional structural component. This is important relating to the realization of adaptronics in civil engineering structures (tall buildings, industrial plants, bridges etc.). The placement of the active members at locations of maximum strain energy using colocated sensors and actuators provides a redundant and robust control system and appears very attractive for civil engineering structures. Since only the first few vibration modes are of interest, the locations of the maximum strain energy will be near the base of the structure. The measurement of the forces at the base is similar to using accelerometer data at the top of the structure. The sensors used for control can be member forces, the direct measurement of a parameter related to structure collapse. As Wada and Das (1992) showed [120], the load–carrying member can become the actuator itself; it has the requirement of a large load–carrying system with the potential to dissipate energy when placed at the locations of maximum strain energy. The actuator would be very inexpensive, located near a more attractive location near the base (or pylon in case of bridge control), and will not require additional functional physical space. Actuators developed for space systems are not appropriate because of their force and stroke limitations. An active member may include a hydraulic system that triggers the member force to introduce phase delay and stiffness changes to attenuate the loads in the structure. Adaptive structures may present attractive options for the control civil engineering structures in future.

8.6 Vibration Control of Ropeway Carriers
H. Matsuhisa

Ropeways are popularly used in many places, such as skiing and sightseeing venues. Because of their low cost, the use of ropeways as a means of transportation in cities has attracted increasing interest. However, swing of a ropeway carrier is easily caused by wind, and the ropeway cannot operate if the wind velocity exceeds about 15 m/s. The study of how to reduce the wind–induced swing of ropeway carriers has attracted many researchers. There appear two methods to reduce the swing, a dynamic vibration absorber and a gyroscope; however, it had been said that it was impossible to reduce the vibration of pendulum–type structures such as a ropeway carrier by a dynamic absorber, and a gyroscope is not always practical for ropeway carriers because they do not generally have an electric power supply.

In 1993, Matsuhisa showed that the swing of a ropeway carrier can be reduced by a dynamic absorber if it is located far above or below from the center of oscillation [121]. Based on this finding, a dynamic absorber composed of a moving mass on an arc track was designed for practical use, and it was installed in chair–lift–type carriers and gondola–type carriers in snow skiing sites in Japan in 1995 for the first time in the world. It has been shown that a dynamic absorber with the weight of one tenth of the carrier can reduce the swing to half [122].

The moment induced by a gyroscope can be used to control the swing of a pendulum–type structure. Since the moment is proportional to the spin velocity of rotor, the weight of the gyroscope can be lightened by increasing the spin velocity. The moment is also proportional to the angular velocity of precession of the gyroscope. One method to obtain the moment is an active control of precession of gyroscope, which is called CMG (control moment gyroscope). The second method is a passive gyroscope, in which the precession axis is connected to a rotary spring and a rotary damper and the precession becomes a single–degree–of–freedom vibration system. The swing of the primary system gives the moment to the gyroscope to precess. Then the precession of the gyroscope gives the moment to the primary system to reduce the swing. This passive gyroscope is applicable to the carriers because it does not need a control device. Several design methods of a passive gyroscope were also proposed. The experiments with prototype passive gyroscopes on a gondola for six passengers were carried out and satisfying results were obtained. These studies have been carried out by Nishihara, [123]–[127].

8.6.1 Dynamic Vibration Absorber

As shown in Fig. 8.47, a ropeway carrier can be regarded as a rigid–body pendulum of mass m_1, inertia moment of I and damping of zero. Assume that the distance between the center–of–gravity G and the fulcrum O is l_1, the equivalent length of pendulum OG' is $l'_1 (= I/m_1l_1)$. There are several types of dynamic absorber: a linear mass and spring type, a pendulum type, and so on. Here, an arc–track type of dynamic absorber is chosen for a ropeway carrier because it is easy to tune the natural frequency and its shape is suitable for the ropeway carrier. Assume that the mass of the dynamic absorber is m_2, the radius of the arc track is l_2, the damping constant is c, and the distance between the fulcrum O and the mass of dynamic absorber is l. Letting θ_1 and θ_2 represent the angular displacements of the carrier and the absorber, respectively, and T the external moment acting on the carrier, the linearized equations of motion can be expressed as

$$(I + m_2l^2)\ddot{\theta}_2 + m_2l_2l\ddot{\theta}_2 + (m_1l_1 + m_2l)g\theta_1 + m_2l_2g\theta_2 = T, \tag{8.1}$$

$$m_2l\ddot{\theta}_1 + m_2l_2\ddot{\theta}_2 + cl_2\dot{\theta}_2 + m_2g\theta_1 + m_2g\theta_2 = 0. \tag{8.2}$$

Introducing the dimensionless symbols,

$$\mu = m_2/m_1, \ \gamma = l/l'_1, \ \alpha = l'_1/l_1, \ \omega_1^2 = m_1l_1g/I = g/l'_1,$$

$$\omega_2^2 = g/l_2, \ \zeta = c/(2m_2\omega_1), \ f = \omega_2/\omega_1, \ h = \omega/\omega_1,$$

$$\Theta_{1st} = T/(m_1gl_1), \ \Theta_{2st} = \alpha T/(m_1gl_2),$$

the complex amplitude of the carrier and the dynamic absorber under the harmonic external moment $Te^{i\omega t}$ become

$$\Theta_1 = \Theta_{1st}(A + i2\zeta B)/(C + i2\zeta D) \text{ and} \tag{8.3}$$

$$\Theta_2 = \Theta_{2st}E/(C + i2\zeta D), \tag{8.4}$$

where

$$A = f^2 - h^2,$$

$$B = h,$$

$$C = (1 - h^2)(f^2 - h^2) - \mu\alpha(\gamma f^2 - 1)(\gamma h^2 - 1),$$

$$D = \{1 + \mu\alpha\gamma - (1 + \mu\alpha\gamma^2)h^2\}h, \text{ and}$$

$$E = -(1 - \gamma h^2).$$

Equation 8.3 shows the frequency response of the carrier, and all curves pass through two fixed points P and Q independent of the dimensionless damping coefficient ζ (cf. Fig. 8.48). As the conventional linear system with a linear dynamic absorber [128], by making the two points equal in height and making them the maximum, the optimal natural frequency ratio of the dynamic absorber to the main system, f_{opt}, and the optimal damping of the

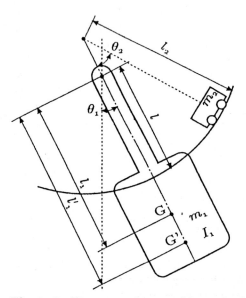

Fig. 8.47. The arc–track type of dynamic absorber attached to a carrier

dynamic absorber, ζ_{opt}, can be obtained as

$$f_{opt} = \frac{\sqrt{1 + 2\mu\alpha\gamma + \mu^2\alpha^2\gamma^3}}{1 + \mu\alpha\gamma^2}, \tag{8.5}$$

$$\zeta_{opt} = \frac{1}{2}\sqrt{\frac{AA' - |\Theta_1/\Theta_{1st}|^2 CC'}{-BB' + |\Theta_1\Theta_{1st}|^2 DD'}}, \tag{8.6}$$

where priming represents $\partial/\partial h$. The optimum damping ζ_{opt} of point P and that of point Q are slightly different from each other. However, since this difference is negligible for practical purposes, we can use the arithmetic average value. The equivalent mass ratio μ_e, which expresses the efficiency of the dynamic absorber, can be obtained from (8.3). Substituting $f = 1, h = 1$ into the real part of the denominator C, the equivalent mass ratio is given as

$$\mu_e = \mu\alpha(1 - \gamma)^2. \tag{8.7}$$

The equivalent mass ratio μ_e is thus proportional to the nominal mass ratio μ and to the second power of $(1 - \gamma)$. This means that if the dynamic absorber is attached to the center of oscillation G' $(\gamma = 1)$, the dynamic absorber does not work at all, and that the dynamic absorber must be attached at a point as far from the center of oscillation as possible. When $\gamma = 1$, the first term (the inertia force acting upon m_2 due to the acceleration of θ_1) and the fourth term (the gravity force due to the decline of θ_2) in (8.1) cancel each other ones, and the external force on m_2 disappears. Then m_2 does not move relative to the primary system and energy of the primary system is not dissipated.

In the case of a conventional linear system, the efficiency of the dynamic absorber is given by μ, and the dynamic absorber must be attached to a point that has a large amplitude of vibration. However, in the case of pendulum system such as a gondola, the dynamic absorber works better when it is attached further from the center of oscillation. Even if the dynamic absorber is attached to the fulcrum, the dynamic absorber reduces the vibration remarkably. Furthermore, it is better to attach the dynamic absorber above the fulcrum $(\gamma < 0)$.

Fig. 8.48 shows the theoretical prediction of the frequency response of a ropeway gondola for six passengers $(l_1 = l_1'=4\,\mathrm{m}, m_1=1{,}000\,\mathrm{kg})$ with an optimally tuned dynamic absorber. In this calculation, the damping ratio of the gondola is assumed to be 1%. In a case of the gondola without absorber, the maximum value of the normalized amplitude $|\Theta_1/\Theta_{1st}|$ is 50. When the gondola has a dynamic absorber with the equivalent mass ratio $\mu_e = 0.025$, the maximum value of the normalized amplitude is 9, and when $\mu_e = 0.05$, the maximum value is 6.4. Fig. 8.49 shows the step responses; Fig. 8.50 shows the random response that is simulated by the wind. These figures show that the dynamic absorber is very effective in reducing the swing of gondola and the effectiveness is represented by the equivalent mass ratio μ_e.

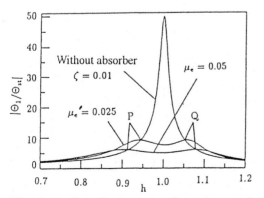

Fig. 8.48. Compliance curves for various equivalent mass ratios

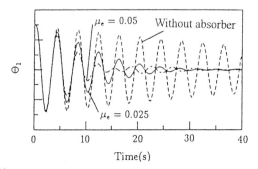

Fig. 8.49. Step responses for various equivalent mass ratios

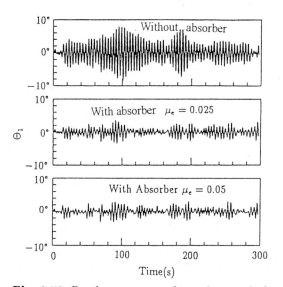

Fig. 8.50. Random responses for various equivalent mass ratios

It is possible to use another type of dynamic absorber for a gondola — for example, a pendulum–type dynamic absorber or a typical dynamic absorber composed of a mass and a spring on a straight track. In the case of a pendulum–type absorber, the equations are the same as for the arc–track type of absorber. In the case of the mass–spring type of absorber, the same results are induced by replacing the natural frequency of the absorber as $\omega_2^2 = k_2/m_2$. For all types of dynamic absorbers, the amplitude of vibration, the optimum tuning, and the equivalent mass ratio are given by (8.3) to (8.7). Thus the theoretical analysis mentioned above appears to be a universal one that can be applied to all kinds of dynamic absorber attached to pendulum–type structures.

Fig. 8.51 shows the prototype dynamic absorber ($m_2 = 48\,\text{kg}$) attached to a gondola ($m_1 = 790\,\text{kg}$) for 10 passengers. The experimental result of step response is shown in Fig. 8.52. In this case the absorber was attached to a very high position and the swing was attenuated very rapidly.

Fig. 8.51. Prototype dynamic absorber on a gondola for ten passengers

Based on the theoretical predictions and the experiments mentioned above, the dynamic absorbers were installed in the practical chairlifts in Japan in 1995. This is the first application of the dynamic absorbers for ropeway carriers in the world. The figure of the lift and the step response are shown in Figs. 8.53 and 8.54. The weight of the carrier was 156 kg and the weight of the moving mass in the arc–shaped aluminum track was 11 kg. The radius of the arc track was 2300 mm and its length was 1400 mm. The damping was induced by the electromagnetic force caused by a permanent magnet attached to the moving mass on the aluminium track. The swing of the lift was decreased from 10° to 1° in six periods.

In 1996, the dynamic absorber was also installed in gondolas for 15 passengers in Japan, as shown in Fig. 8.55. In this case, since the rope span between the neighboring towers was long, the natural frequency of the swing of gondola varied depending on the location. When the gondola was located at the middle of the span the natural period was 4.8 s, and when it was near the tower the natural period was 4.3 s. Thus the gondola had two dynamic ab-

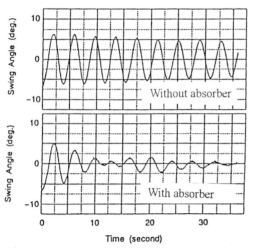

Fig. 8.52. Step responses of a gondola with the prototype dynamic absorber

sorbers whose natural periods were 4.3 s and 4.8 s. The weight of the gondola was 830 kg and the weight of the moving mass was 17.6 kg. The step response is shown in Fig. 8.56. The swing was decreased remarkably by the double dynamic absorber. While the old–type gondola without the dynamic absorber had to stop operating when the wind velocity was more than 15 m/s, the new gondola with the dynamic absorber was able to operate in wind velocities as high as 20 m/s.

8.6.2 Passive Gyroscopic Damper

A gyroscope is used to reduce the swing of gondola, as shown in Fig. 8.57. Assume that the precession axis is x, the rotor axis is z, the gyroscopic torque is exerted about the y–axis and the gondola swings about the y–axis, which passes through the fulcrum O. Letting I_R represent the polar moment of inertia of the rotor, Ω the rotating velocity, $\dot{\theta}_y (= \dot{\theta}_1)$ the angular velocity of the swing of gondola about the y–axis, the gyroscopic torque $I_R \Omega \dot{\theta}_y$ about the x–axis is exerted. The gyroscopic torque $I_R \Omega \dot{\theta}_x$ about the y–axis is also exerted by the precession θ_x, which can be used for control of the swing of gondola. Letting I denote the polar moment of inertia of the gondola about the fulcrum, I_G the polar moment of inertia of the rotor and the gimbal about the x–axis, T the external exciting torque, T_a the external torque acting on gimbal axle, and the precession axle be supported by a rotary spring k and a rotary damper c, the equations of motion for the small swing θ_y become

$$I\ddot{\theta}_y + m_1 l_1' g \theta_y = I_R \Omega \dot{\theta}_x (\cos \theta_x) + T, \tag{8.8}$$

$$I_G \ddot{\theta}_x + c\dot{\theta}_x + k\theta_x = -I_R \Omega \dot{\theta}_y (\cos \theta_x) + T_a. \tag{8.9}$$

Fig. 8.53. The dynamic absorber 'Libra' was attached to chairlifts for the first such use in the world (1995)

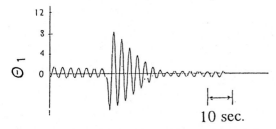

Fig. 8.54. Step responses of the chairlift with the dynamic absorber 'Libra'

Fig. 8.55. Gondola for 15 passengers with two dynamic absorbers 'Libra'

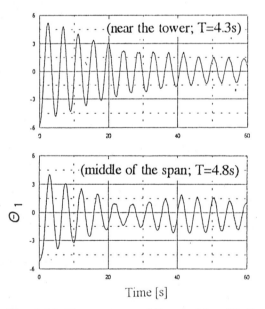

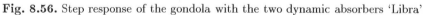

Fig. 8.56. Step response of the gondola with the two dynamic absorbers 'Libra'

In the case of an active gyroscope, the torque T_a is given by a motor connected to the gimbal axle. However, in the case of the passive gyroscope, T_a is zero.

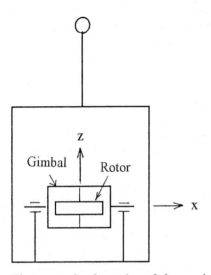

Fig. 8.57. Configuration of the passive gyroscropic damper

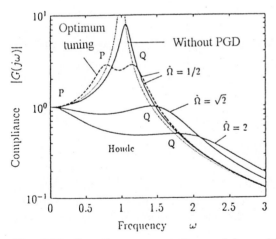

Fig. 8.58. Compliance curves of a pendulum with a passive gyroscropic damper tuned by the fixed–point theory [124].

Since the device on the gondola must be simple and require only a little power, the passive gyroscope is suitable for the gondola. From (8.8) and (8.9), the complex compliance is obtained as

$$\frac{\Theta_{\mathrm{y}}}{T} = \frac{-I_{\mathrm{R}}\omega^2 + k + i\omega c}{(-I\omega^2 + m_1 l_1' g)(-I_{\mathrm{G}}\omega^2 + k + i\omega c) - I_{\mathrm{R}}^2 \Omega^2 \omega^2}, \tag{8.10}$$

where $\cos\theta_x$ is assumed to be 1. This equation is similar to that of the conventional linear dynamic absorber. The swing can be reduced by tuning the spring constant k and damping constant c, and hereafter this gyroscope is called a 'passive gyroscopic damper'.

Here, we introduce nondimensional parameters for the moment of inertia, the rotor velocity and the damping as

$$\alpha_{\mathrm{m}} = I_{\mathrm{G}}/I_{\mathrm{R}}, \ \beta_{\mathrm{m}} = I/I_{\mathrm{R}}, \ \tilde{\Omega} = \frac{|\Omega|}{\omega_1}\sqrt{\frac{I_{\mathrm{R}}^2}{I_{\mathrm{G}}I}}, \ f = \omega_2/\omega_1, \ z = c/(2I_{\mathrm{G}}\omega_1).$$

By making use of these nondimensional parameters, (8.10) is expressed in the following form:

$$G(s) = \frac{\Theta_y(s)}{T(s)} = \frac{s^2 + 2zs + p^2}{(s^2 + 2Zs + 1)(s^2 + 2zs + p^2) + \tilde{\Omega}^2 s^2}. \tag{8.11}$$

There are several methods to determine the optimum values of the spring constant k and the damping constant c: fixed point theory [124], minimum variance criteria [125], numerical optimization by the nonlinear vibrational analysis [125], Monte Carlo optimization by genetic algorithms [126] and stability maximization [127]. Here the fixed–point theory and the stability–maximization theory will be explained because these methods have closed–form simple design formulas.

When $\tilde{\Omega} \leq \sqrt{2}$, the theory of the fixed points on the compliance curve can be used. The optimum natural frequency and the damping of the gimbal support become

$$f^2 = 1 - \frac{\tilde{\Omega}^2}{2}, \tag{8.12}$$

$$z = \frac{1}{2}\left(\frac{\tilde{\Omega}}{2}\sqrt{\frac{3}{2} + \frac{\tilde{\Omega}}{2\sqrt{2}}} + \frac{\tilde{\Omega}}{2}\sqrt{\frac{3}{2} - \frac{\tilde{\Omega}}{2\sqrt{2}}}\right). \tag{8.13}$$

The optimum damping is obtained by the arithmetical average of the two optimum values, one is for the point P and the other is for the point Q. The equivalent inertia moment ratio, which represents the effectiveness of the gyroscope, is

$$\mu_{\mathrm{m}} = \tilde{\Omega}^2 = \frac{I_{\mathrm{R}}^2 \Omega^2 \omega^2}{m_1 l_1' gk} \cong \frac{I_{\mathrm{R}}^2 \Omega^2}{I I_{\mathrm{R}}\omega^2}. \tag{8.14}$$

The compliance curves of the primary system with the tuned passive gyroscopic damper are shown in Fig. 8.58.

When $\tilde{\Omega} \geq \sqrt{2}$, it is not possible to place points P and Q at equal compliance because the natural frequency of the gimbal becomes negative, as (8.12) shows. In this case, we can choose the Houde damper ($k = 0$). The fixed point P disappears and only point Q exists, and the optimum damping is that which gives the horizontal tangent at point Q. The optimum damping is

$$z = \frac{1}{2}\sqrt{1 + \frac{\tilde{\Omega}^2}{2}}. \tag{8.15}$$

The compliance curves for the Houde damper are also shown in Fig. 8.58. The degree of stability is defined as the absolute value of the maximum real part of the root of the characteristic equation, which is given by the denominator of (8.11). When $\tilde{\Omega} \leq 2(1 - Z)$, the condition for the maximum stability $\lambda = Z + \tilde{\Omega}/2$ is given by $p = 1$ and $z = \tilde{\Omega} + Z$. The stability increases with $\tilde{\Omega}$ and it reaches value 1 at $\tilde{\Omega} = 2(1 - Z)$. When $\tilde{\Omega} \geq 2(1 - Z)$, the stability is less than 1, and then the maximum stability is given by $\tilde{\Omega} = 2(1 - Z)$.

Two prototype gyroscopes, as shown in Fig. 8.59, were installed on an actual gondola for six passengers to carry out the experiment. It is possible to diminish the torques about the x–axis and the z–axis, which may cause pitching and yawing by using two gyroscopes that have reverse rotating directions to each other. The weight of gondola was 500 kg, the natural period of it was 3.5 s. The dimension of the gyroscope was 700 mm (height) and 450 mm (width and depth). The moment of inertia of rotor was 0.434 kg/m^2, the diameter 390 mm and the weight 17.6 kg. The gimbal axis was connected to a linear spring with a pulley and wire, and had a rotary–type viscous damper. The gyroscopes were set such that the gimbal axes were vertical to the floor of the gondola and parallel to the pitch axis. The moment of inertia ratios were $\alpha_m = 0.96$ and $\beta_m = 5200$. The rotor was driven by a 24–volt DC motor rotating at 2400 rpm ($\tilde{\Omega} = 2$). The parameters p and z were determined by the maximum stability theory in order to make use of a high rotational speed of the rotor.

The step response of the gondola displacement is shown in Fig. 8.60. The swing decreased from 5° to zero in two periods.

8.6.3 Conclusions and Outlook for Future Research

The wind–induced vibration of ropeway carriers used to be inevitable. However, in 1993 it was found that the vibration can be reduced easily by a dynamic absorber, and one has been in use since 1995 in Japan. Since the theory can be applied for ships and other structures, research into the dynamic absorber will be accelerated and other types of dynamic absorber may appear.

Fig. 8.59. The passive gyroscopic damper

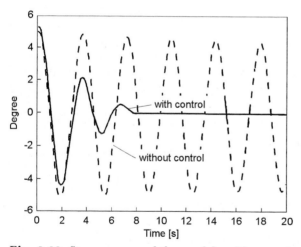

Fig. 8.60. Step response of the gondola with a passive gyroscopic damper tuned by the stability–maximization theory [127].

The gyroscope is very effective in reducing the vibration — not only the passive type but also the active type. Such gyroscopes could be applied for many structures: rope suspended bridges, ships and robot arms.

References

1. Forward, R.L.: Electronic Damping of Orthogonal Bending Modes in a Cylindrical Mast—Experimental. Journal of Spacecraft and Rockets, **18**, Jan. 1981.
2. Wada, B.K.: Adaptive Structures: An Overview. Journal of Spacecraft and Rockets, **27** 3, May–June, 1990.
3. Wada, B.K., Editor: Adaptive Structures. AD–Vol. 15, ASME Winter Annual Meeting, San Francisco, CA, Nov. 1989.
4. Wada, B.K.; Fanson, J.L. and Miura, K, Editors: First Joint US/Japan Conference on Adaptive Structure. Technomics Publishing Company, Lancaster, PA, 1990.
5. Wada, B.K.: Robust Structures Through the Incorporation of Adaptive Structures. IAF–94–I.4.199, 45th Congress of the International Astronautical Federation, Jerusalem, Israel, Oct. 1994.
6. Wada, B.K.; Fanson, J. and Chen, G–S: Adaptive Structure to Enable Missions by Relaxing Ground Test Requirements. Journal of Spacecraft and Rockets, **28** 6, Nov.—Dec. 1991.
7. Wada, B.K.: Adaptive Structures for Deployment/Construction of Precision Space Structures. Proceedings of the 18th International Space And Astronautical Science Meeting, Kagoshima, Japan, March 1992.
8. Das, S.K.; Utku, S.; Chen, G–S and Wada, B.K.: Optimal Actuator Placement in Adaptive Precision Trusses. *Intelligent Structural Systems*, Editors Tzou and Anderson, Kluwer Academic Publishers, 1992.
9. Chen, J.C. and Fanson, J.L.: System Identification Tset Using Active Members. AIAA 89–1290, Proceedings of the 30th AIAA Structures, Structural Dynamics and Materials Conference, Mobile, AL, 1989.
10. Kuo, C.P.; Chen, G–S; Pham, P. and Wada, B.K.: On–Orbit System Identification Using Active Members. AIAA 90–1129, Proceedings of the 31st AIAA Structures, Structural and Materials Conference, Long Beach, CA, 1990.
11. Wada, B.K.: Structural Design for Control. Flight–Vehicle Materials, Structures, and Dynamics–Assessment and Future Direction, **5** Chapter 10, Editors Noor and Venneri, ASME, New York, NY, 1993.
12. Bruno, R.; Salama, M.A. and Garba, J.: Actuator Placement for Static Shape Control of Nonlinear Truss Structures. Proceedings of the Third International Conference on Adaptive Structures, Technomics Publishing, Lancaster, PA, San Diego, CA 1992.
13. Chen, G–S. and Lurie, B.J.: Active Member Bridge Feedback Control for Damping Augmentation. Journal of Guidance, Control and Dynamics, **15** 5, Sept.–Oct. 1992.
14. Neat, G.W.; O'Brien, J.F.; Lurie, B.J.; Garnica, A.; Belvin, W.K.; Sula, J. and Won, J.: Joint Langley Research Center/Jet Propulsion Laboratory CSI Experiment. 15th Annual AAS Guidance and Control Conference, Keystone, CO, Feb. 8–12, 1992.
15. Rahman, Z.; Spanos, J.; O'Brien, J. and Chu, C.: Optical Pathlength Control Experiment on JPL Phase B Testbed. AIAA 93–1695, Proceedings of the 34th

Structures, Structural Dynamics and Materials Conference, La Jolla, CA, April 1993.

16. Lawrence, C.; Lurie, B.; Chen, G–S and Swanson, A.: Active Member Vibration Control Experiment in a KC–135 Reduced Gravity Environment. Proceedings of First US/Japan Conference on Adaptive Structures, Technomics Publishing Co., Lancaster, PA, May 1991.

17. Bousquet, P.W.; Guay, P. and Mercier, F.: Evaluation of Active Damping Performances in Orbit. Proceedings of the Seventh International Conference on Adaptive Structures Technology, Technomics Publishing Co., Lancaster, PA, Rome, Italy, Oct. 1996.

18. Manning, R.A.; Wyse, R.E. and Schubert, S.R.: Development of an Active Structure Flight Experiment. AIAA 93–1114, AIAA Aerospace Design Conference, Irvine, CA, Feb. 16–19, 1993.

19. Geng, Z.J.; Pan, G.G.; Haynes, L.S.; Wada, B.K. and Garba, J.A.: An Intelligent Control System for Multiple Degree–of–Freedom Vibration Isolation. Journal of Intelligent Material Systems and Structure, **6** 6, 1995.

20. Glaser, R.; Garba, J.A. and Obal, M.: STRV–1b Cryo–cooler Vibration Suppression. AIAA 95–1122, Proceedings of the 36th AIAA Structures, Structural Dynamics and Materials Conference, New Orleans, LA, 1995.

21. Wada, B.K. and Rahman, Z.: Vibration Isolation, Suppression, Steering and Pointing (VISSP). ESA Conference on Spacecraft Structures, Materials, and Mechanical Testing, Noordwijk, the Netherlands, March 27–29, 1996.

22. Agnes, G.S. and K. Silva: Aircraft Smart Structures Research in the USAF Wright Laboratory. AGARD–CP–531, Paper 27, 1992

23. Schmidt, W. and Chr. Boller: Smart Structures – A Technology for Next Generation Aircraft. AGARD–CP–531, Paper 1, 1992

24. Kudva, J.N. et al.: Overview of the ARPA/WL, Smart Structures and Material Development – Smart Wing, Contract. Paper No. 2721–02, SPIE North American Conference on Smart Structures and Materials, San Diego CA, 1996

25. Rogers, C.A. (Ed.): Smart Materials, Structures, and Mathematical Issues. Technomic Publ. Co., 1988

26. AGARD: Smart Structures for Aircraft and Spacecraft. AGARD–CP–531, 1992

27. AIAA/ASME: Adaptive Structures Forum. AIAA Proceedings, 1994

28. ICAST,1991 – 1995: Proc. Internat. Conf. on Adaptive Struct.; Technomic Publ. Co.

29. SPIE: Proc. of: Fiber Optic Smart Structures and Skins II. The Internat. Soc. for Optical Engineering (SPIE), 1988 ff.

30. Crawley, E.F. and E.H. Anderson: Detailed Models of Piezoceramic Actuation of Beams. AIAA 89–1388–CP, pp. 2000–2010, 1989

31. Wada, B.K. and J.A. Garba: Advances in Adaptive Structures at Jet Propulsion Laboratory. AGARD–CP–531, Paper 28, 1992

32. Fanson, J.: Articulating Fold Mirror for the Wide Field and Planetary Camera. Proc. of the 4th Internat. Conf. on Adaptive Structures, Technomic, pp. 278–302, 1993

33. Varadan, V.K. and V.V. Varadan: Smart Materials, Smart Skins, and Composites for Aerospace Applications. Proc.: ECCM Smart Composites Workshop, pp. 17–22, 1993

34. Obal, M., et al.: The Satellite Attack Warning and Assessment Flight Experiment (SAWAFE). AGARD CP–531, Paper 7, 1992

35. Varadan, V.K. and V.V. Varadan: Smart Materials, MEMS and Electronics Integration for Aerospace Applications. Proc.: Internat. Aerospace Symp. (IAS), Nagoya/Japan, pp. 115–123, 1994

36. Boller, Chr.: Fundamentals on Damage Monitoring. AGARD LS–205, Paper 4, 1996
37. Measures, R.: Fibre Optic Sensing for Composite Smart Structures. AGARD CP–531, Paper 11, 1992
38. Tutton, P.A. and F.M. Underwood: Structural Health Monitoring Using Embedded Fibre Optic Sensors; AGARD CP–531, Paper 18, 1992
39. Hickman, G.A.; J.J. Gerardi and Y. Feng: Application of Smart Structures to Aircraft Health Monitoring. J. of Intell. Mat. Syst. and Struct., **2**, pp. 411 – 430, 1991
40. Tracy, J.J. and G.C. Pardoen: Effect of Delamination on the natural Frequencies of Composite Laminates. J. of Composite Material, **23**, pp. 1200 – 1215, 1989
41. Carlyle, J.M.: Acoustic Emission Testing the F–111. NDT–Internat., **22** (2), pp. 67 – 73, 1989
42. McBride, S.; M. Viner and M. Pollard: Acoustic Emission Monitoring of a Ground Durability and Damage Tolerance Test. in: D.O. Thompson & D.E. Chimenti: Review of Progress in Quantitative NDE; **10B**, Plenum Press, pp. 1913–1919, 1991
43. Vary, A. and R.F. Lark: Correlations of Fiber Composite Tensile Strength with the Ultrasonic Stress Wave Factor. J. of Testing and Evaluation; pp. 185 – 191, 1979
44. Keilers, C.H. and F.-K. Chang: Identifying Delamination in Composite Beams Using Built–In Piezoelectrics. J. of. Int. Mat. Syst. & Struct., **6**, pp. 649–672, 1995
45. Victorov, I.A.: Rayleigh and Lamb Wave; Plenium, New York, 1967
46. Kaczmarek, H.; C. Simon and C. Delebarre: Evaluation of Lamb Wave Performances for the Health Monitoring of Composites Using Bonded Piezoelectric Transducers. Proc. of ICIM'96 and ECSSM 96, SPIE **2779**, pp. 130 – 135, 1996
47. Blaha, F.A. and S.L. McBride: Fiber–Optic Sensor Systems for Measuring Strain and the Detection of Acoustic Emissions in Smart Struct. AGARD CP–531, Paper 21, 1992
48. Fürstenau, N.; D.D. Jantzen and W. Schmidt: Fiber–Optic Interferometric Strain Gauge for Smart Structures Applications: First Flight Tests. AGARD CP–531, Paper 24, 1992
49. Tomlinson, G.R.: Use of Neural Networks/Genetic Algorithms for Fault Detection and Sensor Location. AGARD LS–205, 1996
50. Varadan, V.K. and V.V. Varadan: Smart Structures, MEMS and Smart Electronics for Aircraft. Paper 8; AGARD LS–205, 1996
51. Varadan, V.V., et al.: Image Reconstruction of Flaws using Ramp Response Signature. J. Of Wave–Mat. Interact., **10** (1), pp. 67–78, 1995
52. MIL–STD–1530A: Military Standard, Aircraft Structural Integrity Program, Airplane Requirements, 1975
53. Kudva, J.N.; A.J. Lockyer and C.B. Van Way: Structural Health Monitoring of Aircraft Components. Paper 9; AGARD LS–205, 1996
54. DeCamp, R.W.; R. Hardy and D.K. Gould: SAE Internat Pacific Air and Space Technology Conf., Melbourne /Australia, Nov. 13–17, 1987
55. Wadley, H.N.G.: Characteristics and Processing of Smart Materials. Paper 1, AGARD LS–205, 1996
56. Breitbach, E.J.: Research Status on Adaptive Structures in Europe. Proc. of 2nd Joint Japan/U.S. Conf. on Adaptive Structures; Technomic Sc. Publ., 1991
57. Charon, W. and H. Baier: Active Mechanical Components as a Step Towards Adaptive Structures in Space. Proc.: 4th Int. Conf. on Adaptive Structures; Technomic, 1993

58. Herold–Schmidt, U.; W. Schäfer and H.W. Zaglauer: Piezoceramics/CFRP composites for active vibration control and shape control of aerospace structures; SPIE **2779**, pp. 718 – 723, 1996
59. Misra, M.S.; B. Carpenter and B. Maclean: Adaptive Structure Design Employing Shape Memory Actuators. AGARD–CP 531, paper 15, 1992
60. Barrett, R.; R.S. Gross and F. Borozoski: Missile Flight Control Using Active Felxspar Actuators. Smart Mater. & Struct., **5**, pp. 121–128, 1996
61. Barrett, R.: Active Aeroelastic Tailoring of an Adaptive Flexspar Stabilator. Smart Mater. & Struct., vol. 5, pp. 723–730, 1996
62. Lazarus, K.B. and E.F. Crawley: Multivariable High–Authority Control of Plate–like Active Structures; AIAA Paper No. 92–2529; 33rd AIAA Conf. on Structures, Structural Dynamics, and Materials, Dallas/TX, 1992
63. Borchers, I.U., et al.: Selected Flight Test Data and Control System Results of the CEC BRITE/EURAM ASANCA Study. in: Proc. of Internoise 93, pp. 59–64, 1993
64. Fuller, C.R., et al. : Active Control of Interior Noise in Model Aircraft Fuselages Using Piezoceramic Actuators. AIAA Journal, **30** (11), pp. 2613–2617, 1992
65. Strehlow, H. and H. Rapp: Smart Materials for Helicopter Rotor Active Control; AGARD CP–531, Paper 5, 1992
66. Spangler, R.L. Jr. and S.R. Hall: Piezoelectric Actuators for Helicopter Rotor Control; Report #SSL 1–89, SERC 14–90, MIT Cambridge/USA, 1989
67. Hall, S.R. and E.F. Prechtl: Development of a Piezoelectric Servoflap for Helicopter Rotor Control. Smart Mater. & Struct., **5**, pp. 26–34, 1996
68. Straub, F.K.: A Feasibility Study of Using Smart Materials for Rotor Control; Smart Mater. & Struct, **5**, pp. 1 – 10, 1996
69. Straub, F.K. and D.J. Merkley: Design of a Servo–Flap Rotor for Reduced Control Loads; ibid, pp. 68–75, 1996
70. Ben–Zeev, O. and I. Chopra: Advances in the Development of an Intelligent Heli–copter Rotor Employing Smart Trailing–Edge Flaps. ibid, pp. 11–25, 1996
71. Fenn, R.C., et al.: Terfenol–D Driven Flaps for Helicopter Vibration Reduction. ibid, pp. 49–57, 1996
72. Samak, D.K. and I. Chopra: Design of High Force, High Displacement Actuators for Helicopter Rotors. ibid, pp. 58–67, 1996
73. Roglin, R.L. and S.V. Hanagud: A Helicopter with Adaptive Rotor Blades for Collective Control; ibid, pp.76–88, 1996
74. Das, A.; G. Ombrek and M. Obal: Adaptive Structures for Spacecraft – A USAF Perspective; AGARD CP–531, Paper 3, 1992
75. Obal, M. and J.M. Sater: Adaptive Structures Programs for the Strategic Defense Initiative Organization. Proc.: 33rd Struct., Struct. Dynam., & Mat.ls Conf., Dallas/TX, 1992
76. Clark, R.L. and C.R. Fuller: Control of Sound Radiation with Adaptive Structures. J. of Intell. Mater. Syst. and Struct., **2**, pp. 431 – 452, 1991
77. Priou, A.: Electromagnetic Antenna and Smart Structures. AGARD LS–205, Paper 11, 1996
78. Lockyer, A.J. et al.: Development of a Conformal Load Carrying Smart Skin Antenna for Military Aircraft. SPIE **2448/53**, 1995
79. Altandal, K.H.: Smart Skin Structure Technology Demonstration. SPIE Meeting on Smart Structures and Materials; San Diego/CA, 1996
80. Howard, B.M. et al.. Thermoadaptive Antennas; SPIE North American Conference on Smart Structures and Materials, San Diego/CA, 1996
81. Howard, B.M. et al.: Electrochromic Adaptive Antennas. ibid, 1996
82. Varadan, V.K.. Design and Development of a Conformal Spiral Antenna. ibid, 1996

83. Huang, J–B. and C–M. Ho: Micro Riblets for Drag Reduction. SPIE Proceedings, edited by V.K. Varadan, **2448**, pp. 245–250, 1995
84. Varadan, V.K. and V.V. Varadan: Drag Reduction in Aircraft Structures. SPIE Conference, San Diego/CA, 1995
85. Varadan, V.V. and V.K. Varadan: Microriblets for Drag Reduction using MEMS Technology. SPIE Conf. San Diego, 1995
86. Varadan, V.K. and V.V. Varadan: 3D MEMS Structures and their Applications. Proc. of the Internat. Symposium on Microsystems, Intelligent Materials and Robots; Tohoku Univ., Sendai/Japan, 1995
87. N.N.: Advanced Materials and Processes. 9, p. 9, 1995
88. Schrey, U.; Ulm, M.: Verteilte Aktorsysteme im Kraftfahrzeug. Mikroelektronik **7** 4, pp. 214–217 (1993).
89. Menge, W.; Hobein, D.: Integration eines E–Gas Systems in eine modulare Antriebsstrang Management Architektur. VDI–Berichte Nr. 1170, pp. 253–278 (1994).
90. van Zanten, A.; Erhard, R.; Pfaff, G.: FDR — Die Fahrdynamikregelung von Bosch. ATZ Automobiltechnische Zeitschrift 96 **11**, pp. 674–689 (1994).
91. Kücükay, F.; Renoth, F.: Intelligente Steuerung von Automatikgetrieben durch den Einsatz der Elektronik. ATZ Automobiltechnische Zeitschrift 96 **4**, pp. 228–235 (1994).
92. Graf, F.; Rauner, H.: Integration der Steuerung ins Getriebe – Getriebelektronik als Produktionsmodul: Chancen und Risiken. VDI Report No. 1175, pp. 339–349 (1995).
93. Ulm, M.; Lacher, F.; Graf, F.: Die elektronische Getriebesteuerung 2000 – Markt und technische Trends –. VDI Report No. 1170, pp. 201–218 (1994).
94. Bergholz, R.; Rech, B.: Anforderungen und Chancen der Adaptronik im Kraftfahrzeug. Adaptronic Congress Berlin 96, 20–21 November 1996 (Sauer Marketing–Service, Göttingen), pp. 1–21.
95. Spanner, K.; Wolny, W.W.: Trends and Challenges in New Piezoelectric Actuator Applications (Review). Proc. Actuator 96, 5th Int. Conf. New Actuators, 26–28 June 1996, Bremen, Germany
96. Kuhnen, K.; Janocha, H.: Compensation of the Creep and Hysteresis Effects on Piezoelectric Transducers with Inverse Systems. Proc. Actuator 98, 6th Int. Conf. New Actuators, 17–19 June 1998, Bremen, Germany
97. Jungnickel, G.; Wunderlich, B.: Korrektursystem für thermisch bedingte Neigungen an Werkzeugmaschinen. In: wt Werkstatttechnik, **88** 3, p.92 (1998)
98. Rech, B.: Aktoren mit elektrorheologischen Flüssigkeiten. PhD thesis, Universität des Saarlandes 1996. Verlag Mainz, Wissenschaftsverlag, Aachen
99. Janocha, H.; Rech, B.; Bölter, R.: Practice–Relevant Aspects of Constructing ER Fluid Actuators; Proc. 5th Int. Conf. Electro–Rheological Fluids, Magneto–Rheological Suspensions and Associated Technology. W.A. Bullough (ed.), 10–14 July 1995, Sheffield, UK, pp.435–447
100. Morishita, Sh.; Ura T.: ER Fluid Applications to Vibration Control Devices and an Adaptive Neural–Net Controller. J. of Intelligent Material Systems and Structures 4, pp.366–372 (1993)
101. Gosebruch, H.: Rundschleifen im geschlossenen Regelkreis. VDI–Verlag, Düsseldorf 1990
102. Gaul, L.; Nitsche, R.; Sachau, D.: Semi–Active Vibration Control of Flexible Structures. Proc. Euromec 373 Colloquium, Modelling and Control of Adaptive Mech. Structures, 11–13 March 1998, Magdeburg, Germany
103. Köhler, E.; Kunzmann, J.: Schnelle Pneumatikventile mit piezoelektrischen Aktoren. In: O+P Ölhydraulik und Pneumatik **42** 1, pp.43–46 (1998)

368   References

104. Bölter, R.; Janocha, H.: Aktoren mit elektrorheologischen und magnetorheol-
     ogischen Flüssigkeiten. In: atp–Automatisierungstechnische Praxis **39** 5, pp.18–
     26 (1997)
105. Lord Corp, Cary, North Carolina: Product information, 1996
106. Kobori, T.; Minai, R.: Analytical Study on Active Seismic Response Control.
     Trans. Arch. Inst. Japan. **66**, 257–260 (1960)
107. Yao, J.T.B.: Concept of Structural Control. ASCE. J. Structural Div. **98** (7),
     1567–1574 (1972)
108. Leipholz, H.H.E.: Structural Control. North–Holland Publ. Comp., Amster-
     dam, New York, Oxford (1980)
109. Leipholz, H.H.E.: Structural Control. Martinus Nijhoff Publ., Amsterdam
     (1985)
110. Soong, T.T.: *Active Structural Control.* Longman Scientific & Technical, Essex
     (1990)
111. Housner, G.W.; Masri, S.F.: International Workshop on Structural Control,
     Hawaii USC Publ. Number CE–9311, Los Angeles (1993)
112. Housner, G.W.; Masri, S.F. and Chassiakos, A.G.: First World Conference
     on Structural Control. Proc. International Association for Structural Control,
     USC, Los Angeles (1994)
113. Wada, B.K.; Fanson, J.; Crawley, E.: Adaptive structures. Journ. of Spacecraft
     and Rockets **27** (3). pp. 157–174 (1990)
114. Bachmann, H. et al.: *Vibration Problems in Structures.* Birkhäuser, Basel,
     Boston, Berlin (1995)
115. Sockel, H. et al.: *Wind–excited Vibrations of Structures.* CISM Courses and
     Lectures No. 335, Springer, Wien, New York (1994)
116. Koshika, N. et al.: Research, development and application of active–passive
     composite tuned mass dampers. Proc. 4th Int. Conf. on Adaptive Structures,
     Technomic Publ. Co.Inc. (1993)
117. Tamura, K.: Technology of active control systems for structural vibration. Int.
     Post–SMiRT Conference Seminar, Capri (1993)
118. Reinhorn, A.M. et al.: Active bracing system — A full scale implementation
     of active control. Technical Report NCEER-92-0020 (1992)
119. Traina, M.I. et al.: An experimental study of the earthquake response of build-
     ing models provided with active damping devices. Proc. Ninth World Conf. on
     Earthquake Eng., VIII 447–4522 (1988)
120. Wada, B.K. and Das, S.: *Application of adaptive structures concepts to civil
     structures.* Intelligent Structures–2, Ed. Wen, Y.K., Elsvier Publishing (1992)
121. Matsuhisa, H.; Gu, R.; Wang, Y.; Nishihara, O. and Sato, S.: Vibration Con-
     trol of a Ropeway Carrier by Passive Dynamic Vibration Absorbers. Jpn. Soc.
     Mech. Eng. International Journal (C), **38** (4), pp. 657–662 (1995). [Japanese
     original: Trans Jpn, Soc. Mech. Eng., **59** (56), C, pp. 1717–1722 (1993)]
122. Matsuhisa, H.; Nishihara, O.; Sato, K.; Otake, Y. and Yasuda, M.: Design of
     Dynamic Absorber for a Gondola Lift. Proc. of Asia–Pacific Vibration Confer-
     ence, pp. 215–220 (1995)
123. Nishihara, O.; Matsuhisa, H. and Sato, S.: Methods for Designing Vibration
     Control Mechanisms with Gyroscopic Moment. Proc. of Asia–Pacific Vibration
     Conference **1**, pp. 3.56–3.61 (1991)
124. Nishihara, O.; Matsuhisa, H. and Sato, S.: Optimum Design of Vibration Con-
     trol Mechanisms with Gyroscopic Moment for Harmonic and Stationary Ran-
     dom Excitations. Proc. of the 1st International Conf. on Motion and Vibration
     Control **1**, pp. 321–326 (1992)

125. Nishihara, O.; Matsuhisa, H. and Sato, S.: Passive Gyroscopic Damper for Stabilization of rigid Body Pendulum. Proc. of Asia–Pacific Vibration Conference **3**, pp. 889–894 (1993)
126. Nishihara, O.; Ishihara, H.; Matsuhisa, H. and Sato, S.: Design Optimization of Passive Gyroscopic Damper by Genetic Algorithms – Monte Carlo Optimization under Random Excitations. (in Japanese), Trans. Jpn Soc Mech Eng, No. 640–26(I), pp. 31–34 (1994)
127. Nishihara, O.; Yasuda, M.; Kanki, H; Nekomoto, Y.; Sato, K.; Otake, Y. and Matsuhisa, H.: Stability Maximization of Passive Gyroscopic Damper for Ropeway Gondola. Proc. of Asia–Pacific Vibration Conference, pp. 864–869 (1995)
128. Den Hartog, J.P.: Mechanical Vibration. 4th ed., McGraw–Hill, pp. 87–106 (1956)

9. Adaptronic Systems in Biology and Medicine

9.1 The Muscle as a Biological Universal Actuator in the Animal Kingdom
W. Nachtigall

Active movement is one of the signs of life. The universal actuator in the animal kingdom is the muscle. The striated skeletal muscle — which is the focus of the following discussion — is the evolutionary highlight of biological actuators. Early stages functioning as contractile elements can even be seen in protozoa such as amoebas.

Actuators took shape on the molecular scale as thread–like protein molecules formed that they could attach themselves to other similar molecules or biological surfaces with a type of crossbridge, and that the angle of attachment of these crossbridges can be changed by applying chemical energy. The 'carrier molecules' of the crossbridges must move actively relative to the point of attachment. This discovery took place very early in biological evolution, surely more than 600 million years ago. The discovery proved so useful that it not only still forms the basis for movement and mobility today, but it also made way, with the appearance of multicellular creatures, for a highly specialised cellular differentiation: muscle fiber. Muscle fiber is the functional unit upon which all actuators in the animal kingdom are based.

Muscle types of very different nature — from slow, smooth muscles, as found in the intestines of vertebrates, to extremely quick, oscillating fibrillary wing muscles of small insects — have developed in response to the various demands placed on such actuators (quick/slow contraction, large/small force, sustained contraction/brief twitching, etc.). The striated skeletal muscle exhibits the broadest range of application. Despite numerous modifications, this muscle type has maintained a uniform construction and functionality. The following summary concentrates on the striated skeletal muscle. Special biological and physiological details will be left out; instead, an attempt will be made to elaborate on the mechanics of contraction, the mechanical aspects of producing force and extension, the way in which such biological actuators interact with skeletal elements and, finally, their incorporation into complete systems with feedback control. This description should provide the engineer, technician and physicist with a direct analogy to technical actuators, perhaps spurring initiative for autonomous technological developments or improvements — i.e. aspects of 'bionics'.

At its Bionics Conference in Düsseldorf, Germany in 1993, the VDI (German engineering association) and the author presented the following definition: 'As a scientific discipline, bionics deals with the technical realisation and application of principles involving construction, process and development found in biological systems'. Actuators would fall into the categorisation 'constructional bionics'.

9.1.1 Principles of Construction and Function

Coarse and Fine Structures. Characteristic, categorizable elements can be seen upon cutting across any skeletal muscle, such as *musculus biceps brachii* (the 'biceps'). A closer look reveals finer and finer structures. Each hatched or shaded element is presented in more detail in a hierarchical fashion, as shown in Fig. 9.1.

The muscle is several centimetres in diameter. It works within a relatively stiff sheath of connective tissue, the fascia, from which the muscle is separated by a layer of loose connective tissue. The muscle is divided into bundles of muscle fibers by internal boundaries (perimysium).

Such a muscle fiber bundle has a diameter of about one centimetre. The fiber bundle consists of a series of muscle fibers, each enveloped in a fiber sheath and held apart from the others by a loose connective tissue, permitting relative motion during muscle contraction.

The muscle fiber is formed in ontogeny (individual development) as a fusion of single cells to form a sort of giant polynucleate cell. Its thickness is no greater than $100\,\mu m$. The partially aqueous interior medium contains bundles of myofibrils in addition to nuclei and mitochondria (cellular power plants).

The myofibril (approximately $1\,\mu m$ in diameter) is composed in cross–section of a regular hexagonal arrangement of interfaces between molecular filaments, of which the myosin filament is about twice as thick as the actin filament. Each myosin filament is surrounded by 6 actin filaments, resulting in a submicroscopic regularity nearly resembling a crystalline structure.

Motion. A longitudinal section of the muscle reveals that the thin actin filaments are connected from both sides to a common anchoring membrane (the Z–membrane). The thick myosin elements are located in between, and their ends maintain a certain distance from the Z–membranes. The thick and thin elements glide along each other in a telescopic fashion until the ends of the myosin elements hit the Z–membranes. (Holes in the Z–membranes of super contractile muscles allow the myosin ends to travel a bit further.)

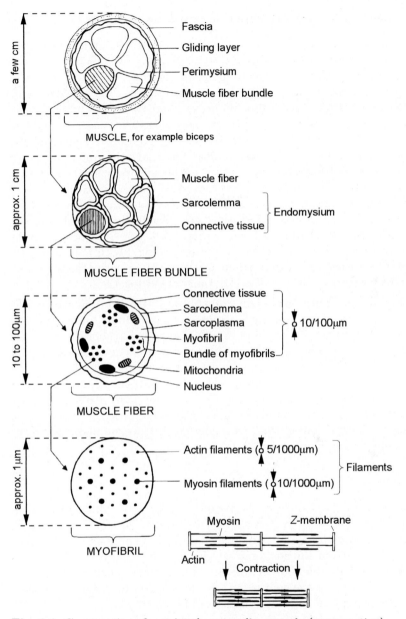

Fig. 9.1. Construction of a striated mammalian muscle (cross–section)

9.1.2 Analogies to Muscle Function and Fine Structure

The Boulder Analogy. Analogies often help to clarify complex structures. Let us assume that a stone–age man is trying to bring two boulders closer to one another (see Fig. 9.2). When he pulls on one rope, he is too weak and slips (Fig. 9.2a). Should he pull on two ropes connected to the boulders lying opposite one another, he won't slip, but he is still too weak (Fig. 9.2b). Several men pulling next to each other on long ropes (Fig. 9.2c) cause the system to cant, but they are still too weak. By adding more ropes and arranging themselves diagonally opposite one another (Fig. 9.2d), the system no longer cants — *summa summarum* — but there is not enough room for the necessary team of men.

The men now divide themselves into two groups, connect themselves by a bracing in the centre and, alternating, grip diagonally outward onto one of two ropes connected to each boulder (Fig. 9.2e). They can now slide the boulders a bit closer to each other by pulling strongly with both arms. By alternately releasing, gripping and pulling at other places along the two ropes, they manage to bring the boulders closer and closer together.

Submicroscopic Fine Structure of the Longitudinal Section. The following associations can be made to the biological model for the analogy presented. Boulders = Z–membranes; ropes connected to boulders = actin filaments; central bracing = myosin filaments; men = myosin heads; arms directed diagonally outward = actin–myosin crossbridges (Fig. 9.2e).

A photo taken with an electron microscope with the same orientation (longitudinal section) is displayed in Fig. 9.2f. The actin filaments run toward each other from the net–like Z–membrane structures; the crossbridges, radiating outward from the thick, centrally located myosin filaments to the actin filaments, are plainly visible.

9.1.3 Muscle Contraction

Filaments and Elementary Contraction. An actin filament is about $1\,\mu$m long (see Fig. 9.3). It consists of two threads (F) wound about each other and composed of globular monomers of actin maintaining their polarity. Fine tropomyosin threads, onto which troponin molecules are attached at regular intervals, run into the niches. The entire configuration is functionally significant for the connection and detachment of crossbridges; their peculiarities cannot, however, be discussed in detail here.

The myosin filaments, which are approximately twice as thick as actin filaments, are about $1.5\,\mu$m long and consist of several hundred myosin molecules connected in parallel, each with a braced head at the end. These heads are located at a distance of 426 Å from one another in whorls of three.

Due to its charge and geometry, a myosin head is able to combine with a monomer from the thread–like F–actin. Upon supplying it with energy (reaction and disintegration of an adenosine triphosphate molecule (ATP)), the

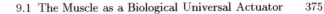

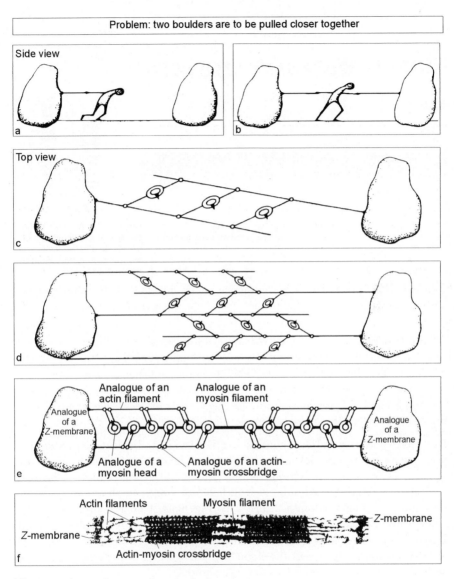

Fig. 9.2. An analogy to the muscle function

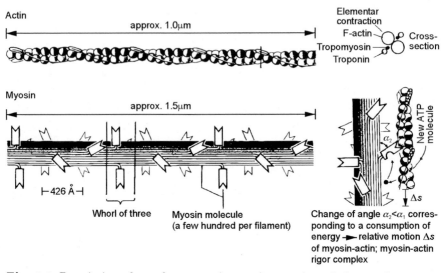

Fig. 9.3. Description of muscle contraction: actin, myosin and elemental contraction

F–actin changes its angle with the longitudinal axis of the myosin filament through a complex series of reactions. When many such heads take hold, the myosin and actin filaments move relative to one another over a certain elemental distance Δs (elemental contraction). The precise processes (reaction, chemomechanical transduction, detachment, reattachment, aspects concerning energy) cannot be described in detail here.

Increase of Stress with Progressing Extension. Excitation of a muscle fiber (or an entire muscle) under increasing extension results in a maximum of the relative stress (maximum stress set to 100%) when plotted against the relative length (unstimulated length set to 100%) in the vicinity of the unstimulated length (points 2 and 3 in Fig. 9.4). Under compression, the configuration is distorted and the stress that can be developed reduces (point 1). During passive elongation, the number of crossbridge possibilities reduces and so also does the stress developed by excitation. The stress reduces to zero when no more crossbridges can take hold.

Single Twitches and Tetanus. A typical skeletal muscle reacts to an artificially induced electrical impulse with a brief contraction, resulting in a mechanical deflection under suitable experimental conditions (see Fig. 9.5). An increase in the excitation results in a superposition of contractions when the successive impulse takes place before the muscle has dilated completely: incomplete tetanus. Single twitches are no longer detectable at an excitation frequency of about 50 impulses per second: complete tetanus.

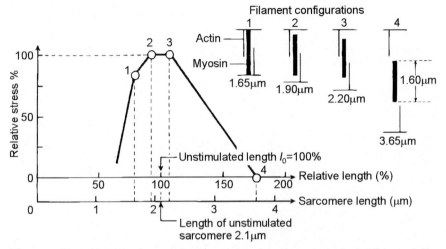

Fig. 9.4. Descriptions of a muscle contraction: length–stress relationship and filament configuration

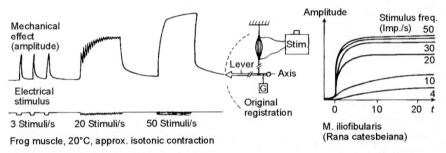

Fig. 9.5. Description of a muscle contraction: from single twitch to tetanus

Muscle operation often approximates to complete tetanus under natural conditions. The amplitude of contraction increases for increasing motor impulse frequency.

9.1.4 Aspects of Muscle Mechanics

Losses in Muscle Work During the Elongation and Contraction Cycle. Different curves (elongation and relaxation curves) result when plotting the load against the length during passive elongation and successive relaxation of muscle fiber (without electrical or neural excitation) — see Fig. 9.6. Since the area in a force–length diagram has the dimension 'work', the area enclosed by the elongation and relaxation branches of the curve corresponds to the losses per cycle of passive stretching. In a similar fashion (not shown here), losses also result per contraction cycle of an active muscle. The elastic efficiency of the system undergoing passive elongation can be determined as indicated in the diagram. The efficiency of skeletal muscles typically lies around 85%. The mechanical efficiency can be determined in a similar fashion for active contraction. In this case, the values vary due to a strong dependency on the boundary conditions.

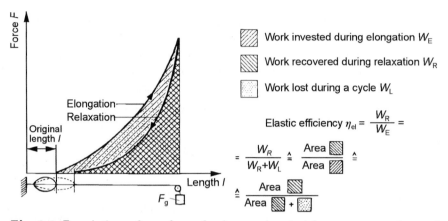

Fig. 9.6. Descriptions of muscle mechanics: passive stretch curve, energy loss and elastic efficiency

Force–Speed Relationships. It is clear that an actuator cannot develop a large force and a high speed of contraction simultaneously; the two parameters counteract each other. A hyperbolic relationship (the Hill's equation) results when plotting the speed of contraction of a muscle against the load (see Fig. 9.7a).

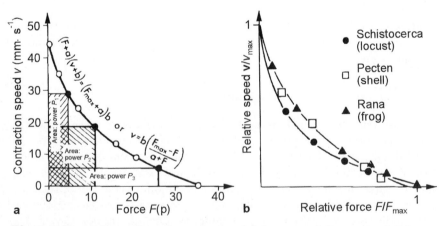

Fig. 9.7. Descriptions of muscle mechanics: **(a)** force–speed diagram and **(b)** standardized force–speed diagram

The area of a speed–force diagram has the dimension 'power'. As can be seen, a muscle is capable of producing its greatest power at medium values of speed and force; the expendable power sinks in the vicinity of the extremes (by either high contraction rates or large loads).

The characteristic curves of diverse muscles can in practice be represented by a single curve when normalised with respect to the maximum speed and maximum force. This applies to muscular systems functioning in different ways, including the wing muscles of insects, shell–closing muscles specialised to tonic contraction, or the leg muscles of cold–blooded animals. This indicates an inherent constructional principle among all of these greatly varying muscles (Fig. 9.7b).

Possibilities of Contraction. An isotonic type of contraction (reduction in length under constant stress) is exemplified by an isolated muscle excited to lift a weight hanging on one end — see Fig. 9.8. (The cross–sectional area experiencing loading is assumed to remain approximately constant.) By fixing both ends and applying an excitation, the stress increases without a change in the length: an isometric type of contraction.

A muscle works under isometric conditions followed by isotonic conditions when lifting a weight from a supporting surface ('supporting contraction'). Upon reaching a limit near completion of an isotonic contraction, the isotonic work of the muscle changes to isometric work ('limited contraction'). Isotonic and isometric behavior is combined when letting the muscle work, for example, against a strong elastic spring; this type of work is referred to in the field of physiology as auxotonic contraction and, among the experimental cases mentioned, exhibits the closest resemblance to contraction occurring in nature.

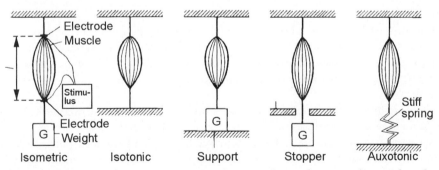

Fig. 9.8. Descriptions of muscle mechanics: experimental contraction modes of a skeletal muscle

9.1.5 Principal Types of Motion Achievable by a Muscle and its Antagonists

Muscles are always arranged so as to interact with an antagonist. This antagonist is typically an opposing muscle. However, mechanically elastic elements can also fulfil the function of an antagonist. In principle, an angle, a distance, an area or a volume can be changed by the contraction of a muscle.

Change of Angle. A contraction of our *musculus biceps brachii* leads to a reduction of the angle α between the upper arm and the forearm (see Fig. 9.9). The opposing *musculus triceps brachii* undergoes a simultaneous extension. A contraction of the triceps leads to an increase in the angle α and to an extension of the biceps. Biceps and triceps are muscular antagonists.

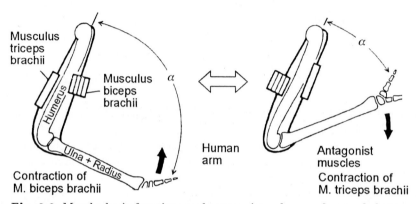

Fig. 9.9. Muscle: basic functions and antagonists: change of an angle between two skeletal elements

Change of Distance. The downstroke and upstroke muscles that are used to drive the wings found in a dragonfly also operate as antagonistic muscle pairs. These muscles tilt the wing about its basal joint, causing the angle of the longitudinal wing axis to change relative to a reference axis (see Fig. 9.10a).

More highly developed flies and bees function in a different manner. Their thoracic capsule oscillates quickly (up to several hundred strokes per second!) through a more automatic indirect drive. In a sort of lid–and–pan system, the 'lid' (the upper side of the thoracic capsule, displayed a bit smaller in the oversimplified model) is pulled into the larger 'pan' (remaining portion of the capsule) by dorsoventral muscles fixed between the two thoracic pieces (see Fig. 9.10b). This action results in an upward motion of the connected wings. Dorsal muscles running longitudinally through the capsule deform it in the other way, causing the wings to effect a downward stroke. The dorsoventral muscles and their antagonists don't effect a change of an angle but rather the distance between their attachment points.

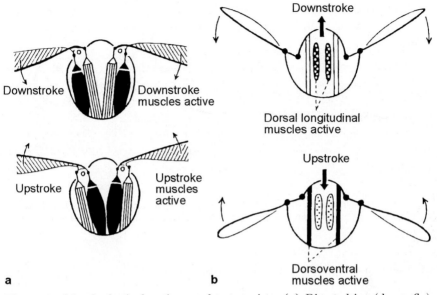

Fig. 9.10. Muscle: basic functions and antagonists. (**a**) Direct drive (dragonfly); (**b**) indirect drive (fly, bee)

Changes in Area and Volume. The cuttlefish can change its lightness and colouring within fractions of a second. This is due to quickly contracting, fine radial muscles that are capable of pulling pigment–filled cells apart: the result is dark colouring. When the muscle contraction releases, the cell pulls together due to its elasticity, the pigments are condensed point by point, and the light background appears: light colouring (Fig. 9.11a).

Segmentally arranged so–called wing muscles (having nothing to do with insect wings) work in a rhythmic sequence in an insect heart. Valve flaps on the inside prevent the circulatory fluid (hemolymph), sucked out of lateral openings, from flowing in the reverse direction. A unidirectional flow of circulatory fluid results. The elasticity of the entire system and, to some degree, the muscles undergoing subsequent contraction function together as antagonists (Fig. 9.11 b).

To reduce the volume of our small arteries and thereby control the speed of flow of our blood, we involuntarily place under tension smooth muscle fibers that enclose the vessel in a circular or spiral fashion. By reducing the free lumen, the speed of blood flow is changed dramatically, as described by the Hagen–Poiseuille law of capillary flow. The blood pressure functions as antagonist, causing the vessel to expand when the muscle contraction ceases (Fig. 9.11c).

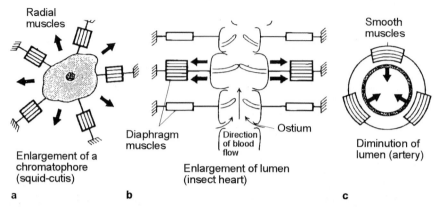

Fig. 9.11. Muscle, basic functions and antagonists. Change of areas and lumina:
(a) enlargement of chromatophone; (b) enlargement of lumen; (c) diminution of
lumen

9.1.6 Force and Position of Muscular Levers

Balance and Location of the Muscle Attachment Points. Of two conceivable possibilities — weak actuators that contract over great distances, or strong actuators that generate large displacements via translation (muscular levers) — nature generally chooses the latter. The biceps muscle is fixed relatively close to the joint. An angular motion of the forearm therefore requires large forces but small displacements. Fig. 9.12 illustrates an additional concept for fixation, which could also be used to raise the hand, but an unseemly supply system would be necessary and the arm would hardly be a practical tool for daily activities. Working with powerful actuators, small displacements and large forces leave room for free motion of the lever arm.

Principle of Catapulting Twitch. Beginning with a wide angle between the upper arm and the forearm, the biceps works with a small leverage and therefore must generate a large force (see Fig. 9.13). Through its angular motion, the horizontal distance between the weight (in the hand) and the elbow joint becomes smaller; and so the moment created by the load becomes smaller. At the same time, the perpendicular distance between the muscle tendon and the joint increases; so (for a constant muscular stress) the force moment increases. Both tendencies are favorable and accommodate the characteristics of the muscular actuator: it can operate initially with a large force (nearly isometric) and the needed force decreases as the arm angle is reduced more and more. As soon as the weight has been set in motion, the muscle can stop contracting even before reaching the end position of the movement: 'a catapulting twitch'. (The opposing muscle must simultaneously begin its contraction before the end position of the movement has been reached in order to achieve early braking.)

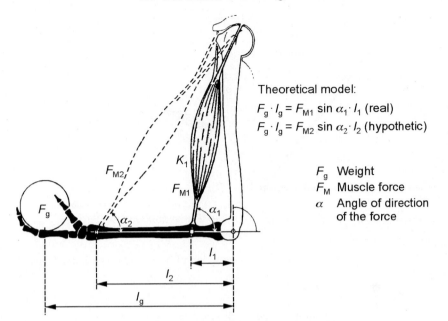

Theoretical model:

$F_g \cdot l_g = F_{M1} \sin \alpha_1 \cdot l_1$ (real)
$F_g \cdot l_g = F_{M2} \sin \alpha_2 \cdot l_2$ (hypothetic)

F_g Weight
F_M Muscle force
α Angle of direction
    of the force

Fig. 9.12. Aspects of muscular levers. Balance of unilateral lever

Tensor Muscles. Not all muscular actuators function in the sense of inducing motion as presented here so far. Some muscles place mechanical systems under tension — systems that first become capable of motion upon being driven by other muscles. An example is found in the click mechanism of flies. A double leverage allows a central joint (over schematised in Fig. 9.14) to toggle after it has pressed the periphery joints apart. In this fashion, the wing is torn upwards or downwards (as applied in the cap of a can of shoe polish). However, the system only operates when the outer joints are drawn toward each other by a mechanical load. This loading is effected by tensor muscles represented in the drawing by the pleurotergal muscles (which pull between the sides and the top) and the pleurosternal muscles (which pull between the sides and the basal piece).

9.1.7 Cooperation of Unequal Actuators

Model calculations show that the jumping motion of animals can be considerably more efficient than their running motion (assuming the same average speed of locomotion). Several muscles work together to create jumps. One muscle performs the main drive and at least one secondary muscle assumes auxiliary functions. Jumping motion is a good example of the cooperation of unequal biological actuators.

Jump of a Locust. Fig. 9.15a shows phases of jumping of a large locust. Phase 3 is presented in more detail in Fig. 9.15b. Clearly, the angle between

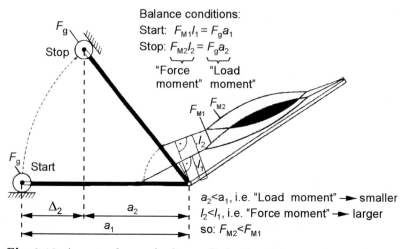

Fig. 9.13. Aspects of muscular levers. Reduction of force under continuous change of angle

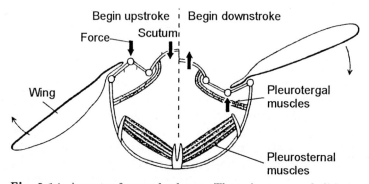

Fig. 9.14. Aspects of muscular levers. Thoracic stress and click mechanics in flies

the femur and the tibia is increased during the jump, while the tarsus located on the ground is unrolled until shortly before losing contact with the ground. The main driving muscle is the *musculus extensor tibiae* (which increases the angle between the tibia and the femur). As shown in Fig. 9.15b, this muscle works at a very small leverage distance from the joint, implying short contraction and large force. The instantaneous muscle force can be determined based on the geometric configuration, the mass of the body, and the acceleration during the jump. The force is about 5 N, corresponding to a mechanical stress in the muscle of $140\,\mathrm{kNm^{-2}}$. For a cross–sectional area of the muscle apodem of $0.01\,\mathrm{mm^2}$, the stress reaches $500\,\mathrm{Nmm^{-2}}$. The experimentally determined ultimate tensile strength of the biological apodem is about $600\,\mathrm{Nmm^{-2}}$. (This strength is comparable to that of structural steels, which starts at $450\,\mathrm{Nmm^{-2}}$.) The factor of safety against breakage of the apodem is only 1.2, and the locust jumps near its biomechanical limit.

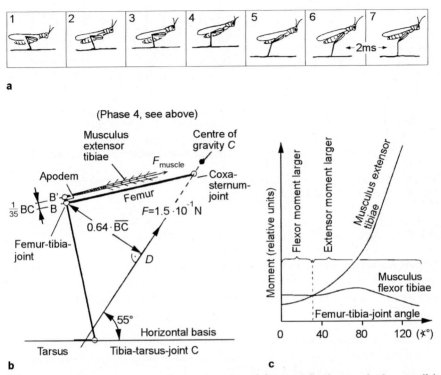

Fig. 9.15. Unequal actuators during jumping: **(a)** jumping phases of a locust; **(b)** jump mechanics; **(c)** flexor and extensor moments (From Brown)

As shown in Fig. 9.15c, a flexor muscle (*musculus flexor tibiae*) works against the jumping muscle (*musculus extensor tibiae*). At the beginning of a jump, the moment generated by the flexor exceeds that of the extensor until an angle of about 30° is reached. A sort of auxiliary catapult is prestressed and unloaded after reaching the 30° angle, thereby reducing the launch time.

Jump of a Flea. In a similar fashion, a flea jumps by contracting the main jump muscle (which is affixed to the 'reverse side') and deforming a highly elastic biological polymer (resilin) in compression. After storing this energy, an auxiliary muscle pulls the tendon–like apodem of the main jump muscle to the 'correct side' allowing the leg to spring out and catapult the flea — see Fig. 9.16.

This is a biological catapult, i.e. a translation of power, where the energy necessary for the jump is stored slowly and released over a shorter period than is possible with direct muscle contraction.

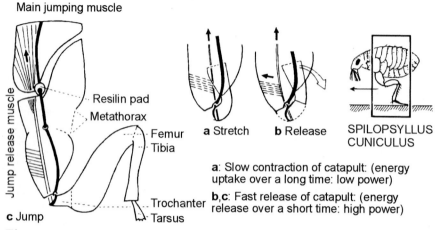

Fig. 9.16. Unequal actuators during jumping. Jump of a flea (From Bennet–Clark):
(a) slow contraction; **(b)** fast release; **(c)** the jump

9.1.8 Muscles as Actuators in Controlled Systems

Servo Assistance in the Control of Extension. Numerous control mechanisms take place in the arm musculature when we guide a glass of water to our mouth. Muscle contraction receptors (muscle spindles) effect servo assistance. The process can be described by four steps (Fig. 9.17).

1. The extrafusal fibers EF of the muscle M and the intrafusal fibers IF of the muscle spindles MS contract simultaneously when commanded by the central nervous system G.
2. The feedback signal sent via nerve fibers (so–called Ia–afference) remains constant for equivalent reduction in length of EF and IF; the spindle control loop is inactive.
3. Disturbances induced for example by unexpected increases or reductions in the load L cause the sensor ends SE in the muscle spindles to stretch or become compressed.
4. As a result, the Ia–feedback signal changes, effecting a corresponding change in the excitation of the so–called α–motoneurons. The spindle control loop is active. This process can be described as 'conditioned feedback'. The disturbance is essentially compensated (servo assistance within the γ–loop).

Final Control of the Stress.

1. The extrafusal fibers EF of the muscle M contract on command from the central nervous system G, as shown in Fig. 9.18. (The muscle spindle control loop is ignored here.)
2. Tendon and tendon organs experience increased stretching.

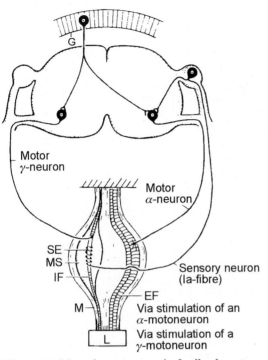

Fig. 9.17. Muscular actuators in feedback systems. Servo support of movement via muscle spindles (length regulation)

3. Due to the geometry of the tendon organs (see insert in Fig. 9.18), the sensitive nerve endings are squeezed and thereby excited.
4. This excitation is fed back negatively to the α–motoneuron via the so–called Ib–fibers and an inhibitory interneuron I.
5. The resulting effect is a reduction of muscle tension. The tendon organ practically functions as a limit switch that shuts off muscle activity before the threat of tearing the tendon becomes real.

Crossed Extensor Reflex. Fig. 9.19 is self–explanatory: the dilemma of having to react extremely quickly by reflex without upsetting a stable position of the body is accomplished in nature through positive and negative control of the bending and stretching muscle groups located in the left and right legs. A negative interneuron is necessary to switch from positive to negative control.

9.1.9 Control Loops in Biology: Similarities Within Biology and Engineering

Control of Muscle Length. The control of length with servo assistance via muscle spindles, as described in Sect. 9.1.8, is displayed again in Fig. 9.20. The

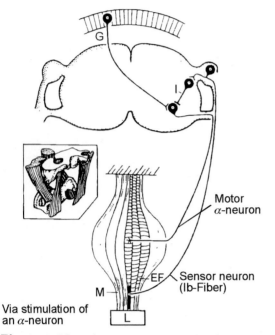

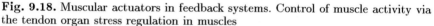

Motor
α-neuron

EF — Sensor neuron
(Ib-Fiber)

M

Via stimulation of
an α-neuron

L

Fig. 9.18. Muscular actuators in feedback systems. Control of muscle activity via
the tendon organ stress regulation in muscles

Example: stepping on a sharp stone triggers a reflex:
a antagonistic influence on the stretching (S) and bending (B)
muscles in both the left (l) and right (r) legs
b antagonistic movement of both legs

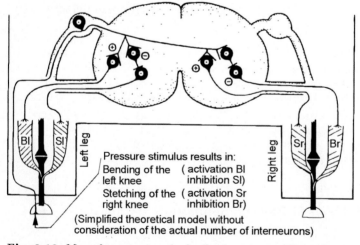

Left leg

Bl   Sl

Pressure stimulus results in:
Bending of the   (activation Bl
left knee   inhibition Sl)
Stetching of the   (activation Sr
right knee   inhibition Br)
(Simplified theoretical model without
consideration of the actual number of interneurons)

Right leg

Sr   Br

Fig. 9.19. Muscular actuators in feedback systems. 'Crossed extensor reflex' — a
more complicated reflex

control variable x is the muscle length. The control involves adaptive sensors: the sensitivity of the muscle spindle as a sensor is set by commands from the set–point adjuster before or during the preliminary phase of contraction.

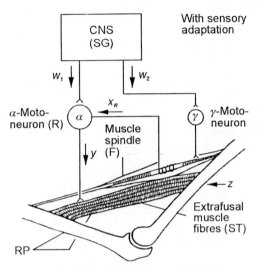

Control variable x: muscle length

Fig. 9.20. Muscular–cybernetic analogy. Feedback control of muscle length

Control of a Furnace. The control variable in the example in Fig. 9.21 is the temperature. A camshaft can influence the sensitivity (via a second cam in the model) of the thermoactuator (by changing the fill volume). The analogy between the two examples (control of length and control of temperature) is given in detail in Figs. 9.20 and 9.21.

Common Control Loop: Control with Adaptive Sensing. The control loop presented in Fig. 9.22 is just as applicable to biological as to technical control. The set–point adjuster informs not only the controller (via reference input 1) but also the sensor (via reference input 2) of its intentions. The sensor is therefore more capable of reacting with the proper signal at the right time. The transient response can be accelerated thereby for small differences between the command variable and the control variable.

Outlook. The similarity of processing within biological and engineering systems is apparent although the morphological designs of their actuators differ. Biology and engineering can be viewed as final components within a joint continuum governed by the laws of nature. By applying this viewpoint, boundaries that have developed between scientific disciplines through traditional dividing strategies could be dissolved.

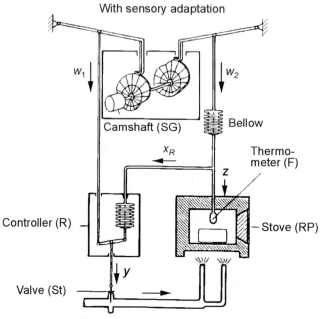

With sensory adaptation

w_1 w_2

Camshaft (SG)   Bellow

x_R Thermo-
meter (F)

z

Controller (R)   Stove (RP)

y

Valve (St)

Control variable x: temperature

Fig. 9.21. Muscular–cybernetic analogy. Feedback control of furnace temperatur

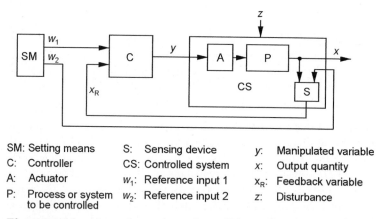

z

SM w_1
        w_2 C y A   P x

        x_R CS   S

SM: Setting means	S: Sensing device	y: Manipulated variable
C: Controller	CS: Controlled system	x: Output quantity
A: Actuator	w_1: Reference input 1	x_R: Feedback variable
P: Process or system to be controlled	w_2: Reference input 2	z: Disturbance

Fig. 9.22. Muscular–cybernetic analogy. Cybernetic control scheme, valid for Fig.
9.20 and Fig. 9.21

9.2 Adaptronic Systems in Medicine —
Applications and Prospects
J.-U. Meyer

Adaptronic technical systems and structures in medicine are characterized by sensing the biological system and adjusting their performance using multi-functional elements in order to accomplish a beneficial effect for the subject. Medical devices generate a demand for adaptive features since most biological systems exhibit time–invariant and nonlinear properties. The controller design has to adapt to the varying system dynamics caused by the variability of the biological system and the lack of a complete description of state–variables. The incomplete description of the system is compensated for by employing artificial neural networks (ANN) or fuzzy based regulation algorithms. Open– and closed–loop feedback and feed–forward adaptive systems are applied.

Adaptronic systems imply adaptive technical systems that heavily rely on microelectronics and microelectromechanical components integrated in one element for improved performance and enhanced functionality per device size. Our body is a perfect example of a highly complex, well tuned and adaptive system — for example, the proper moment to moment functioning of our motor systems depends on a continuous inflow of sensory information. The cortical motor system extracts the information necessary to guide the movements and transmits the signals via the brain stem and spinal cord to the skeletal muscles.

Three possible adaptronic systems for biomedical applications are conceived and illustrated in Fig. 9.23. In Fig. 9.23a, an open–loop adaptronic system with closed–loop options is illustrated, e.g. for applications in the neural/muscular control of a mechanical prosthesis (open–loop) or for a drug delivery system (closed–loop) [1]. In the latter, the biological system is altered in response to the effector performance and therefore adopts the function of a regulator. The signal controller is decoding the signals that are generated by the biological system. In Fig. 9.23b, the adaptronic system is characterized by regulation that depends on two variables, namely the biological system and the ambiance sensor input. The ambiance sensor is located outside the body, e.g. in a sensorized prosthesis. (The adaptronic system of Fig. 9.25 differs from the ones above in respect of encoding the signal from an ambiance input to achieve an effector response in the biological system.) Biosensing is used for regulation. One example is an implantable stimulator for grasping, with feedback response from an implanted sensor that is measuring motion.

In the following, adaptronic systems for biomedical applications have been selected that illustrate recent activities in the field. An emphasis is given on systems that employ microtechnologies. The microsystems described comprise electronic implants, as advanced pacemakers and neural prostheses, as well as adapting diagnostic devices, as tactile sensors and self–adjusting blood–flow monitors.

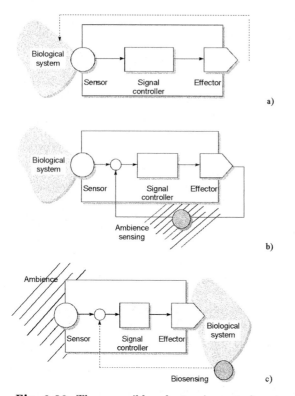

Fig. 9.23. Three possible adaptronic control systems for biomedical application: (a) open–loop (solid line) adaptronic system with closed–loop option (dashed line); (b) regulation depends on two variables, namely the biological system and the ambiance sensor input; (c) signal is encoded from an ambiance input to achieve an effector response in the biological system. Biosensing is used for regulation

9.2.1 Adaptive Implants

Microelectronic implants have gained increasing interest in the bioengineering research community [2] and the medical device industry. Heart pacemakers are the most prominent example of an implantable microsystem that exhibits adaptive properties [3]. In the following section, adaptive properties of advanced pacemaker systems are described. Experience gained from heart stimulation devices has led to the development of a new class of implantable neural stimulators and sensors systems, namely neural prosthetic devices. Fabrication and the envisioned adaptive control mechanisms are described here.

Advanced Pacemakers. Common pacemakers stimulate the heart with a fixed rate. For the benefit of the patient, it is desired to adjust the cardiac output to the physiological state of the patient. Among other factors, the physiological condition of a patient is influenced by body motion, posture, metabolism, ambient conditions, and emotional states. It is the objective of advanced pacemaker systems to adjust cardiac output by altering either the cardiac stroke volume or the cardiac rate. The regulation of stroke volume is refined by its inherently limited range. Current activities focus on rate–adaptive heart–pacing systems. Cardiac and non–cardiac state variables can be used for the regulation of cardiac output. Variables include central venous oxygen saturation, oxygen uptake and ventilation, mean arterial blood pressure, parasympathetic and sympathetic activity, which are difficult to monitor in an integrative manner [3]. Applied are corporeal control parameters, such as motion and central venous temperature, as well as cardiac control parameters that include intracardiac impedance measurements.

The pacemaker as an adaptronic system is shown in Fig. 9.24. Sensory inputs may comprise parameters for body motion, central venous temperature, and intracardiac impedance. Body motion is measured using piezoelectric or micromachined accelerometers that are located in the housing of the pacemaker. Implantable electrodes are used, sensing intracardiac impedance and for heart pacing.

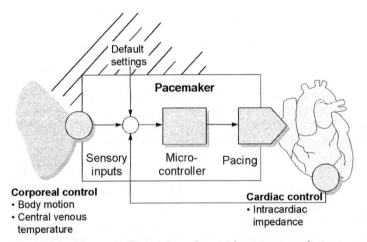

Fig. 9.24. Schematic illustration of control parameters that are used to adjust the rate of pacing in a pacemaker device

Neural Prostheses. Neuroprosthetic microimplants are considered an emerging field in rehabilitation engineering [4]. The neural prostheses are designed to compensate for lost or impaired nervous functions. Close interfaces are established to the neural tissue for sensing or for generating neural signals. Detection and decoding of peripheral neural signals are needed when decoding neural information, e.g. to control an artificial limb. Microimplants for encoding and stimulating neural structures are applied when restoring motion in paralyzed limbs or mimic sensory functions. The same implantable electrode elements may serve as sensors for nerve signal registrations and as actuators for nerve stimulation.

Neural Interfaces for Amputees. Fig. 9.25 illustrates a simplified open–loop adaptronic system for controlling an artificial limb with muscular or nervous signals that are sensed in the stump of an amputee.

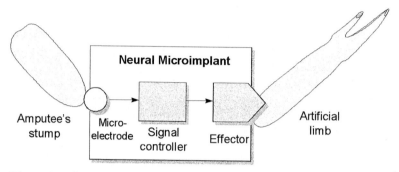

Fig. 9.25. Open–loop adaptronic system for controlling the movements of a mechanical prosthesis

The main challenge of the approach shown in Fig. 9.25 in terms of adaptive signal processing is the design of the controller. The controller's task is to decode the extracellular compound nerve potentials and identify the ones that are related to efferent signals containing information about motion. Artificial neural networks are applied for the purpose of separating the various signal components of the neural activity. Subsequently, meaningful information about the neural activity is converted into electric signals that control the artificial limb.

Micromachined, flexible, neural electrodes have been designed and manufactured in the laboratory for contacting the tightly packed neural elements in the nerve trunk [5]. Fig. 9.26 illustrates the design of multichannel sieve electrodes with integrated interconnects that are interfacing regenerating nerve axons of the stump. Fig. 9.27 is a micrograph of the polyimide sieve structure with platinum electrodes for neural signal detection.

Implants for Neural Stimulation. Neural stimulation is most commonly applied in the field of functional electrical stimulation (FES) of the neuromuscular apparatus to restore motion in paralyzed extremities. Stimulation is achieved through rather bulky surface electrodes that are part of an orthosis or individually attached to the skin. Investigators have applied self adaptive neuro–fuzzy algorithms to control an actuated orthosis [6]. The controller utilizes a closed–loop, supervised, learning–adaptive network controller. Inputs comprise knee and hip angles, from which hip torque and the pulse width of the stimulation are generated as outputs. The generation of appropriate

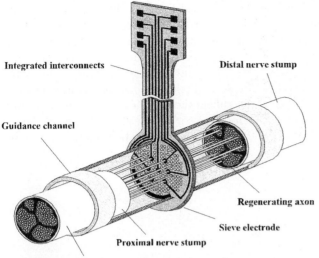

Integrated interconnects

Distal nerve stump

Guidance channel

Regenerating axon

Sieve electrode

Proximal nerve stump

Fascicle with axons

Fig. 9.26. Flexible, polyimide–based sieve electrodes interfacing regenerating axons in peripheral nerves

stimulation parameters is the major challenge for controlling motor activity. A major drawback of functional electrical stimulation with surface electrodes is the rather inconvenient handling of electrodes and the adjustment of supporting structures. Therefore, recent research in rehabilitation engineering has focused on implantable neuromuscular stimulators with adaptronic functions.

Various microelectrode designs have been constructed for contacting nerve trunks and neural tissue. A new breed of flexible, light–weight multi–channel electrodes have been developed for interfacing nerves using micromachining

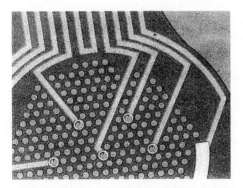

Fig. 9.27. Micrograph of the micromachined sieve structure with circular platinum electrodes. The diameter of the holes is about $40\,\mu$m

technologies [7]. The electrodes are designed to allow the integration of microchips on the electrode substrate in order to obtain an adaptronic microsystem particularly suited for nerve stimulation. Fig. 9.28 depicts an adaptronic microsystem of the future for restoring grasping in paralyzed arms and hands. The system comprises motion sensors as inputs, a neural net for encoding the information into stimulation pulses, several telemetric units, a transcutaneous signal and energy transmitter, the subcutaneous microelectrodes with integrated chips, and an implantable force sensor for tactile feedback. Materials such as piezoelectric foils or conducting polymers can be used as smart skins for slip detection.

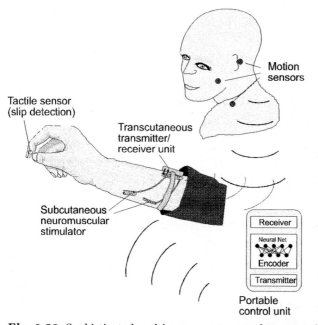

Fig. 9.28. Sophisticated multicomponent control system with adaptronic elements (slip detector) for controlling grasping in a paralyzed arm

A major challenge in biomedical engineering has been the development of microelectronic systems for substituting impaired or lost sensory functions. Cochlear implants have demonstrated that only eight to twenty electrodes are needed to stimulate sensory nerves in the cochlea so as to give a hearing perception. Cochlea systems consist of a halter device–sized speech processor and a transcutaneous inductive link to a microprocessor implant that generates the appropriate signals for the stimulating electrodes in the cochlea. Their function corresponds to the control system, as shown in Fig. 9.23a. During the training phase, parameters of the speech processor are adjusted for gaining an optimal performance of hearing. In general, cochlea systems are open–loop regulating systems that perceive feedback of perception only during training and readjustment of the system. Even more ambitious are research programs in the U.S. and Europe that are aimed at developing a neural prosthesis for the blind. Present activities focus on the design of flexible, multi–channel microelectrodes for stimulating retinal nervous tissue. First retinal contact microstructures have been developed in the laboratory (Fig. 9.29). The polyimide–based material has been functionalized by embedding an array of platinum electrodes. The retina contact structure is particular shaped to adjust and adhere to the particular geometry of the retina.

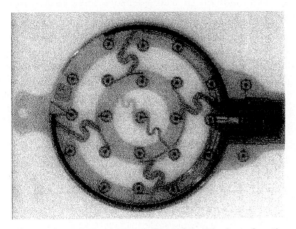

Fig. 9.29. Micrograph of the polyimide–based retina contact structure

The retinal implant system is being jointly developed by a team of twelve investigators, who are experts in ophthalmology, neuroinformatics, and microtechnology. The project is sponsored by the German research ministry BMBF. One emphasis is given on the design of optimal, appropriate, adaptive stimulation. An adaptive retinal encoder is currently being developed. Adaptive visual fields are employed that are adjusted to the function of retinal ganglion cells [8]. The implant is composed of a micromachined polyimide in which an array of platinum electrodes is embedded. The design of the contact structure provides a snug fit to the retina. The arrangement of electrodes allows sophisticated stimulation procedures. The described retinal prosthesis is considered one of the most challenging adaptronic systems of the future.

9.2.2 Adaptive Diagnostic Systems

Adaptive diagnostic systems are characterized by their enhanced performance when adapting to the varying dynamics of the biological system under inspection. Enhanced performance includes additional feedback information received from a sensory front–end or is expressed by adaptive tracking of the biological system. The devices described below represent two implementations of adaptronic systems that have recently been developed.

Tactile Sensing and Feedback. Endoscopic diagnostics and therapy have gained worldwide recognition as a method for minimally invasive interventions. However, endoscopic procedures are restricted to instrumentation that does not allow for a direct tactile contact or palpation of the operator's hand or fingers with the tissue under observation. Recent research is directed to investigate sensor and actuator principles that will provide tactile information for the surgeon. An endoscopic forceps has been developed into the mouth of which an array of piezoresistive silicon pressure sensors have been integrated [9]. The pressure sensors are covered by a thin steel foil to withstand rugged handling and sterilization of the endoscopic instrument. Currently, the pressure signals are displayed on a monitor; the next generation of adaptronic instruments will be equipped with an actuated handle that will house an array of cushions that are filled with electrorheological fluid. The fluids will change their viscosity in an electric field that is produced by individual electrodes. The size of the cushion corresponds to the size of the pressure elements. Addressable fluid channels and valves will control filling of the cushions. The applied electrical field will control the hardness of the cushion by changing the viscosity of the fluid. The necessary microelectronic circuits and the battery will be housed in a round steel case that is mounted close to the handles. A major obstacle to overcome is the mounting, assembly, and interconnection technology so as to fit all the components in the specialized forceps without impairing its mechanical properties.

Neonatal Blood Flow Monitoring. Ultrasonic monitoring of blood flow is a well established diagnostic tool in medicine. Despite recent advances in Doppler blood flow measurements, long–term, operator–independent monitoring of blood flow is still facing the problem of tracking the blood vessel under observation, particularly when movements occur. Recent developments have investigated the potentials of mechanical or electrical ultrasonic beam steering to accomplish continuous tracking of the blood vessels under investigation. Electrical beam steering is achieved utilizing phased arrays. Phased arrays comprise an assembly of single piezoelectric transducers arranged in a line (a 1–D array) or in rows and columns (a 2–D array) and electronics for properly delaying the signals going to the elements for transmission or signals arriving at the elements for receiving [10]. Appropriate, adaptive algorithms are needed to apply ultrasonic beam steering to a moving target, e.g. a cerebral blood vessel. The following describes a cerebral blood–flow monitor for pre–term infants, with automatic tracking of the sample volume under observation.

Pre–term infants face the risk of intraventricular hemorrhage that may lead to neurological disorders. Most commonly, the hemorrhage is a consequence of disturbed cerebral blood flow. The author's institute has developed a Doppler flow system that is capable of long–term cerebral blood–flow tracking in premature newborns. Major components of the system are board–based plug–ins, namely a scan converter, a microprocessor–based controller board, and up to sixteen beam formers. Each beam former contains twelve piezo-electric elements, which perform the functions of beam steering and Doppler signal detection. The system is working with a dynamic, adaptive aperture control performed in the scan converter. Signals are received from a center point and four points whose positions rotate around the center. The recording at multiple sites is achieved without spatial readjustment of the array position. A new center position is determined as a function of a vector that points to an expected new signal maximum. A schematic illustration of the cerebral blood–flow system is shown in Fig. 9.30.

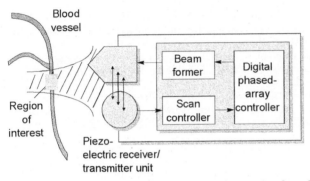

Fig. 9.30. Schematic illustration of an ultrasonic phased–array system for tracking blood vessels and recording blood flow in the cerebral arteries of premature newborns

9.2.3 Conclusions and Outlook

The examples above have given a brief insight on the possibilities of adaptronic systems in medicine. Adaptronics in medicine is an evolving field that is currently in its infancy. A recent conference on minimally invasive surgery has highlighted the potentials of adaptronic systems. Microgrippers actuated by shape memory alloys (SMAs) have been demonstrated whose dimensions are smaller than 0.5 mm [11]. Several research teams are working on SMA actuated endoscopes that can be steered through cavities, lumina, and blood vessels inside the human body. Implantable drug delivery systems will comprise glucose sensors that control delivery of insulin. Newest research is directed towards the design of glucose–sensitive gels that actuate valve and pumping systems in a self–regulating manner.

Despite the recent advances in the design functionalized or biomimetic materials, nature itself is providing the most mature and sophisticated adaptronic materials and systems. This is reflected by research in the field of bioartificial organs. For example, pancreas cells possess the capability of glucose–level–dependent production of insulin; the insulin production is inherently regulated inside the cell. It will be a major challenge of the future to employ the adaptability and multifunctionalty of living cells to design the bioadaptronic systems of the future.

References

1. Woodruff, E.A.: Clinical care of patients with closed–loop drug delivery systems. *The Biomedical Engineering Handbook* (ed. Bronzino, J.D. Boca Raton, FL: CRC Press and IEEE Press, 1995) pp. 2447–2458
2. Edell, D.J.: Tomorrow's implantable electronic systems. eds. Edell, D.J.; Kuzma, J. and Petraitis, D., 18th Annual International Conference of the IEEE EMBS, Minisymposia; IEEE Press, Amsterdam pp. 1–3 (1996)
3. Schaldach, M.: Reestablishment of physiological regulation, a challenge to technology. *Electrotherapy of the Heart* (ed. Schaldach, M.; Springer, Berlin, Heidelberg 1992) pp. 209–221
4. Edell, D.J.: Chronically Implantable Neural Information Transducers eds. Meyer, J.U.; Edell, D.J.; Rutten, W..C. and Pine J. 18th Annual International Conference of the IEEE EMBS, Minisymposia. IEEE Press. Amsterdam, pp. 1–3 (1996)
5. Stieglitz, T.; Beutel, H. and Meyer, J.–U.: A flexible, light–weight multichannel sieve electrode with integrated cables for interfacing regenerating peripheral nerves. Sensors and Actuators A 60, pp. 240–243 (1997)
6. Trasher, A.; Wang, F. and Andrews, B.: Self adaptive neuro–fuzzy control of neural prostheses using reinforcement learning. 18. IEEE EMBS Conference. IEEE Press. Amsterdam (1996), pp. (CD–ROM version)
7. Stieglitz, T.; Navarro, X.; Calvet, S.; Blau, C. and Meyer, J.U.: Interfacing regenerating peripheral nerves with a micromachined polyimide sieve electrode. 18. IEEE EMBS Conference. IEEE Press. Amsterdam (1996), pp. (CD–ROM version)
8. Eckmiller, R. and Napp–Zinn, H.: Information processing in biology–inspired pulse coded neural networks. IJCNN'93, Nagoya. Heidelberg (1993), pp. 643–648. VDI/VDE
9. Flemming, E.: Tactile Sense in Minimal Invasive Surgery: The TAMIC–Project mst news, **19**, February 1997, p. 13
10. Shung, K.K. and Zipparo, M.: Ultrasonic transducers and arrays. IEEE Engineering in Medine and Biology Magazine, **15** (6), pp. 20–30, 1996. IEEE Press, New York
11. Heyn, S.P.: Microsystem technologies and allied technologies in medicine: spotlights on recent activities in the USA, Canada, and Europe mst news, vol. June 96, no. Special Issue, pp. 134–139, 1996. VDI/VDE

10. Future Perspectives: Opportunities, Risks and Requirements in Adaptronics
B. Culshaw

10.1 Introduction

We live in a world which changes frighteningly quickly — a well worn cliché perhaps, but a simple statement that encapsulates the potential and the problems not only in adaptronics but in countless other areas of science and technology.

Most — possibly all — of the rate of change can be ascribed to the availability of increasingly complex, sometimes elegant, technological changes. The consequential availability of tools and the artifacts that these tools can create has benefited humanity enormously and at the same time created some far from insignificant problems. Sanitation engineers and pharmaceutical companies minimise infant mortality and maximise lifespan. The world population now increases at the rate of 100 million per annum worldwide, although it drops in the so–called affluent countries. Faxes, e–mails and web pages inundate us all with information, yet we all agree that even the standard postal system is very capable of ensuring that junk mail is in the majority. However, we can now phone our friends in China, India or South America in seconds rather than weeks — or even not at all.

This chapter is being prepared on a computer that is far more powerful than the machine I shared with hundreds of other users as a student in the late 1960s and which at that time employed a staff of 20 and cost half a million pounds sterling. It all seems like overkill, but the same computer has given us access to many of the exciting possibilities put forward in this volume. Without it we may all have been less confused and frequently less frustrated, but we would also have missed out on an enormous range of benefits and opportunities.

Scientific and technological advances would be lost without it. Chaos has been controlled to allow us to model evolution. The cellular automaton replicates natural selection and neural networks mimic human thought processes [1, 2, 3]. We can now design complex molecules, modify proteins [4, 5], predict the weather with a reasonable accuracy, and model the quirks of the stock exchange. The computer has metamorphosed the unpredictable into the defined. The only prototypes to the Boeing 777 were built, assessed, evaluated and corrected on a computer[1]. A huge machine has found the last quark[2], a much smaller machine produced a Bose–Einstein condensate[3].

So the tools are there and for most of the time they work. The tools, however, cannot be fully exploited since all but the few who design the tools can never appreciate their complete potential. Moreover, the information filters that are required to feed the tools in the right way are, at best, crude. The garbage out remains no better than the garbage in — though it might look different through a wavelet transform [6]. This leads into considerable challenges in defining the education process for engineering and technological professionals in an era in which, in some subjects, the knowledge half–life is at most a couple of years. The educators and the educated can no longer assume that the approach to life can be 'teach to use', it must be 'teach to learn'.

There is the additional challenge of how to get the best from the engineering resources available and how the engineer should serve the economic, environmental and social infrastructure in which he or she operates. Environmental needs supported by an increasingly complex legislative framework tend to impose a repair–and–retain philosophy rather than one of replace–and–dispose. Adaptronics is really about optimizing this process in its most general sense. Philosophically these comments also expose one of the most important intellectual hurdles in exploiting these ideas: the major infrastructure investments in transport systems, highways and buildings are designed and realized by civil, structural and mechanical engineers in an environment where 'design to code'rather than 'design to purpose' is the norm. Approvals for new codes take years rather than weeks, and so these professions — quite correctly since so much is at stake — are culturally unaccustomed to change. Compare this with computing and communications.

The Industrial Revolution sought to create material artifacts to improve the material well–being and prosperity of the populace, within which those far off technologies emerged. These material artifacts inevitably produce material garbage (Fig 10.1) that is well characterized [7] and most directly attributable to the material benefits that preceded it. Furthermore, the greater the material benefit, the greater still the garbage. The past half–century has seen the beginnings of another revolution — this time taking us into the Information Age that produces information artifacts for which, unfortunately, we have yet to characterize the garbage. However the same rule still applies: the greater the information, the greater still the irrelevance.

---

[1] The 777 design process has featured in TV documentaries (The BBC's *Horizon* series, among others) and in professional journals.

[2] The top quark was discovered in 1995/1996 at Fermilab, Illinois by colliding protons and antiprotons, each accelerated to 900 GeV. Even so, less than one collision in a billion (10^9) produced the quarks.

[3] E. Cornell and C. Wiemann of NIST reported the 'superatom' in *Nature* in summer 1995. This needed a relatively modest cryogenic vacuum system.

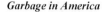

Garbage in America

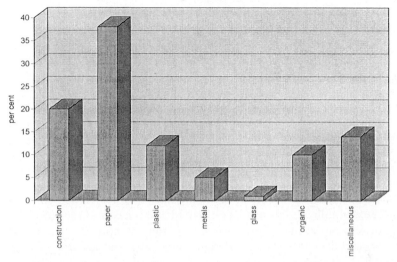

Fig. 10.1. The consequences of the first Industrial Revolution are now well characterized [7]. But what about the second, leading us to the Information Age?

Adaptronics is also about the interface between these two industrial revolutions. There will be benefits in the interface but the two cultures must learn each other's language and the educators must endeavor to produce linguists. There will also be mistakes and these will produce the garbage. What is interesting is that we are back at the beginning of a new learning curve. The rest of this article will present a largely personal view of where this learning curve may lead.

10.2 Adaptronics — the Generic Enabler?

I was pleased our editors chose a new term. Even though I personally view adaptronics as synonymous with smart structures and materials, this latter label is one with which I have never felt comfortable despite its currency. Maybe 'adaptronics' will catch on.

There have been countless attempts to define the subject area. With many others, I too have contributed to this intriguing, arguably futile, debate. I concluded by observing that adaptronics, smart structures or whatever we call it, is nothing other than a synonym for good engineering and surely good engineering (Fig. 10.2) is what the profession aspires towards?

Adaptronics is arguably a universal panacea. It applies the tools available from material science, mathematics, computing and numerical analysis to the optimization of material artifacts. The history of spectacular failure and less spectacular but far more costly decay in the infrastructure that uses these

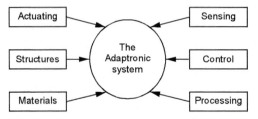

Fig. 10.2. A sense of adaptronics — or is this just good engineering?

artifacts provides compelling evidence that any potentially useful optimization process, whether applied to the design, to the use, to the maintenance,
or preferably to all three, must offer very substantial benefits. These benefits can be quantified: for example, if the highway had not needed so much
repairing the cost differentials can be estimated and the cost of the social disruption can also be estimated. The same logic may be applied to grounded
aircraft, automobiles called back for design faults, potential savings in energy
consumption, in transport, industrial processes and heating and cooling —
indeed, across the entire sector. But the take up is, at best, leisurely. The
principal reason seems to be that the universal enabler (Fig. 10.3) has to
cope with not only engineering integration but also (and especially for major
infrastructure) the conflicts between the political and economic aspirations of
the various sectors within the society in which an adaptronic structure must
operate.

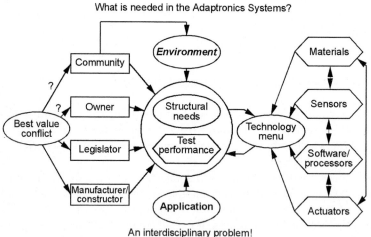

Fig. 10.3. The needs and ideals for the adaptronic engineer. But to whom should
the engineer listen?

It is, however, far from a totally gloomy picture. At one extreme, civil and military aircraft and extraterrestrial vehicles epitomize the intelligence and adaptability that could be built into current mechanical engineering concepts. They are also an excellent model from which to expose the needs in the future. The global telecommunications network remains a remarkable example of the fact that an extremely complex system can be made remarkably reliable and deceptively easy to use. The civil engineering infrastructure, on the other hand, presents the paradox of an extremely conservative design to code targeted at building nominally permanent structures with almost totally unproven materials — most notably concrete. It seems that the extrapolation that the survival of medieval masterpieces was evidence that anything derived from stone was automatically permanent. What is more, if you reinforced it with steel it would be more permanent still. The potential for adaptronics to cope with anomalous loads, to detect deterioration and decay, to identify tests and prove new materials and new material systems is only now becoming appreciated within a civil infrastructure that accounts typically (Table 10.1) for 5–10% of GDP (Gross Domestic Product) in most developed economies.

Table 10.1. How some countries spend their money

Country	GDP ($bn)	Agri-culture %	Mining %	Manu-facture %	Construction %	Transport and communication %
US	6738	1.9	1.1	17.6	3.7	5.9
Japan	4693	2.2	0.3	30.4	8.8	6.4
UK	1042	2.0	2.2	20.9	5.4	8.5
India	81	30.3	2.4	17.3	5.8	8.1
Brazil	471	12.4	1.8	24.9	7.4	6.2

Note: GDP for the U.S. and Japan is for 1994, and for 1993 otherwise.

Of course, any additional complexity in any of these systems implies inevitably additional cost. This additional cost is only tolerated if either there is a demonstrable benefit to the customer within the timescale relevant to the customer or society puts in place legislation that compels complexity. The aerospace industry is an excellent example of the latter. The dilemma in the former may be exemplified by a hypothetical discussion of a new concrete bridge structure to be purchased by a local or national government driven by politicians who wish to continue to be elected and therefore wish to be seen to be saving taxpayers' money during their term of office. The instrumented bridge costs an extra £5 000 000 but the instrumentation will direct a repair program 15 years after it is installed and well before the condition of the bridge becomes critical. The bridge would be repaired then for 10% of the original cost. Without instrumentation, the need for repairs will not be detected for 25 years but the cost of the repairs will be half of the cost of the original bridge. The social disruption of this major repair is immense: the bridge is restricted to one lane of traffic in each direction instead of five, and a city full of commuters is frustrated. The politician will, however, be determined to save money during the relevant term of office unless the technical arguments can be extremely convincingly put forward and agreed.

Engineers and technologists thrive on the satisfaction of the application of their art and design. The design–code philosophy that dominates most branches of large–project engineering does, however, mitigate against the risk–taking adventurer, but this conservatism has produced a whole catalogue of spectacular mistakes. There is then a need to take risks, to be adventurous, to gamble on the prospects of spectacular successes to demonstrate that the integration of the diverse engineering professions can produce hitherto unrealised benefits. Even now it is far from impossible to cite some examples of both the risks of conservatism and the benefits stemming from the adventurous spirit.

10.3 Engineering Conservatism: The Concealed Risks

The conservative engineer designs to established codes. These codes are themselves conceived to ensure that any artifact designed to them will be safe, long lived, secure and functional. In the vast majority of cases this approach works very effectively, and indeed the result is over–design. But even within the design codes there are the known acceptable risks and, more important, there are surprises. When these surprises are life–threatening or result in fatalities, they make international headlines:

- There are on average about 25 air crashes per annum — usually caused by natural events outside the design envelope but sometimes through often mysterious design failure.
- On 3 December 1984 Union Carbine's pesticide plant in Bhopal, India, leaked poisonous gas into the surrounding hutments. 2,500 people died and countless more were maimed. The litigation continues.
- The Channel Tunnel between England and France suffered its first fire on a train carrying freight trucks on 19 November 1996. The tunnel was closed for several weeks — to the delight of the ferry operators who had seen the tunnel rise from a novelty to the conduit for about half the cross–channel traffic. The fire safety procedures were found to be lacking and have since been modified. Cross–channel traffic tolerated temporary but extreme delays.
- On 28 January 1986 the space shuttle 'Challenger' exploded a little more than a minute after lift–off. The shuttle programme was delayed for over two years whilst the cause was identified. Richard Feynman performed his

now famous experiment, shattering an O–ring after immersing it in liquid nitrogen to illustrate the point.
– In Seoul, South Korea, on 21 October 1994 a section of a bridge over the River Han sheered neatly and fell into the river. Miraculously the bridge section floated but over 30 people died.

The list is, of course, endless but spans all sections of the human endeavor and grabs international headlines. The fact that these and countless other similar accidents all involved conservatively designed engineering systems does perhaps explain the reticence within the community towards the nominally unproven.

Especially when the more adventurous system can also make the headlines. The most remarkable of these in recent times was the delay, exceeding a year, in the opening of the Denver International Airport. It appeared that the highly automated baggage–handling system would certainly deliver the baggage — but quite possibly shredded. Most who read this book will no doubt have driven automobiles with the early version of the 'new' electronic ignition and been stranded somewhere. Intel's problems with the launch of the Pentium processor chip are rumoured to have cost half a billion dollars to correct. Worldwide, that solid reliable material concrete is cracking, flaking and deteriorating, threatening the infrastructure and compromising the mobility which, whilst once a luxury, is now essential to our way of life.

So there are risks in conservatism, risks in designing to accepted code and risks in the tried–and–trusted system. The consequence of these highly publicized and well documented risks is that those who invest the large sums of money in major projects, many of which could benefit from the application of adaptronics, need serious convincing that these new systems have something to offer. These principal prospective beneficiaries seem to be within the civil infrastructure, aerospace, ground transportation and manufacturing systems.

However, all of these industries are under pressure from environmental legislation, from the ever increasing demands to improve user efficiency, from the exposed fallibilities in 'design to code', and from the need to conserve and recycle the store of the earth's resources. Adaptronics surely has a role to play and the risk in ignoring it must exceed the risks of some experimentation and, by yet more, the risks in ignoring the fallibility of current practice.

10.4 The Adventurous Engineer: Some of the Benefits

The engineer is not necessarily the conservative soul, and when operating in areas that are not hidebound by codes and legislation can even be adventurous and speculative. He can even be stimulated to meet the impossible specification. There are many examples:

– The compact disk player that we all take for granted nowadays is a miracle of high–volume, very high–precision, very low–cost electromechanical

engineering. An initial problem statement defining a rapidly spinning disc within which a track has to be found using a non–contact method with radial definition of submicron accuracy presents an obvious technical challenge. To meet this challenge in systems which cost no more than a few pounds to assemble is a technical miracle and in my view exemplifies the engineer achieving the apparently impossible.
– The global communications network has only recently (Table 10.2) agreed its international standards for extremely high–speed links. Paradoxically the agreement came well after most of the developed world and a significant fraction of the developing world had been linked together using efficient (often optical fiber–based) links. The real cost per bit–kilometre of communications transport had dropped by at least eight orders of magnitude in thirty years. The transport costs will soon be negligible compared with switching and handling costs, so the world will be accessible through a local phone call (indeed, it already is for Internet users). Throughout much of this spectacular development, the system was designed, implemented and operated more rapidly than the codes and standards bureaucracies could finalize the documentation that described it. And already there is serious talk of yet higher channel data rates and yet greater capacity for multiple channels funnelled into a single optical fiber using a diversity of optical wavelengths.

Table 10.2. Standardized transmission rate in the digital hierarchy[d]

SDH[a] System	SONET[b] System	OC[c]	Rate	Maximum Voice Channels
	STS-1	OC-1	51.84 Mbps	783
STM-1	STS-3	OC-3	155.52 Mbps	2 349
STM-4	STS-12	OC-12	622.08 Mbps	9 396
		OC-24	1244.16 Mbps	18 792
STM-16	STS-48	OC-48	2.4888 Gbps	37 584

[a] Synchronised digital hierarchy (CCITT: European Standard)
[b] SONET — (ANSI: US Standard)
[c] Optical Carrier (ANSI: US Standard)
[d] These standards, now globally recognized, emerged from cable transmission systems, originally nationally defined. But in interim interfaces have been built to convert from one domain to another. TV broadcast standards are still defined locally, but digital conversion is imperceptible. In communications, the nationally based (relatively) adventurous engineer in his national monopoly evolved systems to meet local needs before addressing the international implications. The globalization of the PTT industry inhibits future independence.

– The personal computer industry sets its standards through the dominant product line both in hardware and software. This is an industry that continues to be a miracle in marketing, promoted by the technical enthusiasm of a few who dominate the needs of the many. Most computers are used for word processing, spreadsheets, databases and elementary graphics — needs that can be more than adequately met by state–of–the–art machines from the early 1990s, all of which are now obsolescent and impossible to maintain. The needs for current super machines (soon to be yet more super) are far less evident, but here the total absence of legislation and the power of marketing muscle assure us of continued innovation stimulated by the technocrats rather than the consumer. After all, who really wants to play clips from Braveheart on their computer system?

- Space exploration catches the imagination and operates at the limit of our current technological ability. Even though shuttle flights no longer make news, they are still annually listed in the Britannica Year Book so that, at least by Britannica's criteria, they are significant global events. The whole programme gained substantial credibility thanks to the launch of a defective telescope. The Hubble Repair Programme was one of the most impressive achievements using the space shuttle technology. The space shuttle is another example of a situation where technical imagination is allowed at least some dominance over history and precedent.

All these examples, and there are very many more, combine a technical risk, often initially stimulated by the adventurous engineer, with proven success. They all also involve very significant financial backing, often from major investment sources — either a multinational organisation gambling for profit or a taxpayer gambling for kudos. The adventurous engineer who ignores or is entirely oblivious of precedent, codes, regulations and standards is also the one who initiates the new small company ploughing an often lone furrow, developing new markets and new technologies simultaneously. Thankfully he starts with a good chance of making substantial financial returns and also benefiting society through creating opportunities for others. This, of course, contrasts with the calculating conservative engineer involved with the technical management of large operations and providing the necessary technical direction to please financial masters.

But the prognosis for adventurous engineering remains excellent. There are both needs and opportunities for socially and economically profitable technology developments, sometimes correcting former historical errors (for example, in the environmental control industry) but more often expanding infrastructure and the overall quality of life.

A moment's glance will confirm that there is minimal overlap between the operational environment of the conservative engineer and that of the adventurous engineer. Whilst the conservative engineer contemplates the limits of his codes of practice, the adventurous engineer carries on in a culture that essentially ignores their presence. Perhaps the most intriguing feature of adaptronics is that the conservative engineer must work with the adventurous engineer, with very significant potential benefits for both. The intrinsic cultures, though, are very different.

10.5 Opportunities in Adaptronics —
the Adventurous Conservative?

The art of adaptronics brings together different complementary engineering worlds. The materials scientist has historically sought to improve the structural qualities of structural materials such as steel and carbon fiber composites and the responsivity of transducer materials such as magnetostrictives and piezoelectrics (the latter group often referred to, for reasons that seem totally incomprehensible, as 'intelligent materials'). In more recent times biological materials have come to the fore but predominantly the materials scientist is here the observer and interpreter, though as time progresses he becomes more the modifier and manipulator. Slowly the structural and transduction cultures have come just a little together, but the biologists remain aloof. Adaptronics also involves structural engineering, an area more traditionally responsive to needs than technologically innovative, although the needs have often stimulated innovative solutions of which perhaps the currently most public is the cable–stay bridge. In another domain, structural engineering to pre–ordained criteria explains why most automobiles are now the same shape — optimizing wind–tunnel tests and simulations to reduce air drag seems to converge on nominally identical solutions. It makes for less interesting highways but lower fuel bills. The vast majority, though, of current structures are only based on structural materials. Transducer materials and (far less) biological materials are a different world.

Sensor technologies have advanced apace and some are described in other chapters in this book. The principal conceptual development lies in the ability to address long lengths and/or large areas of a structure using relatively simple technologies. But most of the potential users of these systems have yet to ascertain exactly how to use all the data. However the measurement industry is historically very niche–oriented, very application–specific and very dependent on emerging integrated technological approaches for its success. It is a fascinating paradox: initially the domain of the adventurous engineer but in its operation the domain of the conservative, who must meet the precise demands of his/her own sector and maintain the technology introduced by an adventurous colleague (or maybe even by a former self). Finally in data processing both the tools and the hardware change almost daily. Adaptronics seeks to bring these diverse worlds together.

In doing so, it offers much. Conceptually it is simply the integration of diverse disciplines. Through this integration, the abilities to recognize and correct for changes in the operational environment may be built into a huge variety of engineering artifacts. In many ways the potential offered by the application of adaptronics is the greater the more alien the cultures that are brought together. However, the major developments will inevitably be evolutionary rather than revolutionary.

They will need some arranged relationships between the disciplines in order to optimize the overall engineering potential. It is here that the responsibilities of the matchmaker come to the fore. Whilst in industry assembling cross–disciplinary teams is a relatively common practice, in academia, where at least in principle the research activity is more free–ranging and adventurous, tradition encourages the loner.

The benefits, though, of structural artifacts that can recognize and correct damage, deterioration and the onset of anomalous loading conditions and are flexible in use are potentially huge. Combining the contributory technologies promises very substantial rewards. These rewards have their most obvious possibilities in established engineering disciplines, many of which are already pursuing an adaptronic pathway. The principal examples of such traditional engineering disciplines include:

- Air and space transportation systems, for which the tradition of adaptronics is already well established. The future opportunities lie in part in gilding the lily by enhancing the control of the gas turbine engine, by improving the accuracy and stability of navigation systems and stabilization control, by ensuring more responsive and reliable safety systems, and by introducing more reliable surveillance and detection. There is also the obvious scope for improved health monitoring, especially within load–bearing structures. There are new opportunities too, of which the most obvious is the adaptive aerofoil and overall more responsive aerodynamics. For space transportation the obvious barriers are the cost of hardware and of the necessary energy to accelerate to escape velocity. Here adaptive structures operating in a weightless environment have already gained their spurs (see Sect. 8.1) and, whilst intergalactic travel is still a fantasy due to the immense velocities required, some even now view the starship *Enterprise* as a documentary of the future.
- Ground transportation systems that respond very much to market forces. Trains breakdown sufficiently frequently to be noticed, and condition-monitoring systems that provide a useful early warning have obviously some potential. Similarly, safety systems that nip the much–feared fire in the tunnel in the bud would be welcome. Road vehicles already have significant adaptronic content, ranging from responsive shock absorbers to complex plug–in engine monitoring systems, to dashboard alarms that politely request an oil change, to engine control systems that have significantly improved fuel consumption in the past couple of decades. In ground transportation, the principal demand appears to lie in condition monitoring. Most drivers still prefer to take themselves to work (otherwise they would take the bus) but there is no doubt that a complete self–guided vehicle that has excellent fuel consumption and is fully collision–proofed through its sensory systems could be extremely attractive. Such vehicles have already been proved on the test track: all that is required is confidence — and money. There are already vehicles on the market that will tell you the route through Tokyo (in your choice of language)[8].

- Marine transportation, which again highlights the need for effective condition monitoring, especially in large and difficult–to–inspect cargo transporters. Autonavigation is already a reality and, apart from military vehicles (especially submarines), enhancing adaptive hydodynamics is probably just a luxury.
- Civil and structural engineering, which appear at present to offer by far the greatest scope. This is an area in which there is minimal tradition of monitoring, and even less of responsive structures. It is an area in which every nation invests very substantially and an area in which there is a proven track record of structural deterioration and extremely expensive repair schedules, in both direct and social costs. Here, monitoring must have much to offer but here also the 'best value' conflict, highlighted in Fig. 10.3, comes to the fore. With the arrival of 'design, build and operate' contracts the 'best value' conflict will be slightly resolved though to date the handover period seems to arrive somewhat before any anticipated structural difficulties. For condition monitoring systems, some progress has been made in offshore structures over the past twenty years and for mainland structures during the past ten. There is still a great deal to learn, especially in the optimization of multiple–parameter and multiple–technology sensing networks, both in specifying the best measurement process and interpreting the results therefrom [9]. Responsive structures that actively respond to wind and earth tremors are relatively common in Japan, with a few in the U.S. The tuned passive response, especially for damping, in such structures can also be most impressive [10].
- Manufacturing systems exist in any automated production plant even though they already rely upon a significant presence from monitoring and control systems. Whilst there is always scope for technical improvement and cost reduction, at a conceptual level the current systems appear to be very well suited to the task. Of course, we could speculate towards the genetic control and for artifacts with a large biological content, and this too is already beginning to happen. Applying this approach to automobiles and washing machines stretches even the most liberal imagination.

So much for mainstream engineering. Most of the above, with the possible exception of the civil sector, is essentially more of the same but better. Opportunities, though, for an innovative contribution probably lie principally within the broad confines of the leisure and entertainment industries. EPCOT, Orlando's 'theme park of the future', and Disneyland impress even the most sceptical hard–nosed engineer, and every prize that Disney's 'imagineers' have been awarded has been well warranted. Within the mass markets, home entertainment systems have become more complex (do you know anyone over the age of 35 who can operate their video recorder?). Those of us with daughters have been pestered by talking dolls and everyone sneaks a quiet session on the computer games. These are all extremely profitable spin–offs from a very sophisticated technological base. Where this will lead in the future depends upon entrepreneurial imagination and technological availability. I wish I could predict, but if I could with confidence identify a successor to the Cabbage Patch doll I certainly would not publish it here.

So adaptronics undeniably offers the adventurous engineer the opportunity to shine. He can opt for the safe path of refining more of the same or he can venture into relatively unexplored territory, where the pickings can be profitable but the opportunities for straying from the correct path abound. Over the gamut of conventional engineering, the civil sector appears to offer the best untrodden territories not only for safety and condition monitoring but also in the general area of responsive structures. For those who wish to flirt with the whims of the public, the leisure industry has much to offer and is particularly attractive since it is totally undefined. Depending upon your bank balance and your aspiration, you can take your choice.

10.6 Adaptronics Research

We have already hinted at this. The principal research focus has to be on integrated adaptive systems, which in order to demonstrate their prowess must have contributions from a variety of different teams. The contributory technologies must also be then identified to ensure that they too progress outside the systems context. However the overall component interactions within the adaptronics system are critical to its success and research into a whole which is far greater than the sum of the parts is essential.

Most of the contributory technologies have been discussed and identified within the chapters in this work. Even so, the direct contribution made through the availability of low–cost computing cannot be underestimated. The techniques for signal processing and feature extraction are central to the application of any of these contributory technologies within an adaptronic system. The application of progress within the signal–processing community into the adaptronic environment is an essential feature. There is already a substantial body of evidence that highlights both the possibilities and the pitfalls inherent in the exploitation of genetic algorithms, neural networks, novelty extraction and the cellular automaton [11]. Undoubtedly there is scope for significantly more research in this area: it is all very well to create data but the data is useless unless properly interpreted.

The industrial development teams will approach these requirements on an as–needed basis but in doing so this precludes the speculative and inhibits the realization of generalized analytical and design approaches. It is here that the academic researcher must come to the fore, but it is also here that the paradoxes of the academic research environment become evident. In the early 1970s as an unexperienced postdoctoral fellow I encountered an experienced researcher from the then Bell Laboratories who was the first to define research for me as 'learning more and more about less and less'. This is the obvious antithesis of any philosophy that supports progress in adaptronics but it is still one that is consistently embraced throughout the international academic community. There are pressures on the interdisciplinary academic researcher to maintain excellence in his own discipline whilst simultaneously applying his experience in the interdisciplinary forum. Whilst doing the latter he usually makes but very modest progress in the former and the risks are obvious — even more so when it is recognized that world–class biochemistry is only feasible when world–class biology and world–class chemistry are combined.

The various national and international funding agencies also contribute. Most are organized on a subject–specific sub–group with individual budgets. Most if not all of these budget are oversubscribed, so interdisciplinary proposals are often passed to the next committee. To an extent this is being recognized and corrected, particularly by the European Union agencies which require interdisciplinary teams to operate together. However at the national level the prognosis is less optimistic.

There are also the management issues. In the UK university departments go through a research assessment exercise on a four–or five–yearly basis. Applications are judged by peers within subject groups. Publications in biological journals when judged by a mechanical engineering panel cause wrinkled brows and confusion. There are similar processes, at local, regional or national level, in other countries. Academic promotions are also based upon local or national/international peer opinion. The optics researcher who applies his art (but does not advance it) in structural monitoring is vulnerable to the accusation that the optics is pedestrian. Again, this is an international rather than a local phenomenon. The international academic community is really only most comfortable when encouraging extreme specialization. I believe that the paradox inherent within this and the consequential impact upon the exploitation of ideas and technologies is slowly being recognized, so the situation should change. However those of us in universities are all too aware of system inertia.

My principal source for optimism is that there certainly is the will among many senior academics with distinguished specialized research records to see their knowledge successfully applied, even exploited, within a complementary discipline so as to enable the realization of conceptual advances within the global systems and applications arena. Some of the initial results which have combined the efforts of multidisciplinary teams from totally different backgrounds have proved extremely encouraging [12]. The area of adaptronics — and, indeed, many more disciplines in the engineering and technology umbrella — need to encourage much more cross–discipline fertilisation, through which the advanced tools can meet those who may use them to best effect.

10.7 Adaptronics and Education

Engineering education has changed radically in the past quarter of a century. Initially in most undergraduate schools worldwide the introductory courses at university would be common over most if not all engineering disciplines. Every graduate emerged with the rudiments of thermodynamics, electronic circuits and structural mechanics. Subsequently every discipline has succumbed to the information overload in its particular sector and eliminated all but material perceived to be relevant. The pressures on time to learn the new tools have become so intense that, much to everyone's chagrin, there are now electrical engineering graduates, and very many of them, with little knowledge of the electromagnetic fields that many regard as fundamental to the art.

Many countries, not least the UK, have also seen a change in the philosophy of the school curriculum, from one targeted to training for entry to university to one seeking to educate the majority. Consequently, school courses are less intense, broader and more diverse, presenting inevitable dilution of the academic rigor within the undergraduate course structure. The intriguing paradox is that the educational experience in most undergraduate institutions has progressed from a discipline–oriented specialization on a very general and academically rigorous engineering foundation to a within–discipline subject specialization covering a narrower subject area, on a less rigorous foundation but with a much broader subject content. In particular, the wide–ranging intellectual mathematical and conceptual toolbox has become much less available. The graduate certainly receives a broader education within his own subject discipline than was possible but is also less equipped to communicate across the disciplinary boundaries.

In the UK the situation is somewhat exacerbated by the institutional accreditation process. The institutions involved within the adaptronics envelope include all the major engineering authorities (mechanical, electrical, civil and chemical), the Institute of Measurement and Control, the British Computer Society, the Royal Institution of British Architects and the Royal Aeronautical Society — and there are probably more. Within the present academic structure it is unlikely that all of these would be content to accredit the same degree course content.

But adaptronics needs the specialists — and good ones — namely, those who have been through the specialist degree mill. The educational fraternity feels that it is comfortable and knows how to produce such competent trained individuals of the right caliber and background. Even the needs of the specialist are changing rapidly. The engineer's responsibilities (see Fig. 10.4) are moving gradually up the design chain. He no longer needs to solve routine technical problems: the computer package designs the circuit, checks the loading distribution, or decides whether the cable structure is sound and complies with international standards. The emphasis is now on framing the problem and assessing the solution, currently against essentially technical criteria. The evolution of even more capable tools will produce a revolution in the engineer's responsibilities, rapidly encompassing social and aesthetic criteria. Even the specialist will need a general education. Addressing the liberal

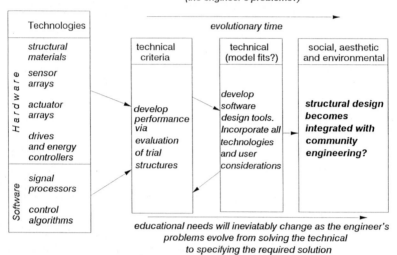

Fig. 10.4. Engineering will change, as the design aids become more complete. The engineer's priorities will evolve too, and educators must adapt

arts in parallel with technology and science will progress from the optional to the mandatory. Society will need the socialized engineer.

The educational process also needs to educate the advanced generalist, the systems integrator who not only knows and appreciates the basic available tools but (more importantly) can synthesise these tools into an innovative conceptual engineering framework. Some of the very few remaining general engineering courses could perhaps serve as a basis, but it is at least equally important to encourage the interdisciplinary philosophy into the thought process of senior academics responsible for course definition. Small steps, for example organizing interdisciplinary group undergraduate projects, can contribute much and some of these small steps have already been taken. All this has to take place against a general educational revolution [13], where interactive computer–aided learning will augment, (and sometimes replace) conventional educational processes.

Planning a course to educate engineers is a very difficult compromise, imposed through the ever broadening (but paradoxically academically narrow) needs within the specialist disciplines and the pressure to ensure that education processes are complete within a three– or four–year undergraduate curriculum. The interdisciplinary socialized engineer will undoubtably need to be educated for a longer period of time. Within the UK the MEng degree provision (at least in one of its many manifestations) seeks to address some of these issues through a 4–5 year undergraduate course. Meanwhile we have conferences in engineering education, institutional committees fretting about engineering education, potential employers bemoaning the shortage (but when pressed not being really sure in answering 'a shortage of what?') and yet behind it all, the prime responsibility of undergraduate education must be to inspire interest, enthusiasm and commitment. When equipped with these attributes, a new graduate can be assured of both an ability to flit from discipline to discipline and, in a much broader sense, a challenging and rewarding career. All too easy to define and all too difficult to implement!

10.8 Some Conclusions

Adaptronics could certainly be argued to be little more than the application of good sound engineering practice. In order to exploit the technology to its best and establish the conceptual framework from which adaptronic designs can evolve, a structured interdisciplinary approach is essential. At present this approach is a rarity, both within the engineering profession and the educational process. However an increasing awareness within the policy making community, particularly in publicly funded research, that interdisciplinary is a Very Good Thing is slowly addressing this issue through directed funding. For specialist researchers to contribute their state–of–the–art knowledge into advancing the art of interdisciplinary integration is at least as valid as continuing to advance the specialist discipline. Thankfully a few senior research figures have ventured along this 'less and less about more and more' path.

The potential benefits stemming from thorough and complete integration of adaptronics concepts into engineering are wide–ranging, especially so in the disciplines where a passive system has been the norm, including for example civil engineering, structural engineering and architecture. Adaptronics is already there and running in aerospace and, to various degrees, in marine transportation and ground systems.

The principal challenges for the practitioner and the enthusiast lie in converting the wary through a combination of effective communication between disciplines, identification and quantification of the benefits and risks applied to a particular potential application and — most important of all — enthusiasm and commitment about the technological, social and economic benefits. There is a slowly expanding community converted to and representing this cause, and there is no doubt that these engineering techniques and the underlying philosophies will have much to contribute in the future.

References

1. Deutsch, S. and Deutsch, A.: *Understanding the Nervous System; An Engineering Perspective* (IEEE Press, New York) 1993.
2. Mandelbrot, B.B.: *The Fractal Geometry of Nature* (Freeman, New York) 1983.
3. Cohen, J. and Stewart, I.: *The Collapse of Chaos* (Penguin, London) 1995.

4. Joyce, G.F.: Directed Molecular Evolution. Scientific American, pp. 90–97 (December 1992)
5. Stemmer, W.P.C.: Searching Sequence Space. Bio–Technology, pp. 549–553 (June 1995)
6. Chui, Ch.K.: *An Introduction to Wavelets, Wavelet Analysis and its Applications* Vol 1, (Academic Press Inc.) Boston 1992.
7. Rarhje, W.L. and Murphy, C.: *Rubbish! The Archeology of Garbage* (Harper Perennial) 1993.
8. Hotate, K.: *The fibre optic gyroscope* Culshaw, B. and Dakin J.P., eds.: Optical Fiber Sensors Vol. IV, (Artech House, Norwood, Mass) 1997
9. Culshaw, B.: *Smart Structures and Materials* (Artech House, Norwood, Mass) 1996.
10. Sakamoto, M. and Kobori, T.: Control Effect During Actual Earthquake and Strong Winds of Active Structural Response Control of Buildings. Fifth Int. Conference on Adaptive Structures, Sendai, Japan, December 1994.
11. SPIE Proceedings Annual Smart Structures Symposium. San Diego (SPIE, Bellingham Washington, annually from 1992.)
12. Staszewski, W.J.; Pierce, S.G.; Worden, K.; Philp, W.R.; Tomlinson, G.R.; Culshaw, B.: Wavelet Signal Processing for Enhanced Lamb Wave Defect Detection in Composite Plates using Optical Fibre Detection. submitted to Optical Engineering, 1997.
13. Snyder, D.P.: High tech and higher education: a wave of creative destruction is rolling toward the halls of academia. in On The Horizon. September/October 1996, Jassey Bass Publishers, San Francisco, USA

11. About the Authors

Horst Baier Horst Baier was born in 1950 near Frankfurt. He gained his degree in mechanical engineering at Technische Universität Darmstadt in 1972. From 1972 until 1977 he was research assistant at the institute of light-weight structures in Darmstadt, with emphasis on research into finite elements and multi-criteria optmization methods. In 1997 he joined the company Dornier Satellitensysteme, where he soon became responsible for structural analysis and technology of mechanical systems. He became more and more involved in the interaction of mechanical with control systems, and one of his focuses was on adaptive structures. Apart from considerations in active damping in general, a major field of application was in very high precision systems, which he pursued with his coauthor Dr. Döngi. In summer 1997 he was appointed to the chair of light-weight structures and became co-director of the Aerospace Institute of Technische Universität München, with research activities in multidisciplinary problems, and especially in adaptive structures.

Prof. Dr.-Ing. Horst Baier, Lehrstuhl für Leichtbau, TU München, 85747 Garching, Germany.
e-mail: baier@llb.mw.tu-muenchen.de

Christian Boller Christian Boller was born in Jevnaker/Norway in 1954. He studied structural engineering at Technische Hochschule (TH) Darmstadt/Germany from 1974 and received his diploma in 1980. From 1981 to 1986 he joined the materials mechanics group of Prof. T. Seeger at TH Darmstadt as a scientific co-worker, working in experimental and theoretical evaluation of structural fatigue life and received a doctoral degree in 1987 from that institute. In 1984/85 he accepted a fellowship from the Japanese Science and Technology Agency, working at the Nat. Res. Inst. for Metals (NRIM) in Tokyo/Japan. From 1987 to 1990 he worked on materials science at Battelle, before taking over a position at MBB (now Daimler–Benz Aerospace) Military Aircraft. There he is the chief engineer for aircraft structures with major interests in structural health monitoring and adaptronics in general.

Dr.-Ing. C.Boller, DaimlerChrysler Aerospace MT2, D-81663 München, Germany.
e-mail: christian.boller@m.dasa.de

John Brignell Born in London, John Brignell began his career as an apprentice at STC. He took the degrees of BSc(Eng) and Ph.D. of London University. He joined the staff at City University, where he worked in a number of areas including dielectric liquids and computer-aided measurement, co-authoring a book 'Laboratory on-line computing' in 1975. He was Reader in Electronics at City University for ten years and has held the Chair in Industrial Instrumentation at Southampton since 1980. He has researched and written extensively in the area of sensors and their applications, and in 1994 co-authored a book with Neil White on 'Intelligent sensor systems'. He has had an extensive private consultancy practice for many years and has advised some of the larger companies in the UK, as well as many small ones, on all aspects of industrial instrumentation. He is a Fellow of IOP, InstMC, IEE and RSA. In 1994 he was awarded the Callendar Silver Medal by InstMC.
Prof. J.E. Brignell, Department of Electronics and Computer Science, University of Southampton, Highfield, Southampton, SO17 1BJ, UK.
e-mail: jeb@ecs.soton.ac.uk

W. A. Bullough Bill Bullough was born in Westhoughton, Lancashire in 1939. He was trade apprentice, student engineer and assistant hydraulics engineer with the De Havilland Aircraft Company. Following his M.Sc. in Thermo Fluid Mechanics at the University of Birmingham he worked as Systems Analyst with Hawker Siddely Dynamics Limited and was latterly research associate in the Osborne Reynolds Hydraulics Laboratory, University of Manchester. Since 1967 he has worked as Lecturer, Senior Lecturer and Reader at the University of Sheffield, UK where he is currently Head of Thermodynamics and Fluid Mechanics Groups and Chairman of SMMART (Smart Machines, Material and Related Technologies Unit). He is, among other things, Chartered Engineer, Fellow of Institution of Mechanical Engineers and Member of the Royal Institution.
Prof. W. A. Bullough, Department of Mechanical Engineering, The University of Sheffield, Mappin Street, Sheffield, S1 3JD, Great Britain.
e-mail: w.bullough@sheffield.ac.uk

Wenwu Cao Wenwu Cao was born in Shulan, Jilin, China in 1957. He earned a B.S. degree in physics from Jilin University in 1982 and a Ph.D. degree in condensed matter physics from the Pennsylvania State University in 1987. In 1990 he became a faculty member in the Materials Research Laboratory, Pennsylvania State University after a year of postdoc experience at the Laboratory of Atomic and Solid State Physics, Cornell University. In 1995, he began a joint appointment between the Department of Mathematics and the Materials Research Laboratory as an Associate Professor of Mathematics and Materials Science. While his teaching is mainly on mathematics, his research interests are highly interdisciplinary, including fundamental theory of phase transitions in ferroic systems, domain microstructure studies, physical property characterization of solid materials as well as transducer and actuator designs using computer simulations.

Prof.Dr. W. Cao, The Pennsylvania State University, University Park, PA 16802, USA.
e-mail: cao@math.psu.edu

Dave Carlson David Carlson was born in 1946 in Erie, Pennsylvania, USA. He received a BS in Physics from Case Western Reserve University in 1968 and a Ph.D. in Physics from the University of Colorado in 1972. Since 1976 he has been a senior research scientist with Lord Corporation, a manufacturer of noise, vibration and motion control systems and specialty materials, where he holds the position of Engineering Fellow in the Materials Division. He has been involved with controllable ER and MR fluids since 1984. His work has included basic studies of the physics and chemistry of ER and MR fluids, development of proprietary fluid formulations and the development of a wide variety of devices and systems which use ER and MR fluids. He holds 26 US patents relating to controllable fluids and is the author of over 70 technical papers and books.

Dr. J. David Carlson, Lord Corporation, 110 Lord Drive, Cary, NC 27511, USA.
e-mail: jdcarlson@lord.com

Frank Claeyssen Frank Claeyssen was born in Lille, France, in 1962. He was awarded an engineering degree from the Institut Supérieur d'Electronique du Nord (ISEN Lille) in 1985 and a Ph.D. in mechanics and acoustics from the Institut National des Sciences Appliquées de Lyon (INSA Lyon) in 1989. Then he joined Cedrat Research, a high tech company involved in Electric Engineering, where he founded a department specializing in Active Material Activities. His current interests are research and industrial development of piezoelectric, magnetostrictive and electromagnetic actuators, motors, transducers and sensors.

Dr. F. Claeyssen, Cedrat Recherche, ZIRST 4301, F-38943 Meylan Cedex, France.
e-mail: claeys@cedrat-grenoble.fr

Bernd Clephas Bernd Clephas was born in 1965 in Herford (Germany) and studied electrical engineering at the University of Hanover where he earned his degree in 1992. Afterwards he worked at the AEG Aktiengesellschaft (Böblingen) in the development and project management of test stands for the automotive sector. Since 1995 he has been a scientific collaborator at the Laboratory of Process Automation at the University of Saarland. His main research topics are the combination of piezoelectric and magnetostrictive actuators and the exploitation of inherent sensory effects (smart actuators).

Dipl.-Ing. Bernd Clephas, Universität des Saarlandes, Gebäude 13, Lehrstuhl für Prozeßautomatisierung, 66123 Saarbrücken, Germany.
e-mail: clephas@lpa.uni-sb.de

Brian Culshaw Brian Culshaw was born in Ormskirk, England in 1945. He was awarded the B.Sc. in Physics (1966) and Ph.D. in Electronic Engineering (1970) from University College London. His first appointment was at Cornell University, followed by Bell Northern Research Ottawa working on microwave semiconductor devices. In 1975 he was appointed to a lectureship at UCL and his research interests shifted into optical fibre sensors. In 1982 he spent a year at Stanford University researching on optical fibre gyroscopes. He joined Strathclyde University as a professor in 1983, extending his interests in fibre sensors to embrace smart structures technologies, particularly sensing and detection in fibre based structural composites.
Professor B Culshaw, Department of Electronic and Electrical Engineering, University of Strathclyde, 204 George Street, Glasgow G1 1 XW, UK.
e-mail: b.culshaw@eee.strath.ac.uk

Rajendra R. Damle Rajendra Damle was born in Poona, India in 1968. He was awarded the Bachelor of Electronics and Telecommunications Engineering degree from the University of Poona in 1985, M.Sc. and Ph.D. degrees from the University of Missouri-Rolla in 1993 and 1997 respectively. Then he joined US Sprint as a Senior Network R&D Engineer. His current research interests are neural networks, fiber optic sensors, and communication networks.
Dr. Raju Damle, Senior Network Engineer, Advanced Technology Labs, Sprint, 1 Adrain Ct.,Burlingame, CA 94010, USA.
e-mail: rdamle@sprintlabs.com

Frank Döngi Frank Döngi was born in Worms, Germany, in 1966. He studied aerospace engineering at the University of Stuttgart, Germany, and at the Cranfield Institute of Technology, England, where he was awarded an engineering diploma in 1991 and M.Phil. in 1993, respectively. In 1996 he recieved his doctorate in engineering from the University of Stuttgart for a dissertation on active flutter suppression by means of smart structures. He then joined Daimler–Benz Aerospace/Dornier Satellitensysteme GmbH as a structural mechanics engineer, where he is now responsible for smart structures and active structural control. Since 1998, he has also lectured on smart structures at the University of Stuttgart.
Dr. Frank Döngi, Daimler–Benz Aerospace/Dornier Satellitensysteme GmbH, 88039 Friedrichshafen, Germany.
e-mail: frank.doengi@dss.dornier.dasa.de

Victor Giurgiutiu Victor Giurgiutiu was born in Bucharest, Romania, in 1949. He received his B.Sc. (1972) and Ph.D. (1977) in Aeronautical Engineering from the Imperial College for Science, Technology and Medicine of

London University, UK. In 1984, he received an M.Sc. in Mathematics from the University of Bucharest, Romania. After a successful scientific career in the Romanian Aviation Institute, he moved to USA in 1992. Until 1996, he was Visiting Professor, then Research Professor and Associate Director in the Center for Intelligent Material Systems and Structures of Virginia Polytechnic Institute and State University. At present, he is Associate Professor in the Mechanical Engineering Department of the University of South Carolina. He has authored and co-authored two books, two book chapters, and over 20 journal papers. His research focuses on adaptronic structures with applications to health monitoring, induced-strain actuation, active sensors, smart active composites, and NDE.

Dr. V. Giurgiutiu, Department of Mechanical Engineering, 300 S. Main St., University of South Carolina, Columbia, SC 29208, USA.
e-mail: victorg@sc.edu

Wolfgang R. Habel Wolfgang R. Habel, was born in Stralsund in 1949. He received his diploma in theoretical methods of electrical engineering/information technology from the Technical University of Ilmenau, Germany, in 1972. After research on high-power energy transmission through superconducting cables, in 1981 he joined an institute of the German Academy of Building. Since 1985 he has been engaged in research on fibre optic sensors for monitoring engineering structures and for measurements of the properties of building materials. From 1992 to 1997 he managed the fibre optic group of the Institute for Renovation and Modernisation of Structures at TU of Berlin. Since autumn 1997 he has continued his research in fiber sensor technique at the Federal Institute for Materials Research and Testing (BAM) Berlin.

Dipl.-Ing. Wolfgang R. Habel, BAM Berlin, Laboratory S.12, Unter den Eichen 87, 12205 Berlin, Germany.
e-mail: wolfgang.habel@bam.de

Jürgen Hesselbach Jürgen Hesselbach was born 1949 in Stuttgart. From 1968 to 1975 he studied mechanical engineering at the University of Stuttgart and completed his Ph.D. thesis at the Institute of Control Engineering of Machine Tools in 1980. Afterwards he joined the department 'Industrial Equipment' of the Robert Bosch company. Since 1990 he has headed of the Institute of Production Automation and Handling Technology at the Technical University of Braunschweig. The institute is working in the field of robotics, microassembly, disassembly and new actuators.

Prof. Dr.-Ing. Hesselbach, Technische Universiät Braunschweig, Postfach 3329, D-38023 Braunschweig, Germany.
e-mail: j.hesselbach@tu-bs.de

Gerhard Hirsch Gerhard Hirsch was born in Pogegen, Lithuania, in 1924. He received his Engineering diploma from the Aachen University of Technology 1954. From 1954 to 1972 he was assistant and senior assistant. Lecturing and research work on structural dynamics and vibration control of light-weight structures were the basis for the senior lecturership that he held from 1972 onwards. He retired in 1996 and works as a consultant in the TMM Ltd. Eschweiler.

Dipl.-Ing. G. Hirsch, TMM GmbH, Königsbenden 38, 52249 Eschweiler, Germany

Hartmut Janocha Hartmut Janocha was born in 1944 in Oppeln and studied electrical and mechanical engineering at the University of Hanover. In 1969 he graduated in high-frequency technology, finished his doctorate in 1973 and his habilitation in 1979. Since 1989 he has been professor at the University of Saarland, where he is chair holder at the Laboratory for Process Automation. Between 1992 and 1994 he was president of the German Assembly of electrical engineering departments and between 1995 and 1997 vice president of the University of Saarland. His main fields of work include new actuators, 'intelligent' structures with system and signal processing concepts, calibration procedures to improve the positioning accuracy of industrial robots and the measuring of 3D geometries using methods of image processing.

Prof. Dr.-Ing. habil. Hartmut Janocha, Universität des Saarlandes, Gebäude 13, Lehrstuhl für Prozeßautomatisierung, 66123 Saarbrücken, Germany.
e-mail: janocha@lpa.uni-sb.de

Matthias Kallenbach Matthias Kallenbach was born in Leipzig, Germany, in 1972. From 1991 to 1996 he studied mechanical engineering with the special field of microsystem technology at the Technical University of Ilmenau. During his study, he spent one year at the Linköping University in Sweden, where he worked on the microtechnical realization of polymer micro-actuators based on polypyrrole. He graduated in 1996 under the supervision of Dr. Smela (Linköping University) and Prof. Wurmus.

Dipl.-Ing. Matthias Kallenbach, Technische Universität Ilmenau, Postfach 100565, D-98684 Ilmenau, Germany.
e-mail: mkalle@maschinenbau.tu-ilmenau.de

Hiroshi Matsuhisa Hiroshi Matsuhisa was born in Osaka, Japan in 1947. He studied mechanical engineering on Kyoto University from 1966 to 1970 and 1972 to 1976. From 1970 to 1972, he studied industrial engineering and received M.Sc. degree from Georgia Institute of Technology, USA. In 1982, he received his doctoral degree from Kyoto University for the study of vibration and noise of train wheels. He became an instructor in 1976, an associate professor in 1987, and a professor in 1994 at Kyoto University, Japan. His current research interests are vibration control, noise control and human dynamics.

Prof. Dr. H. Matsuhisa, Dept. of Precision Engineering, Kyoto University, Kyoto, 520-8501, Japan.
e-mail : matsu@prec.kyoto-u.ac.jp

Detlef zur Megede Detlef zur Megede, born in 1957 in Frankfurt am Main, Germany, studied chemistry at the university of Frankfurt am Main (Diplom 1980) and gained his Ph.D. in 1985 at the Dechema Institute in Frankfurt with a thesis on corrosion of hard metals. From 1985 to 1988 he worked at DECHEMA (Deutsche Gesellschaft für Chemisches Apparatewesen, Chemische Technik und Biotechnologie e.V.) and was involved in the development of a computer based corrosion system. Since 1988 he has worked in research for Daimler–Benz AG. He was initially responsible for sensor-based exhaust-gas detection and also worked in the development of electrochemical actuators. In 1994 he moved to the research centre in Ulm where he is active in the field of fuel cell development for vehicular applications.

Dr. D. zur Megede, Daimler–Benz AG, Wilhelm-Runge-Straße 11, D-89081 Ulm/Donau, Germany

Jörg-Uwe Meyer Jörg-Uwe Meyer was born in Darmstadt, Germany, in 1956. He received his engineering degree from the Technical University of Gießen, Germany in 1981 and his Ph.D. degree in Biomedical Engineering from the University of California, San Diego, USA, in 1988. In 1989 he was awarded a NRC fellowship as a principal investigator at the NASA–Armes Research Center, Moffett Field, USA. Since 1990, he has headed the Sensor Systems/Microsystems Department of the Fraunhofer Institute for Biomedical Engineering, St. Ingbert, Germany. His expertise is on biomedical and industrial sensor developments as well as on telemedical applications. He holds several patents in the field of microsensors and biomedical instrumentation.

Dr. J.-Uwe Meyer, Fraunhofer-Institut für Biomedizinische Technik, Ensheimerstr. 48, D-66386 St.Ingbert, Germany.
e-mail: uwe.meyer@ibmt.fhg.de

Werner Nachtigall Werner Nachtigall, born 1934 in Saaz, studied biology and technical physics in Munich. After working as an assistant lecturer in the zoology and radiobiology departments of the Munich University and as research associate at the University of California, Berkeley, he was appointed director of the Zoology Department at the University of Saarland, Saarbrücken. His main fields of research are movement physiology and biomechanics. For several years now he has offered a course on technical biology and bionics. Combining biology and physics is an important matter of interest to him.

Prof. Dr. rer. nat. Werner Nachtigall, Zoologie, Universität des Saarlandes, Gebäude 6, 66041 Saarbrücken, Germany.
e-mail: W. Nachtigall@rz.uni-sb.de

Dieter Neumann Dieter Neumann was born in 1955 in Gelnhausen and studied sports and physics in Göttingen. From 1983 to 1985, after passing the first State Examination, he underwent in-service training at the Highschool of Salzgitter and worked as a scientific assistant at the Max-Planck Institute for fluid dynamics in Göttingen. In 1989 he gained his doctorate in physics and received a scholarship from the Max-Planck-Gesellschaft. Since 1991 he has been working as a technology adviser at the VDI-Technologiezentrum in Düsseldorf with a focus on the identification and evaluation of future technologies. In 1995 he published his book 'Bausteine intelligenter Technik von morgen' (Components of intelligent future technology). From 1995 to 1998 he was an IT-manager at VDI and since February 1998 has headed the telecommunications department of the SIG GmbH, Germany.

Dr. Dieter Neumann, Heideäcker 27, D-70771 Leinfelden, Germany.
e-mail: dieter.neumann@s-i-g.de

Vittal Rao Vittal Rao was born in Inupamula, India in 1944. He was awarded the Bachelor of Electrical Engineering degree from the Osmania University, Hyderabad in 1969 , M.Tech. and Ph.D. degrees from the Indian Institute of Technology, New Delhi in 1972 and 1975, respectively. Since 1981, he has been on the faculty at the University of Missouri-Rolla, where he is currently Rutlege–Emerson Distinguished Professor of Electrical & Computer Engineering and the Director of Intelligent Systems Center. His current research interests are intelligent control, smart structural systems, structural health monitoring, and intelligent sensor data fusion.

Prof. Vittal S. Rao, Department of Electrical & Computer Engineering and Intelligent Systems Center, University of Missouri-Rolla, Rolla, MO 65409, USA.
e-mail: raov@umr.edu

Craig A. Rogers Craig Rogers was born in Barre, Vermont, USA, in 1959. He received his Ph.D. (1987), M.S., and B.S. degrees in Mechanical Engineering from Virginia Polytechnic Institute and State University (VPI&SU), Blacksburg, VA. During the period 1979–81, he worked as a physical design engineer with Bell Telephone Labs in Holmdel, NJ. From 1983 to 1996, he was an instructor, visiting assistant professor, assistant, associate, and professor in the Department of Mechanical Engineering at VPI&SU. He also held a joint appointment as professor of education in the College of Education (1994–96). Beginning in 1989, until 1996, he was director for the Center for Intelligent Materials Systems and Structures at VPI&SU. In August 1996, he became Dean of the College of Engineering at the University of South Carolina, Columbia, SC. His current research interests are adaptive material systems and structures, and intelligent materials and smart structures.

Dr. Craig A. Rogers, College of Engineering, University of South Carolina, Columbia, SC 29208, USA.
e-mail: rogers@sc.edu

Ben. K. Wada Ben K. Wada was born in 1936, in Los Angeles, USA. He studied Engineering Mechanics at University of California at Los Angeles (UCLA) from 1954 to 1963 where he was awarded a Master's Degree and studied for his Ph.D.. He has been at the Jet Propulsion Laboratory, California Institute of Technology from 1959 to the present time. His career includes working on spacecraft, research in structures, and management of 100 mechanical engineers. In the last 10 years, his research interest has been in Adaptive Structures.

Prof. Ben K. Wada, Jet Propulsion Laboratory, Bldg 125 Room 214A, 4800 Oak Grove Dr., Pasadena, CA 91109, USA.
e-Mail: ben.k.wada@jpl.nasa.gov

Neil White Neil White was born in 1963 in Nottinghamshire, England. He was awarded a Ph.D. from the University of Southampton in 1988 for a thesis on applications of thick-film strain gauges. In 1990 he was appointed as lecturer within the Department of Electronics and Computer Science, University of Southampton. He co-authored the book 'Intelligent Sensors Systems' with Prof. John Brignell. His publications are in the areas of thick-film sensors and intelligent instrumentation. He is currently chairman of the Instrument Science and Technology (ISAT) group of the Institute of Physics and research director of the University of Southampton Institute of Transducer Technology (USITT). Funded research projects include several on applications of thick-film piezoelectric materials.

Dr. N. White, Department of Electronics and Computer Science University of Southampton, Highfield, Southampton SO17 1BJ, UK.
e-mail: N.M.White@ecs.soton.ac.uk

Helmut Wurmus Helmut Wurmus was born in Graustein, Germany, in 1940. From 1957 to 1963 he studied precision device technology at the Technical University of Ilmenau. In 1972 he received his doctor's degree from the same institute under the supervision of Prof. Bischoff. After collaborating on the development of automatic production units by 'Elektromat Dresden' used in the semiconductor industry and a research visit at the TU Sofia (Bulgaria) from 1979 to 1980, he became assistant professor for microelectronics on the faculty of device techology at the Technical University of Ilmenau. Lecture work and research work on opto-electronic sensor technology were the basis for the professorship in micro-mechanics (1988). Since 1990 he has been responsible for the field of microsystem technology at the University of Ilmenau.

Prof. Dr. H. Wurmus, Technische Universität Ilmenau, Postfach 100565, D-98684 Ilmenau, Germany.
e-mail: helmut.wurmus@maschinenbau.tu-ilmenau.de

Index

Absolutely measuring sensors   268
Academic researcher   413
Accelerometers   26
Acceptance angle   257
Accreditation   415
Accuracy   16
Acoustic attenuation   24
Acoustic behavior   17
Acoustic signature   17
Actin filament   372
Active acoustic attenuation   21
Active aerodynamic measures   340
Active bracing   338
Active control   334
Active control algorithms   16
Active damage control   21
Active dampers   327
Active elasticity   32
Active flaps   302
Active forming   121
Active mass dampers   334
Active material stacks   22
Active structural acoustic control
   (ASAC)   17
Active structural damping   21
Active tagging   27
Active tendons   334
Active type (AMD)   337
Active vibration damping   121
Active–passive composite tuned mass
   damper   337
Actuators   13, 124, 242
Adaptability   5
Adaptive architecture   31
Adaptive composite structures   19
Adaptive control   29
Adaptive functionality   13
Adaptive functions   14
Adaptive materials   84
Adaptive rotorblades   290

Adaptive self-preserving features   30
Adaptive structures   79
Adaptive transmission control (ATC)
   314
Adaptive wings   290
Adaptronic material system   13
Adaptronic skins   290, 292
Adaptronic structures   13, 14
Adhesive bond   27
Advanced fiber reinforced composites
   31
Advanced robotics   1
Aero–servo–elastic control   21
Aerodynamic devices   340
A_f 145, 146
Aging   14
Aging process   15
AIAA–Journal   291
Ailerons   15
Air and space transportation systems
   411
Air conditioning   206
Air crashes   406
Airbag   309, 318
Aircraft   15
Aircraft health and usage monitoring
   290
Aircraft wings   15
Airplanes   206
Align   20
Aloha Airlines   16
Alternating magnetic field   27
Amplification   22
Angular deflection   23
Angular motion   23
Anomalies   35, 37, 38
Antagonist muscles   3
Antagonistic actuator design   158
Antennae   295
Anti slip control systems (ASR)   312

Antiferromagnetic 50
Antilock braking (ABS) 309
Apparent viscosity 24
Area averaging sensors 258
Arm 30
Artificial organs 15
Artificial life 29
Artificial life forms 29
Artificial limbs 15
Artificial neural networks 394
A_s 144
Astronautics 10
Atila 125, 132
ATP 374
Austenite 47
Auto–calibration 248
Automated baggage–handling 407
Automatic distance control (ADC) 321
Automatic headlight range control 322
Automobiles 15, 309
Aviation 10

Beam steering 398
Bending mode 112
Bhopal 406
Biaxial films 26
Bilayer 214
Bimetals 8
Bingham–plastic 180
Bioadaptronic systems 400
Bioengineering 392
Biological structural systems 13
Biological structures 2
Biological system 391
Biological world 3
Biologically inspired materials 13
Biomaterials 15
Biomedical applications 15, 391
Bionics 372
Biosensing 391
Bit–stuffing 252
Blackbody 25
Blind fasteners 19
Blocked force 130
Blood flow 398
Blood–sugar level 15
Boxer configuration 22
Bragg grating sensors 272
Bragg sensors 265
Braille alphabet 27
Brain 30

Brakes 24, 182
Bridge 407
Bridge joint 24
Bridge piers 340
Bubble principles 220
Built–in antennas 292
Burn 30
Bus standards 311
Bus topology 251

Camber 15
Cantilever beam 3
Capacitance 121
Carbon 197
Carbon–based 27
Carbonyl iron 181
Cascaded actuator 213
Catalyst 32
Catalyzing enzyme 27
Catastrophic damage 17
Cell package 201
Central processor architecture 29
Centrifugal 183
Ceramic 16
Ceramics 107
Cerebral blood–flow 398
Channel tunnel 406
Chemical actuators 195
Chemiresistor 243
Chemo–mechanical drive 208
Chemochromic elements 8
Chimneys 336
Civil and structural engineering 412
Civil engineering 189
Civil engineering constructions 31
Civil engineering structure 190
Clutches 24, 184
Cochlear implants 396
Codes 406
Communication 13
Compact disk player 407
Compensating 246
Compensation 243
Complex communication 30
Complex organism 19
Complex systems 18
Complex tasks 30
Composite 35, 36
Composite components 31
Composite cure monitoring 25
Composite transducers 112
Compressive stresses 20
Computational hardware 30

Computational networks　13
Concrete　17
Condition monitoring　292, 293, 304
Conducting polymers　208
Conservative engineer　406
Contracting muscles　30
Control electronics　123
Control of muscle activity　388
Control processors　31
Control system　30
Controllability　80
Controllable exercise equipment　191
Controllable fluids　180
Controlled damping　188
Controls　13
Converse piezoelectric effect　44
Copper–aluminum–nickel　19
Copper–zinc–aluminum　19
Core　257
Cornucopia　18
Corrosion　264
Corrosion resistance　19
Cost–effective　14
Coupling factor　108
Crack sensor　260
Creep　109
Critical factor　30
Cross–sensitivity　242
Crossed extensor reflex　388
Curie temperature　109
Cybernetics　29

d_{31} mode　108
d_{33} mode　108
Damage detection　3, 25
Damage index　25
Damage metric　24
Damage monitoring　290
Damage of composite materials　18
Damage–control　20
Damage–tolerant　16
Damaged　14
Damaged structure　17
Damages　2
Dampers　184
Damping　19, 124, 246
Decomposition　91
Defects　3
Deployment　285
Design adaptronic structures　32
Design chain　415
Design philosophy　54, 56
Design symmetry　243

Design to code　407
Design variables　90
Detailed information　18
Detect edges　27
Diagnosis　311, 318
Diagnostic devices　391
Diaphragms　22, 219
Dielectric breakdown　183
Digital filtering　246
Digital signal processors　101
Dimensional stability　16
Direct shear　184
Direction of loading　270
Disabled aircraft　16
Disneyland　412
Displacement amplification principle　23
Displacement–amplifying system　329
Distributed in–situ sensing　27
Distributed moisture detection　264
Distributed sensor network　2
Domains　40, 47
Drug delivery systems　399
Dry fluids　165
Durability　13
Dynamic absorbers　325, 326
Dynamic loading　334
Dynamic wind loads　339

Earthquakes　334
Economic efficiency　13
Economic viability　2
Eddy current　182, 191
Eddy current probe　27
Effective coupling factor　128
Efficiency　129
Eigensystem Realization Algorithm (ERA)　60
Elasticity　23
Electric field　19
Electrical admittance　25
Electrical dipoles　20
Electrical energy　14
Electrical impedance　24
Electrical or magnetic disturbances　26
Electrical stimulation　395
Electro–active　16
Electro–chemo–mechanical energy converter　208
Electrochemical actuators　195
Electrochemical devices　30
Electrochromic elements　8

Electrode   197
Electrolysis   197
Electrolyte   198
Electromagnetic coil   190
Electromagnetic compatibility   102
Electromechanical impedance   24
Electron–hydraulic time delay   169
Electronic engineers   32
Electronic materials   15
Electronic nose   250
Electronic transmission controls (ETC) 312, 314
Electronically trainable artificial neural network   68
Electrorheological fluid   6
Electrorheological fluids   8, 17, 104, 180
Electrostatic micropump   224
Electrostatic valve   222, 223
Electrostriction   44, 49, 109
Electrostrictive   125
Electrostrictive effect   44
Electrostrictive particles   27
Electrostrictors   8
Electroviscous effect   161
Embedded   270
Embedded piezoelectric transducers 121
Endoscopic diagnostics   398
Energy as a structural component   32
Energy provider   99
Energy storage   32
Energy transducer   99
Energy–to–mass ratio   216
Energy–to–volume ratio   216
Engine mounts   19
Engineering education   414
Engineering solution   17
Engineering structure   4
Engines   17
Environment   15
Environmental legislation   407
EPCOT   412
Equations of motion   85
Equivalent mass ratio   352
ER   23
ER fluid servo drive   331
ER fluid valve   331, 333
ER fluids   180, 325
Euler's equations   130
European Union   414
Evanescent   25
Evanescent fibers   264

Evanescent sensors   257
Evolution   19
Exercise equipment   191
Exercise machine   193
Expansion bellows   200
Extrinsic   256

Fabry–Pérot   265
Failed system   30
Fatigue failures   17
Fatigue–tolerant   16
Fault–tolerant   30
FE codes   94
Feedback   188
Feedback control of furnace temperature   390
Feedback control of muscle length   389
Ferrimagnetic   50
Ferroelastic   48
Ferroelectric   36, 37, 40, 41, 44, 45, 48, 52
Ferroic   48
Ferromagnetic   48, 50
Ferromagnetic materials   124
Fiber cladding   257
Fiber optic   26
Fiber optic sensor system   255
Fibers   31, 209
Field applications   275
Field energy   102
FIELDBUS   253
Fingers   30
Finite Element Method (FEM)   131
Flaps   15
Flexible composite sensors   27
Flow mode   184
Fluid flow rate   19
Force impulses   348
Force of expansion   195
Fragile objects   19
Frequency response curves   23
Friction   17
Frictional devices   24
Full–scale active brace   347
Functional density   13
Functional gels   8
Functional materials   35, 36, 38, 43, 46, 51, 55
Funding agencies   414
Fuselage   24
Fuzzy logic   320

Gas pump   197

Gels   18
Giant Magnetostrictive Alloys   124
Global communications network   408
Golf ball   31
GPIB   253
Grain boundaries   38, 40, 46
Graphite   17
Ground testing   287
Ground transportation systems   411

HAHDIS   22
Hamilton principle   130
Hardware   29
Hardware–in–the–loop   95
HART   254
HDLC   252
Health and usage monitoring   297
Health monitoring   277
Heart pacing   393
Heating systems   204
Helicopter rotor blades   15, 292
Hierarchical approach   30
Hierarchical architecture   30
Hifi speakers   21
High winds   334
High–force linear motors   21
High–level computer languages   320
High–power sonar   21
High–power transducers   135
High–precision   20
High–speed   20
High–speed trains   340
High–torque   21
Home appliances   32
Hubble telescope   16
Human muscles   19
Human user   14
Hybrid actuator   119
Hybrid amplifier   102
Hybrid composite systems   18
Hybrid material systems   30
Hybrid type (HMD)   337
Hydraulic actuators   21
Hydraulic semiconductor   163
Hydraulic valves   24
Hydraulically–Amplified High–
   Displacement Induced–Strain
   22
Hydrochromic elements   8
Hydrostatic amplification   22
Hysteresis   109, 144, 148, 149, 152, 153,
   323, 324
Hysteresis models   106

Hysteresis operators   103

IEEE488   253
Imagineers   412
Immunity to noise   26
Impedance response curve   25
Implantable stimulator   391
In–line fiber etalon   268
In–plane structural impedance   18
In–service health monitoring   27
Inanimate objects   29
Inchworm motor   118, 124
Inchworms   135
Incipient local damage   24
Individual blade control   301
Induced strain actuators   17, 23
Industrial development teams   413
Inflight catastrophe   16
Injection valves   116
Ink–jet printers   21
Insects   29
Inspection robots   158–160
Inspection tasks   156
Instilling life functions   32
Insulin   15
Integrated avionics   290
Intelligence   13
Intelligent actuator   316, 317
Intelligent processing   24
Intelligent sensors   241
Intelligent structures   14
Interconnections   31
Interdisciplinary interactions   9
Interferometric   25
Internal sensoric effect   153, 156
International Journal on Intelligent
   Material Systems and Structures
   291
Intracardiac impedance   393
Intrinsic   256
Intrinsic sensing   24
Isolation   188

Kinetic abilities   19
Knowledge–based systems   320

Large bending deformation   20
Large buildings   17
Large flexible structures   16
LARIS (Large–Amplitude Rotary
   Induced–Strain)   23
Lead zirconate titanate (PZT)   20
Learning from nature   13

Leisure and entertainment industries 412

Length   16

Life cycle   3

Life experiences   29

Life features   13

Life functions   18

Lifespan   30

LIGA technique   218

Light level   15

Light transmission characteristics   24

Lightweight   31

Linear actuators   132, 135

Linear Quadratic (LQ) performance 59

Linearization   285, 288

Living species   19

Living systems   14

Load cell   243

Loads   16

Locus   246

Long life   17

Long–gauge–length sensor   262

Long–span bridges   335

Longer life   16

Look–up tables   242

Low–frequency noise   17

Low–signal operation   108

Low–speed rotating motors   21

LQ optimal control   340

Lubricating systems   226

Magnetic circuit   186, 187, 189

Magnetic domains   21

Magnetic field   19, 180, 182

Magnetic field sensors   26

Magnetic heads   21

Magnetic particles   27

Magnetic saturation   183

Magnetic valve   187

Magnetic–energy   182

Magnetic–hysteresis   191

Magneto–elastic coefficients   125

Magnetorheological fluids   8, 17, 180, 181

Magnetostrain   124

Magnetostriction   50, 124

Magnetostrictive materials   19, 115

Magnetostrictive motor   138

Magnetostrictive particles   27

Magnetostrictive transducers   103

Magnetostrictors   8

Man–machine interface   100

Man–made structural systems   2

Management of energy   121

Maneuverability   16

Manipulate molecules   31

Mankind   27

Manufacturing systems   412

Marine transportation   412

Martensite   46, 47, 54

Martensitic phase transformations   19

Martensitic transformation   143

Masts   336

Material configuration   31

Material properties   18

Material science   29

Maximum control force   345

Measure strain   24

Measurement industry   410

Measuring range   270

Mechanical characteristics   19

Mechanical grippers   156

Mechanical impedance   24

Mechanical muscle   16

Mechanical strength and geometrical stability   269

Mechanical systems   1

Mechanical work   195

Mechanics   29

Mechatronic   182

Mechatronic systems   316

Medical devices   156, 392

Medical rehabilitation   180

Medical ultrasonic transducers   27

Medicine   399

Memory   13

MEMS (Microelectromechanical Systems)   292, 300, 307

Metabolic cost   2

Meter level   31

Methodology   15

M_f 145

Microactuator   140

Microactuators   125, 217

Microanalysis systems   225

Microassembly   156

Microbend sensor   261

Microcracking   110

Microdosing systems   223, 225

Microdrop injector   226

Microelectrode   395

Microelectromechanical   391

Microelectronics   391

Microengineering   254

Micromachined accelerometers   393

Micromachining technologies 396
Micromanipulators 19
Micromechanics 218
Micromixers 225
Micron level 31
Micron–sized particles 24
Micropositioners 21, 124
Microprocessor 241
Micropumps 223
Micropush motor 118
Microstructures 13, 14, 217
Microsystems 391
Microtechnologies 391
Microvalves 160, 229
Microvibrations 92
Milliactuators 228
Miniature grippers 157, 160
Miniaturization 122
Ministry for Research and Technology 11
Ministry of International Trade and Industry (MITI) 10
Mirrored surface 16
Missiles 300
Mobility 19
Modal characteristics 20
Modal domain 25
Model reduction 87
Model Reference Adaptive Control 59
Modeling 125
Modern technology 16
Modern–day alchemy 32
Monitoring 318
Monotube damper 187
Motor control 15
Motor control system 14
Motor systems 391
Motors 13, 124
Movement of an arm 3
Moving vehicles 339
MPB 52, 53
MR 23
MR brakes 191
MR fluid 180, 331, 332
MR fluid brake 332
MR fluid clutches 332
MR fluid damper 192
MR fluid rotary brake 192
MR fluid valves 331
MR seismic damper 189, 191
M_s 145
Multichannel electrodes 395
Multifunctional 3, 37

Multifunctional effects 8
Multifunctional elements 5
Multifunctional materials 104
Multifunctional member 116
Multifunctional properties 218
Multifunctionality 310
Multilayer transducers 110
Multimode fibers 261
Multiplexing 276
Muscle 3, 13, 209
Muscle types 371
Muscle, basic functions 382
Muscular levers 384
Muscular–cybernetic analogy 389
Myofibril 372
Myosin filament 372

Nano level 31
Natural frequency 19
Nature 1
Naval Ordnance Lab 19
Nerve stimulation 396
Nerves 13
Net–linked collagens 208
Neural network–based adaptive control technique 66
Neural networks 18, 295
Neural prostheses 391, 397
Neuro–fuzzy algorithms 394
Neuromuscular stimulators 395
Neurons 30
Neuroprosthetic microimplants 393
New Glass Association 11
New Glass Forum 10, 11
Newtonian liquids 180
Next materials revolution 33
Nickel–titanium alloys 19
Nitinol 19
Nitinol actuators 19
Nitinol wires 19
Noise 16, 242
Noise reduction 301
Noise–control 15
Nondestructive materials evaluation 25
Nonlinearity 242
Notched tensile coupons 20

Observability 80
On–board diagnostics 311
One–way effect 145
Open–tube structure 23
Optical fibers 24

Optical time–domain reflectometry 256

Optical tracking devices   21

Optimal load   129

Optimization   89

OTDR   256, 260, 263, 264

Output power   129

Overload response   14

Pacemakers   391

Paper cups   19

Parameter drift   242

Parametric studies   18

Parametric vibrations   339

Paschen effect   218

Passive dampers   325

Passive type (TMD)   337

Pendulum–type tuned active damper 347

Personal computer industry   408

Phase transformation temperature   19

Phase transition   35, 37, 40, 46–48, 50, 51, 53

Phase transitions   41, 46

Phased arrays   398

Philosophy of engineering design   32

Photochromic   5

Photochromic elements   8

Photochromic glass   7

Photoelastic   20

Physical therapy   195

Piezoactive motors   135

Piezoceramics   104

Piezoelectric   41, 44, 48, 54, 55

Piezoelectric actuators   17, 20, 324, 327, 330

Piezoelectric ceramics   16, 26, 137

Piezoelectric composite materials   27

Piezoelectric drives   218

Piezoelectric driving mechanisms   218

Piezoelectric effect   106

Piezoelectric elements   8

Piezoelectric materials   24

Piezoelectric paint and coatings   27

Piezoelectric particles   27

Piezoelectric polymers   26

Piezoelectric powders   27

Piezoelectric ultrasonic motors   138

Piezoelectric valve   330

Piezomagnetic laws   125

Pipe couplings   19

Plastic   31

Plastic viscosity   186

Plasticity   23

Pneumatic drives   196

Pneumatic valve   220

Polyelectrolyte gel   208

Polymers   15

Polypyrrole   212

Polyvinylidene fluoride (PVDF)   26

Position sensor   24

Power modulation devices   31

Power sources   29

Power supply   183

Precision antenna   16

Preisach   103

Preisach model   104

Pressure   269

Prevent damage   30

Processing   101

Processing algorithms   30

Production modules   316

Propellers   15

Propulsion systems   26

Prosthetic devices   392

PTC   35, 37, 40, 46

Pulse control   334

PVDF films   27

Pyroelectric effect   27

Pyroelectricity   109

PZT   41, 52, 54–56

PZT actuators   21

PZT sensor–actuators   24

PZT transducer   24

Quartz   110

Quasi–distributed measurement   262

Quasi–distributed sensors   258

Quiet commuter aircraft   15

Rabinow   180

Radiating modes   18

Radiating the noise   18

Reaction chambers   225

Real–time   101

Reconfigure   31

Recoverable strains   19

Recovery strains   19

Redundancies   4

Redundancy   30

Refractometric   25

Rehabilitation   194

Rehabilitation therapy   195

Rejuvenation   29

Reliability   30

Repair   17

Research assessment exercise   414
Researchers   1
Resistance feedback   154
Resistance to adverse environments   26
Resistance to damage   17
Resolution   270
Resonance   124
Resonators   110
Response time   186
Retaining systems   321
Retina   397
Retinal implant   397
Retirement   14, 27, 29
Rheological   186
Rheological properties   23
Rheopex fluids   8
Robotic   3
Robotic actuators   19
Robotics applications   24
Robots   21
Robust actuators   343
Robust design   287
Room temperature   15
Ropeway carrier   350, 351
Rotary hydraulic actuator   22
Rotary resistance   191
Rotorblades   301
Runability of vehicles   340

Safety (fatigue)   336
Safety tasks   311
Salt draining   213
Sandys   125
Saturation magnetization   181
School curriculum   415
Science and Technology Agency (STA)   11
Science fiction   32
Scientists   1
Security systems   24
Seismic   180, 188
Seismic activity   17
Seismic damage   188
Self–check   248
Self–diagnostics   318
Self–monitoring   311, 318
Self–repair   13
Semi–active control   187, 188
Semi–active damper   189, 328
Semi–active suspension   187
Semi–active vibration control   180, 182

Semiactive type (SAMD)   337
Sensing   3, 24
Sensing capabilities   24
Sensitivity   270
Sensor   242
Sensor equation   86
Sensor network   31
Sensor system   31
Sensors   13, 116
Serviceability   336
Servo–valves   184
Servovelocity seismometers   345
Shakers   17
Shape control   15, 92, 290, 300
Shape memory   47, 48, 54
Shape memory actuators   152, 154–157, 160
Shape memory alloys   9, 16, 104, 144–146, 148, 150, 155, 399
Shape memory effect   143, 145–147, 152
Shape memory elements   8
Shape recovery   19
Shear stress   184
Shock absorbers   6, 184
Short–gauge–length sensors   258
Signal processing   394
Silicon electronics   30
Silicon microchip   31
Silicon–based   27
Silver   199
Single cell   19
Single–mode fibers   257
Skeletal form   15
Skeletal structure   3
Skin–like sensor   27
Skis   31
Skyhook damping   187
Slender, tower–like structures   335
Smart actuator   107, 121, 324
Smart magnetostrictive transducer   106
Smart materials   13
Smart materials and structures   291
Smart piezoactuator   105
Smart Skin Program   10
Smart skins   25, 292, 295, 307
Smart structures   13, 59
Snalog   101
Socialized engineer   416
Solid oxide electrode/fuel cell   197
Solid–state actuation   21
Solid–state transducers   106

Sound radiated   17
Space exploration   409
Space shuttle   406
Space structures   285, 286
Space truss   16
Spatial resolution   18
Speakers   15
Speed   186
Spillover   88
Spinal cord   30
Sports equipment   31
Squeeze film   184
Squeeze mode   185, 334
Stability   81
Stacks   110
Standards for high–level signal
    characteristics   123
Star topology   251
State equation   86
State of the art   335
State–of–health   14
State–space representation   80
Static adjustment   285, 286
Static yield   163
Steel   17
Stepping motor   135
Stiffen   14
Strain   262, 269
Strain concentrations   16
Strain gauge   244
Strain profiles   263
Strain–field strength response   108
Strengthen   14
Stress–free strain sensor   271
Striated skeletal muscle   371
Structural component   13
Structural composites   184
Structural fatigue   16
Structural health monitoring   24, 25
Structural Health Monitoring System
    297
Structural impedance   3, 17
Structural implications   32
Structural modes   18
Structural requirements   2
Structures communities   29
Structures revolution   32
Sub–systems   123
Submarine   17
Super slip   166
Support brackets   24
Surface application   272
Surface electrodes   395

Surface micromachining   218
Surface–attached sensors   260
Surface–mounting   270
Surfactants   181, 183
Survival   19
Suspension systems   19
Suspensions   187
Sweet spot   31
Switching   101
Switching time   187
Symmetry   37, 42, 43, 45, 48, 50, 52, 54
Synthetic material systems   2
Synthetic materials   209
Synthetic rubber (in solvents)   208
System diagnosis   318
System identification   285, 287

Tactile feedback   396
Tactile information   398
Tactile responses   15
Tactile sensors   27, 391
Tagging particles   24
Tall buildings   335
Technologies of the 21st Century   11
Temperature   19, 269
Temperature distribution   264
Temperature–sensitivity   275
Temperature–sensitivity of fiber grating
    sensors   274
Temporal resolution   18
Tendon organ stress regulation in
    muscles   388
Tendon systems   338
Tennis rackets   31
Terbium   21
Terfenol–D   21, 50, 53, 124
Thermal expansion   190
Thermochromic elements   8
Thermomechanical effects   219
Thermomechanical valve   221, 223
Thermopneumatic effects   219
Thermopneumatic valve   222
Thick–film   244
Thin–film   244
Thixotropic   195
Thixotropic fluids   8
Three–level controller   204
Thrust bearings   17
Time (or frequency) response   242
Time–domain   25
Time–efficient   30
Topology   251
Torque   191

Traffic load   339
Transducers   124, 242
Transfer function   82
Transformation temperatures   144,
    147, 148, 151
Transmission of motion   17
Transverse cracks   16
Traveling wave   117
Tunable dampers   24
Tuned mass dampers   336
Turbulent boundary layers   18
Twin   43, 46, 47
Twins   40
Two–port model   108
Two–way effect   146, 150

Ultrasonic monitoring   398
Ultrasonic motors   117, 137
Ultrasonic piezomotor   138
Undergraduate education   417
Uniaxial films   26
Utility of adaptronic structures   15

Valve   182
Variable geometry trusses (VGTs)   17
Variants   40, 45
Variational principle   130, 131
VDI Technology Centre   5
VDR   37
Vehicle dynamics control system (VDC)
    312
Vehicles   206
Ventilation flaps   9
Vertical–tail   24
Vibration   16
Vibration attenuation   289
Vibration control   3
Vibration dampers   184, 185, 326
Vibration damping   290, 300, 304
Vibration isolation systems   24
Vibration suppression   304, 305
Virtual instrumentation   249
Viscosity   23, 180, 182
Viscous liquids   180
Volume change   212
Volume flow   8
Volumetric fluid displacements   22
Vortex–induced cross–wind vibrations
    336

Warping–torsion coupling   23
Wavelength of light   16
Weldless   19
Wet fluid   165

Wheatstone bridge   243
Wing   15
Winslow   161, 180
Wood   31

Yield strength   180–182
Yield stress   180

Druck: Strauss Offsetdruck, Mörlenbach
Verarbeitung: Schäffer, Grünstadt